"The book *Entrepreneurial Innovation in the International Business of Tourism*, edited by María Jesús Jerez-Jerez and Pantea Foroudi, stands as an exemplar in the realm of tourism studies. Seamlessly interweaving empirical research with profound insights, Jerez-Jerez and Foroudi shed light on the complexities of today's dynamic tourism landscape. Each chapter, penned by experts in the field, delves into diverse geographies, cultural nuances, and innovative approaches to sustainable tourism practices. The book's comprehensive examination of technologies, sustainability concerns, and entrepreneurial strategies offers both academics and industry professionals a roadmap for navigating the intricate terrain of international tourism. I wholeheartedly endorse this meticulously crafted volume as a seminal resource, destined to influence the discourse and practice of global tourism for the foreseeable future."

**Prof. T.C. Melewar**, *Professor of Marketing and Strategy, Middlesex University*

"The book, *Entrepreneurial Innovation in the International Business of Tourism*, edited by María Jesús Jerez-Jerez and Pantea Foroudi, is a comprehensive and forward-looking exploration of entrepreneurial innovation in the ever-evolving tourism and hospitality sector. The authors provide fresh insights into the emerging trends and the multifaceted dynamics of entrepreneurial innovation by taking consideration of factors such as policy, society, sustainability, and technology. This book not only offers a rich theoretical foundation but also provides practical insights through numerous case studies. As a unique and timely addition to the literature, it serves as an invaluable resource and inspiration for students, researchers, academics, and industry practitioners."

**Prof. Yan**, *Professor of Information Systems, University of Bedfordshire*

# Entrepreneurial Innovation in the International Business of Tourism

This international case study book provides 28 expertly curated case studies on entrepreneurship and innovation in tourism, each with detailed implementation instructions for the instructor to maximise student participation and learning.

The dynamic characteristic of the tourism industry under the influence of micro and macro environment factors requires future professionals to be equipped with appropriate skills and competencies to deal with change and development in real-life practices. Curated and developed by industry experts and practitioners, these case studies embody real-world scenarios with the aim of best preparing students for their future careers. This compelling set of case studies explores the dynamics of entrepreneurship in global context, analyses emerging markets and new business models, and elicits the implications of innovation and entrepreneurship in different contexts and within a transdisciplinary perspective. The cases illustrate innovation and entrepreneurship as an accelerator of tourism growth and development, under a sustainable perspective.

With reflective questions throughout to aid both in-class discussion and self-study, this book is an ideal study resource for use in higher and vocational education, and its unique, teaching-led approach positions it as a vital study tool for instructors and students alike.

**María Jesús Jerez-Jerez** is Senior Lecturer in International Business at the University of Bedfordshire. Maria received her PhD in Business Management from Middlesex University, London, UK. She has a Master of Arts degree in International Hotel and Restaurant Management from London Metropolitan University, UK, and a Bachelor of Science degree in Hospitality and Tourism, Madrid, Spain. She worked in the hotel and tourism industry in various functional areas and managerial roles before becoming an educator.

**Pantea Foroudi**, Brunel University, London, UK, is the Business Manager and Solution Architect at Foroudi Consultancy, as well as a member of Brunel University London. Her primary research interest has focused on consumer behaviour from a multidisciplinary approach, with a particular focus on the concept of customer perception and its effect on corporate brand identity, design, and sustainable development goals (SDGs). Pantea has published widely in international academic journals such as the *British Journal of Management*, *Journal of Business Research*, and *European Journal of Marketing*. She is the associate/senior/editor of the *International Journal of Hospitality Management*, *Journal of Business Research*, *International Journal of Hospitality Management*, *International Journal of Management Reviews*, *International Journal of Contemporary Hospitality Management*, and more.

## Routledge Critical Studies in Tourism, Business and Management

Series editors: Tim Coles, University of Exeter, UK and Michael Hall, University of Canterbury, New Zealand

This ground-breaking monograph series deals directly with theoretical and conceptual issues at the interface between business, management and tourism studies. It incorporates research-generated, highly-specialised cutting-edge studies of new and emergent themes, such as knowledge management and innovation, that affect the future business and management of tourism. The books in this series are conceptually-challenging, empirically-rigorous creative, and, above all, capable of driving current thinking and unfolding debate in the business and management of tourism. This monograph series will appeal to researchers, academics and practitioners in the fields of tourism, business and management, and the social sciences.

**Published titles:**

**Entrepreneurial Innovation in the International Business of Tourism (2024)**
*Edited by María Jesús Jerez-Jerez and Pantea Foroudi*

The *Routledge Critical Studies in Tourism, Business and Management* monograph series builds on core concepts explored in the corresponding Routledge International Studies of Tourism, Business and Management book series. Series editors: Tim Coles, University of Exeter, UK and Michael Hall, University of Canterbury, New Zealand.

Books in the series offer upper level undergraduates and masters students, comprehensive, thought-provoking yet accessible books that combine essential theory and international best practice on issues in the business and management of tourism such as HRM, entrepreneurship, service quality management, leadership, CSR, strategy, operations, branding and marketing.

**Published titles:**

**Carbon Management in Tourism (2010)**
*Stefan Gössling*

**Tourism and Social Marketing**
*C. Michael Hall*

# Entrepreneurial Innovation in the International Business of Tourism

Edited by
María Jesús Jerez-Jerez and Pantea Foroudi

Routledge
Taylor & Francis Group
LONDON AND NEW YORK

First published 2024
by Routledge
4 Park Square, Milton Park, Abingdon, Oxon OX14 4RN

and by Routledge
605 Third Avenue, New York, NY 10158

*Routledge is an imprint of the Taylor & Francis Group, an informa business*



*British Library Cataloguing-in-Publication Data*
A catalogue record for this book is available from the British Library

ISBN: 978-1-032-44013-2 (hbk)
ISBN: 978-1-032-44014-9 (pbk)
ISBN: 978-1-003-36996-7 (ebk)

DOI: 10.4324/9781003369967

Typeset in Times New Roman
by codeMantra

# Contents

*List of figures* xi
*List of tables* xiii
*List of contributors* xv

**Introduction: Entrepreneurial Innovation in the International Business of Tourism** 1
MARÍA JESÚS JEREZ-JEREZ AND PANTEA FOROUDI

SECTION 1
**Strategies and Insights for Emerging Markets** 3

1 **Exploring the Tourism Potential and Innovative Contributions of Social Enterprises in South-East Nigeria** 5
MAURICE EKWUGHA, CHARLES OHAM, NKECHI OJIAGU AND ROBERT AMADI

2 **Exploring Destination Managers' Approach to Shock Advertising: The Case of Southeast Asian Countries and Turkey** 28
AUGUSTA EVANS

3 **Post-Brexit Tourism: Customer Perspectives** 48
ISABEL DOLORES JIMENEZ JIMENEZ

SECTION 2
**Sustainable Practices and Environmental Considerations** 65

4 **Environmental Sustainability in the Tourism Sector** 67
CLAUDIA SEVILLA-SEVILLA, AINHOA RODRIGUEZ-OROMENDIA AND JULIO NAVIO-MARCO

**5 Water Use in Tourism** 90
AMELIA PÉREZ ZABALETA AND ESTER MÉNDEZ PÉREZ

**6 Smart Tourism: Advancing Sustainable 'Smart' Tourism by a Critical Analysis of the Disparity between the Factors Driving Consumer Decision-Making and the Offerings and Challenges within the Industry** 106
SAIRA SULTANA

**SECTION 3**
**Innovations, Technologies and Branding in Tourism** 129

**7 Immersive Technology in Tourism: Applications of Immersive Technologies and Impacted Tourism Experience** 131
CARMEN ARROYO

**8 Destination Branding Authenticity: Building Relationship Orientation among Visitor Attractions, Tourism and Millennials** 149
ILMA AULIA ZAIM, ANNISA RAHMANI QASTHARIN AND AGI AGUNG GALUH PURWA

**9 Unleashing Innovation through Internal Branding and Resident Involvement** 170
IOANA S. STOICA

**10 Transport: Digital Trends** 191
PAVLOS ARVANITIS

**SECTION 4**
**Management, Policy and Research Insights in Tourism** 209

**11 Managing Cultural Diversity and Communication Inside the Tourism Industry** 211
LESZEK WYPYCH AND IJAZ AHMAD

**12 Changing Government Attitudes: Development and Management of New Knowledge on Security and Safety in Tourism** 232
MARÍA JESÚS JEREZ-JEREZ

**13 Synergies to Promote Successful PMI in the Tourism Industry** 252
LESZEK WYPYCH, IJAZ AHMAD AND SANDHYA SASTRY

**14 Entrepreneurship in Tourism and Hospitality Research: A Bibliometric Analysis** 276
SANAZ VATANKHAH, VAHIDEH BAMSHAD AND SADAF TALLIA

*Index* *299*

# Figures


1.1 Theory of change – a tourism perspective 7
1.2 Action Research (AR) cycle 11
1.3 M-PESA transaction in simple steps 12
1.4 Conceptual framework for developing maritime tourism in South-East Nigeria 13
2.1 Conceptual framework 33
3.1 Key factors affecting Brexit and their consequences in the travel industry 51
4.1 Smart Tourist Destinations methodology – brief description of each area and the specific components that each one addresses to manage sustainability at the destination 73
5.1 Distribution of travel and tourism water use (water footprint) 96
5.2 Evolution of the water supplied since 2004, compared with the increase in the technical performance of the water network 102
6.1 Developing sustainable 'smart' tourism by analysing the gap between consumer underlying factors of the decision-making process and industry offerings and challenges 118
7.1 Conceptual framework 132
8.1 The relationship between destination authenticity and destination branding towards tourist satisfaction and revisit intention 155
9.1 Internal branding through resident entrepreneurship acts 173
10.1 Conceptual framework 193
11.1 Model of research 213
12.1 Learning organisation conceptual framework 234
12.2 Skills gaps and ability to attract migrant workers (WEF, 2010) 240
13.1 Relationship among IC competence, dynamic capabilities, experiential learning and internalisation (Sastry, 2015) 253
14.1 Conceptual model 277
14.2 Annual scientific production of all documents in "T&H entrepreneurship research" 280
14.3 Average article citation per year 280

14.4 The most influential authors (red line: the author's timeline, bubble size: the number of publications, bubble colour intensity: total citations per year) 281
14.5 Country scientific production in "T&H entrepreneurship research" 282
14.6 Top 10 influential sources 282
14.7 Author's keywords trend 283
14.8 Word cloud of author's keywords 283
14.9 Three-fields plot of countries, author keywords, and sources (from left to right) 284

# Tables


1.1 Action Research Implementation 11
1.2 Seminars 19
1.3 Workshops 19
1.4 Co-operative Society 20
1.5 Seed Business Loan (First Disbursement) 20
1.6 Seed Business Loan (Second Disbursement) 21
1.7 Seed Business Loan (Third Disbursement) 21
1.8 2023 Budget 22
3.1 Top Five European Union Nations and Markets for the United Kingdom in 2018 51
4.1 Meta-Challenges in the Tourism Sector for Global Tourism 77
4.2 Onlife World 78
5.1 Distribution of Water Consumption in Hotels 95
5.2 Average Consumption in Litres per Room per Day 95
14.1 Descriptive Breakdown of Included Data 279

# Contributors

**Agi Agung Galuh Purwa** is the Secretary of Information and Communication Technology (ICT) Department of West Java Province, Indonesia. His educational background and research focus on political sciences and government, human development index and social media and government responsiveness.

**Ijaz Ahmad** is Senior Lecturer in Intercultural Competences at the University of Bedfordshire. Ijaz's research interests include cross-cultural impacts on effective leadership, cross-cultural management with a focus on the tourism and international charity sectors, and marketing practices across cultures.

**Robert Amadi** has a background in mathematics, statistics and data analysis. He blends financial acumen with innovative thinking, using an interdisciplinary approach to guide an innate curiosity about the world's economic intricacies. He holds a Master's Degree in Finance and Risk Management from the University of West London.

**Carmen Arroyo** graduated in the University of Westminster (London) in 2020 with an MSc in Information Management and Finance. She has since worked for different companies in information technology and is an ITIL expert with a successful history of accomplishment in the strategic planning and development of service management policies, processes and procedures in large organisations such as BP and the BBC.

**Pavlos Arvanitis** is an Associate Professor of Tourism and Aviation Management at Edinburgh Napier University, United Kingdom. Pavlos was awarded a PhD in Airport Development and Economic Geography by the University of the Aegean, Greece. He has taught at the University of Aegean, Greece, alongside some UK universities and has extensive experience in EU-funded research projects related to destination management and planning as well as airport development and destination development. Before embarking on teaching Pavlos worked for an incoming tour operator based in Athens and an Italian regional airline that was flying from Italy to Greece as a sales executive. His research interests lie within the area of tourism and airport development, interaction, and co-creation. His work has been published in several journals and international conferences.

**Ilma Aulia Zaim** serves as a full-time faculty member at the School of Business and Management, Institute of Technology Bandung (SBM-ITB), within the Business Strategy and Marketing interest group. Her research interests encompass tourism marketing, place branding, destination image, non-visitors, cross-culture, visual research and video elicitation methods.

**Vahideh Bamshad** is a Lecturer at the Graduate School of Business at the University of Bedfordshire. She holds a Ph.D. in Entrepreneurship and has publications to her name. Her research interests include but are not limited to entrepreneurial teams, the lean startup framework, entrepreneurial cognition, entrepreneurial growth and business model innovation.

**Isabel Dolores Jimenez Jimenez** is a qualified PR and Marketing Manager with more than two decades of experience in all aspects of the travel industry. She has been working at the Cuba Tourist Office for more than 20 years.

**Maurice Ekwugha** is a social entrepreneur and researcher with a background in National Health Service (NHS) property management. He is a Visiting Lecturer at the University of Bedfordshire. Maurice's research interests are healthcare management, social enterprise and social innovation.

**Augusta Evans'** research interest is primarily focused on tourism advertising, specifically on the use of shock advertisements by destination managers in discouraging tourists' misbehaviour. She has major conference and book publications and is currently working on journal publications in this area. She has over seven years of teaching experience using up-to-date technology and tools at various levels in higher education such as foundation, undergraduate and post-graduate.

**Pantea Foroudi**, Brunel University, London, UK, is the Business Manager and Solution Architect at Foroudi Consultancy, as well as a member of Brunel University London. Her primary research interest has focused on consumer behaviour from a multidisciplinary approach, with a particular focus on the concept of customer perception and its effect on corporate brand identity, design, sustainable development goals (SDGs). Pantea has published widely in international academic journals such as the *British Journal of Management*, *Journal of Business Research* and *European Journal of Marketing*. She is the associate/senior/editor of the *International Journal of Hospitality Management*, *Journal of Business Research*, *International Journal of Hospitality Management*, *International Journal of Management Reviews*, *International Journal of Contemporary Hospitality Management* and more.

**María Jesús Jerez-Jerez** is Senior in International Business at the University of Bedfordshire. Maria received her PhD in Business Management from Middlesex University, London, UK. She has a Master of Arts degree in International Hotel and Restaurant Management from London Metropolitan University, UK, and a Bachelor of Science degree in Hospitality and Tourism, Madrid, Spain. She worked in the hotel and tourism industry in various functional areas and managerial roles before becoming an educator.

**Ester Méndez Pérez** is Full Professor at the Department of Applied Economics, Faculty of Economics and Business Administration, National Distance Education University-UNED. She was the Head of Department of Applied Economics, Faculty of Economics and Business Administration, UNED (2017–2023). She is Coordinator of the master's degree in teacher training, specialising in Economics, Faculty of Economics and Business Administration, UNED. She carries out investigations related to environmental economics and teaching economics.

**Julio Navio-Marco** holds an MSc in Telecommunications Engineering, a BA and PhD in Economics and Business Administration at the UNED and a postgraduate degree in IESE Business School. He is Professor of Business Organization and Digital Economy and EU Jean Monnet chairholder in Digital Economy. Dr. Navio is also expert for the EC Directorate-General for Regional and Urban Policy (DG REGIO) and Horizon Europe. Dr. Navio was Deputy Vice Chancellor of Technology Innovation at the UNED in Spain. He is Deputy Dean of the Spanish College of Telecommunication Engineers and Vice-President of the Spanish Association of Telecommunication.

**Charles Oham** is Principal Lecturer and MSc Portfolio Lead in the Graduate School of Business, University of Bedfordshire. His specialisation is in the fields of social entrepreneurship, social innovation, community development and entrepreneurship. Charles' career in the public and third sector has included remits in community development, social entrepreneurship and innovation.

**Nkechi Ojiagu** has a Doctorate in Cooperative Economics and Management. She is currently a Senior Lecturer in the Department of Cooperative Economics and Management, Faculty of Management Sciences at the Nnamdi Azikiwe University, Nigeria.

**Amelia Pérez Zabaleta** is Full Professor, Department of Applied Economics, Faculty of Economics and Business Administration, National Distance Education University-UNED. She is Vice-rector of Economy, UNED from 2019, Dean Chairman of the Economics from 2021, Director of the Chair Aquae of Water Economics at Aquae Foundation – Aquadom – UNED from 2013. She carries out investigations related to water, environmental economics and teaches economics. She has participated in 15 collective books, has written more than 40 articles and has directed six doctoral theses. She participates in the European project "Accelerating Water Smartness in Coastal Europe-BWaterSmart."

**Annisa Rahmani Qastharin** is a full-time faculty member at the School of Business and Management, Institute of Technology Bandung (SBM-ITB), within the Business Strategy and Marketing interest group. Her research interests include marketing, social entrepreneurship and business model innovation.

**Ainhoa Rodriguez-Oromendia** holds a PhD and is Professor of Marketing at the Faculty of Economics and Business Administration UNED in Madrid, Spain. She has published several books and articles on business and marketing issues and has also participated in International Congresses. She has been involved in competitive projects related to these research areas. Her main research interest includes the tourism industry and trade show marketing.

**Sandhya Sastry** is Associate Dean in Global Banking School. Previously, Dr Sastry was Faculty Academic Director for Strategic Partnerships at Bristol Business School and Bristol Law School until May 2022. Prior experience in HE spans 15 years of teaching including subject head of Strategy & International Business at the University of Northampton; MBA Programmes Director at the University of Beds and Programme Leader at ARU, Cambridge. Dr Sastry is currently engaged as country investigator on the GLOBE2020 Project for Myanmar Nepal Oman & Sri Lanka.

**Claudia Sevilla-Sevilla** obtained a BA in Geography and History from the University of Barcelona, Postgraduate in Leisure and Tourism at the London Metropolitan University, a master's degree in CSR and Sustainability and a PhD in Economics and Business at the UNED. Dr. Sevilla-Sevilla is Professor of Marketing at UNED in Madrid. She has held various positions as a Marketing Consultant in the tourism and hospitality sector in the past. Since 2018, she has been a full-time professor and her research focuses on marketing and tourism. She is a member of EU Jean Monnet, Chair in Digital Economy and was former Director of Communication and Cabinet at UNED.

**Ioana S. Stoica** is a Senior Lecturer in Digital Marketing at the University of Bedfordshire. Her research explores participatory branding approaches, brand co-creation, and place branding and reputation, with a specific emphasis on the involvement of residents in the co-creation of place brands and place development.

**Saira Sultana** is a Senior Lecturer with an extensive teaching experience in a diversity of units at a variety of HE levels. Disciplinary focus is on qualitative research methods as well as strategy and marketing. Internationally minded educator with a Doctor of Philosophy in Social Media Marketing from University of Bedfordshire. Her research interest is in Social Media Marketing and Consumer Behaviour. She is currently working on journal publications and book chapters in this area while attending major conferences.

**Sadaf Tallia** is Permanent Faculty at the University of Bedfordshire. Her research interest is primarily focused on Entrepreneurship and intention development and has many publications in the same domain. She is an advanced PhD scholar and has worked towards entrepreneurship intention and SME development. She has served as an associate editor for *American Journal*. Her new interest aligns with tourist development through entrepreneurship intention.

**Sanaz Vatankhah** is Senior Lecturer at the School of Aviation, Marketing, and Tourism, specializing in management and marketing. She also serves as the Research Center Leader for the "Tourism, International Business, and Marketing (TIBAM)". In her academic journey, Sanaz has published numerous research papers in esteemed journals and regularly reviews submissions for international journals. Her research interests span business models, strategic management and innovation, sustainable development, Corporate Social Responsibility (CSR) and the critical areas of Equality, Diversity and Inclusion (EDI).

**Leszek Wypych** is Senior Lecturer in Business Communication at the University of Bedfordshire. His research interests include acculturation across international teams with a focus on cross-cultural management.

# Introduction

## Entrepreneurial Innovation in the International Business of Tourism

*María Jesús Jerez-Jerez and Pantea Foroudi*

The scholarly interrogation and critical application of pivotal tourism concepts have underscored an emergent need in both academic discourse and practical realms of management. A discerning assessment of the trajectory of this academic landscape attributes this burgeoning interest to rigorous, paradigm-shifting research contributions emanating from a few elite business schools over the past decades. Seminal researches, some tracing back to the previous millennium and published in journals of repute like the *Journal of Tourism Management*, not only deconstructed essential tourism sectors but also paved the way for a paradigm shift in the industry's structure, thus instigating an intricate research agenda. This agenda encapsulates a panoptic view of tourism business theory, juxtaposed with intricate facets of entrepreneurship, evolving trends, imminent challenges, and novel formats.

Historically, the epistemological roots of this discipline, while robust, were fraught with ambiguities. Antecedent studies manifested a palpable schism in the perspectives of practitioners regarding the bedrock components of tourism theory. This, however, is not to undermine the strides made by subsequent research endeavors that strived for a more unified conceptual framework. Contemporary scholarship has transitioned from merely cataloging global tourism trends to a more nuanced exploration of macroeconomic shifts, geopolitical influences, and the intricate interplay of supply and demand dynamics in the tourism industry. The driving antecedents of propitious entrepreneurial innovation in this sector are discerned to be multifaceted, encapsulating political, technological, environmental, and sociological dimensions. Adherence to legal, regulatory, ethical, and socio-cultural paradigms is no longer a mere compliance requisite but a critical fulcrum upon which entrepreneurial ventures pivot. Emerging research narratives not only amplify our comprehension of the intricate challenges tethered to the business of tourism but also illuminate the way forward.

Despite the voluminous research corpus available, the inherent dynamism and complexity of this discipline perpetually pose novel research conundrums. These range from the nuanced impacts of antecedents—like technological advancements, political oscillations, socio-cultural shifts, and environmental imperatives—to their far-reaching ramifications, encompassing customer loyalty, trust dynamics, and commitment, spanning myriad contexts. This tome endeavors to weave together these multifarious strands, endeavoring to discern not only what consumers

DOI: 10.4324/9781003369967-1

seek but the modalities of their engagement with organizations. With the economic milieu progressively becoming customer-centric and corporate identity being inextricably linked to value co-creation, it's imperative to understand how organizational facets—be it corporate, product-oriented, or service-centric—are being continually redefined through the prism of customer experiences and narratives.

This book positions itself as a crucible for introspection, analyzing both current trajectories and future inflections in the entrepreneurial tourism domain. Submissions that synergize both external (corporate) and internal (organizational) lenses, especially those that provide a harmonious blend of robust theoretical scaffolding with pragmatic insights, are particularly coveted. By foregrounding the inherent complexity and multifaceted nature of corporate disciplines, this book aspires to bridge a conspicuous lacuna in the current academic milieu, proffering fresh research perspectives on the very essence of tourism business paradigms, ranging from strategic management contours, industry structure dynamics, entrepreneurial analyses, to human capital development.

The book is organized into four distinct sections, encompassing a total of 14 chapters. Section 1—Strategies and Insights for Emerging Markets begins with Chapter 1 that, crafted by Ekwugha, Oham, Ojiagu, and Amadi, shifts the focus to South-East Nigeria, emphasizing the novel avenues through which social enterprises can harness tourism to uplift the local communities, especially amid recent global challenges. The conversation is continued by Augusta Evans in Chapter 2, further shedding light on the immense potential that social enterprises present in this context. In Chapter 3, Isabel Dolores Jimenez Jimenez addresses the ramifications of the United Kingdom's exit from the European Union on the nation's travel businesses.

Section 2—Sustainable Practices and Environmental Considerations features Chapter 4 by Sevilla-Sevilla, Rodriguez-Oromendia, and Navio-Marco, who underscore the pivotal role of environmental sustainability in the tourism sector. Chapter 5 by Pérez Zabaleta and Méndez Pérez then elucidates the profound influence of water resource management on the success of tourism destinations. Chapter 6 sees Saira Sultana assessing the evolution and impact of smart tourism.

Section 3—Innovations, Technologies, and Branding in Tourism introduces Chapter 7 by Carmen Arroyo, which illuminates the transformative potential of immersive technologies on the tourism experience. Chapter 8, a collaboration between Zaim, Qastharin, and Purwa, underscores the significance of authenticity in destination branding, while Chapter 9 by Ioana Stoica probes into the entrepreneurial influences of residents on place brands. In Chapter 10, Pavlos Arvanitis offers insights into the digital advancements in 'smart' travel.

Section 4—Management, Policy, and Research Insights in Tourism initiates with Chapter 11 by Wypych and Ahmad, examining the nuances of the multicultural tourist experience. Chapter 12 by María Jesús Jerez-Jerez delves into the challenges and strategies for asset protection in tourism amid socio-political upheavals. Chapter 13, a collaborative effort between Wypych, Ahmad, and Sastry, illuminates the cultural dynamics in mergers and acquisitions within the tourism industry. Finally, Chapter 14, crafted by Vatankhah, Bamshad, and Tallia, presents a pioneering bibliometric analysis on entrepreneurship in tourism and hospitality research.

# Section 1

# Strategies and Insights for Emerging Markets

# 1 Exploring the Tourism Potential and Innovative Contributions of Social Enterprises in South-East Nigeria

*Maurice Ekwugha, Charles Oham, Nkechi Ojiagu and Robert Amadi*

## Introduction

This chapter explores the innovative ways social enterprises can contribute to uplifting the people in South-East Nigeria through tourism. It will consider how the tourism industry can help in mitigating the economic impact of the recent global economic downturn and the social impact UK BAME-led organizations can make in this regard. It will also look at some of the distinctive needs these organizations have and their limitations such as the lack of formal structures, governmental support, economies of scale, etc. Untapped potential is investigated and innovative ways of moving forward will be addressed.

## Background

The South-eastern part of Nigeria is home to over 50 million people with tens of millions dispersed globally through slavery and emigration exacerbated by poverty and a corrupt ruling class. The study is relevant to the economic development of this area and its untapped tourism potential. The region possesses several sites that can potentially be designated as UNESCO world heritage sites such as the Ogbunike Cave, Obubra Salt Lake, Enugu Coal Mines, Arochukwu Slave Routes, etc. Catalysing the development and functioning of several rivers and seaports will boost business activities and increase the export potential of agricultural products, goods and services, while developing its tourism industry.

In South-East Nigeria, the population exhibits aspects of social entrepreneurship possibly quite different from how it is defined in the UK and the Western world in general. The social groups are generous in many ways, using their resources and investments to support widows and extended family members. In one example, Catherine Oham's entrepreneurialism came to the fore when she employed her home economics training gained in England, to run her cake and pastry business by which she supported her family during periods of austerity in 1980s Nigeria (Beugre, 2017; Burns, 2011). However, despite the people's industrious nature, it is impossible to ignore the state of poverty and deprivation in an area considered an extremely resource-rich region.

DOI: 10.4324/9781003369967-3

Travelling along poorly maintained bumpy roads from the oil-rich city of Port Harcourt in the south to the towns of Owerri and Orlu about 100 kilometres to the north, the stark reality of a lack of infrastructure, basic goods and services, and human suffering become increasingly evident. The world-wide financial and leadership turmoil experienced in the last few years has exposed some of the weaknesses of the developmental strategies and structures as they currently stand and further emphasized the need for alternative systems and ways of doing things (Borzaga et al., 2019). Traditional models of economic growth need to be significantly corrected and the hope of achieving a level playing field in these contemporary times requires the intervention of social economy as a developmental tool for galvanizing locals within various communities and their resources (Egorov and Inshakov, 2021). Hence the need to examine the role social entrepreneurs can play in effecting positive change in governance, politics, society, and the economy.

Furthermore, the question of whether thought leaders in the region can upgrade and create systemic change through their world-renowned stakeholder capitalism system known as the Igbo Apprenticeship System (IAS) (Ekekwe, 2021; Osiri, 2020), is a pertinent one. This model system of grooming young professionals is unique in that it does away with the need for any significant capital inlays or equity requirements, but rather is based on filial and relational platforms through a network of apprenticeship social enterprises. It underlines popular thought in the region that the strength and measure of a man is based on how he has made those who work under him better with skills acquisition (Ogoko, 2022).

## Tourism Development and Related Theories

Tourism can be defined as a system that permits the flexible movement and brief relocation of individuals or groups outside of their normal abode for more than one night, except for journeys dedicated primarily to the purpose of earning an income along the way (Leiper, 1979). Another definition of tourism relates it to the legitimate actions of visitors, the attractions and relationships that emanate from these actions, and the facilities developed to cater to them (Cunha, 2012). Netto (2009) concurs stating that, for tourism to exist; the tourist must be present and must of necessity, interact with the host who may be a service supplier or an indigene. Within the context of South-East Nigeria, many people travel to visit relatives or pay respects to their dearly departed but wouldn't settle, rather choosing to remain with their emigrated families on foreign shores. Such people may be classified as tourists since they mostly have foreign passports.

Tourism is worth US$7 trillion to the global economy and supports 292 million people in the world (World Bank, 2018), however, the COVID pandemic led to a 70% decrease in global tourism in 2020. In fact, Africa alone lost almost $55 billion in tourist revenue in April and June 2020 (Reuters, 2020). According to the UN, global tourism may have suffered losses of $0.9–$1.2 trillion (UNWTO, 2020) within that period. Despite this, it is expected that global tourism will bounce back, in part due to new tourism streams such as space tourism and virtual tourism, as well as well-established aspects like cruise tourism and others such as leisure-oriented tourism and pro-poor tourism (ODI, 2023). Leisure-oriented tourists often travel to sea-side resorts, and in the tropics such coastal resorts are popular in both the summer and winter months. The aim of pro-poor tourism is to cater to the indigent populations of mainly less developed nations who seek to attract significant numbers of visitors and make some profit (Ashley and Poultney, 2005). This would be the case in South-East Nigeria if the tourism potential was developed to any degree. A downside is that less than 25% of the revenue gets to the poor (ODI, 2023).

There are many theories for the tourism industry. They include descriptive models which depict tourism systems; explanatory models which show the functioning of systems and subsystems; and predictive models based on causal relationships (Thirumoorthi and Wong, 2015). Development theories underpin this chapter because tourism will boost economic development (Sharpley, 2022) and have informed the premise that capital investment boosts growth.

The World Bank uses the theory of change model to evaluate tourism under, long-term outcomes and assumptions; intermediate outcomes, design intervention patterns, and development indicators (World Bank Group, 2018) (see Figure 1.1). This involves identifying workable long-term outcomes, outlining the conditions, determining the key success indicators required to reach the predicted goals from the onset, and then working backwards to the required interventions and challenges that need to be overcome (Brest, 2010).

Theory of Change is a set of theories used in various projects and aspects of human endeavour to bring about a positive shift in community development. It identifies the main actors as a sequence of stages in the process of specifying theoretical reasons for the changes (Claus and Belcher, 2020). It outlines causal linkages which track the change process backwards from long-term goals to intermediate, and then short-term outcomes linking each pathway in logical relationships with the preceding ones, with reasons why each outcome is a logical step to the next one (Clark and Taplin, 2012). Theories of Change require stakeholders to determine their final outcomes before designing the necessary interventions, and the use of success indicators to track progress.

The beginnings of Theory of Change were conceived through descriptions of Management by Objectives in the book, *The Practice of Management* by Peter

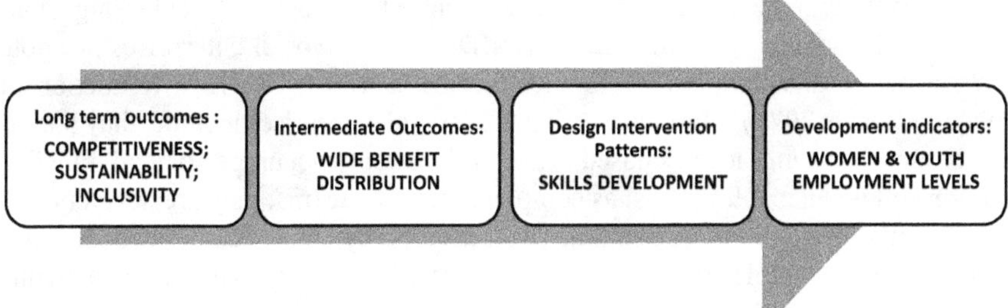


*Figure 1.1* Theory of change – a tourism perspective.

Drucker, from which programme theory evolved in the 1990s (Weiss, 2011). The Theory of Change concept is often given to varied definitions which some find confusing (Belcher and Palenberg, 2018), and this is evidenced by the many ways terminology relating to the subject is applied (Belcher and Halliwell, 2021).

Among other Theory of Change frameworks, Harries et al. (2014) lists the Logic Model as one method of using the Theory of Change to deliver a project. However, Clark and Anderson (2004) differentiate between the two, explaining that though the terms are often used interchangeably and can blend into each other, Theory of Change concentrates on the reasons for the desired change and how it can be achieved through linking specific outcomes and activities, while logic models emphasize the illustration of critical aspects of a programme in a graphic format to keep managers focused on the end results, the necessary actions to be undertaken, and the available resources, in that order. Also, while logic frameworks are able to clearly define the inputs and outcomes of a project, they do not give much traction to the more intricate processes such as the societal, fiscal, leadership, and systemic issues that often drive communal change (Funnel and Rogers, 2011).

One important feature of the logic model, which may also feature in some other Theory of Change concepts, is the 'accountability line' which lies between the intermediate outcomes achieved directly and longer-term goals to which these outcomes may contribute (Harries et al., 2014). In other words, the accountability line enables us to limit outcomes to what is feasible while still being cognizant of the long-term goals not immediately attainable, but which could be achieved in the future (see Figure 1.4).

## Social Enterprise in South-East Nigeria

In the past, the Southeast mainly depended on palm oil as the main stay of the economy. But with the discovery of crude oil, the production and export of palm produce was no longer a priority. However, crude oil has been unable to deliver the masses out of poverty due to mismanagement, government profligacy, and other related factors, thus the need to search for alternative revenue earners.

Organizations like social enterprises are often established with the aim of addressing social problems, and profits realized do not go to individuals but are used to address community needs such as education, poverty, healthcare, etc. They are value-driven organizations with a human-centred approach in addressing challenges in the society (AliAmri, 2019; UNIDO, 2017). Social enterprises promote several of the Sustainable Development Goals such as SDG 10 – Reduction of Inequity (Dave, 2021). The characteristics of social enterprises include, they are often ground-breaking, locally based organizations that encourage societal transformation and inclusion; they exist as complementary concerns; they are development instruments for mobilizing community resources; they create economic opportunities for the vulnerable population; and they work towards goals that benefit the public or community.

Social enterprises in South-East Nigeria mainly involve cooperative societies and community-based organizations (CBOs). Cooperative societies are active and functional organizations that bring a paradigm shift encouraging people to improve their lives by working in teams to maximize services and empower their members.

Cooperatives, a type of social enterprise (Oham and Macdonald, 2016), have seven main principles alongside their core values and ethics that emphasize the importance of people over profit, which makes them one of the understated instruments in the race to achieve the Sustainable Development Goals. They are founded and operated by members, and this gives them the flexibility to deal with localized issues and provide vital remedies in remote locations while also promoting relevant practices which also protect the environment (Cooperatives Europe, 2015; Develtere and Papoutsi, 2021). Some important characteristics of cooperatives are that they are democratic in nature, can be of any size, and are not restricted to any business (Egorov and Inshakov, 2021).

Cooperative organizations usually perform best in a climate of societal reforms where community development programmes are being engineered to deal with issues such as the lack of basic amenities, inclusivity, gender bias, and financial viability. Furthermore, they meet a diverse mix of expectations beyond the usual goals of profit making and ensuring good returns to their shareholders. The ability of cooperatives to reach beyond the scope of group cohesion characterized by their insistence on open and voluntary member involvement, mutual interaction with other cooperatives, and a community development ethos, gives them a coherent social outlook (Dave, 2021 in Ojiagu and Usman, 2023). Cooperative organizations can generate a triple effect as agents of social, economic, and environmental change all in one.

Community-Based Organizations (CBOs) are mainly in the form of women associations, youth organizations, age-grade associations, town unions, social clubs, and church-run charities. Women Organizations are associations of women who come together for their interest and that of their community. This is an exceptional community-founded methodology aimed at rural renovation through group building, self-management, monetary intermediation, and acting as vanguards of social change. The involvement of youth organizations in social enterprises is often underestimated. Youth groups erect health centres, construct town halls for meetings and skills centres, and participate in road construction projects, and a lot more. Age Grade associations are often remodelled to suit current realities. They operate welfare schemes, grant loans to their members for start-up businesses, and provide basic community needs such as electricity, pipe-borne water, housing, etc. (Nwankwo, 2021). Town Unions are a valuable instrument for accomplishing self-reliance and rural community renovation. Social Clubs, apart from social embellishments, engage in community water projects, supplying medicines to village health posts, and donating books to community schools. Church-run charities and churches have been at the forefront of social change and economic development in South-East Nigeria. They have built up a track record of building and running schools, hospitals, motherless babies' homes, and factories that address social issues including rural indigence, gender discrimination, and community advancement (UNIDO, 2017). Comparatively, in the UK formal social enterprises have played a strategic role in shaping government policy towards social value and enabling a thriving and resilient social enterprise ecosystem. However, in South-East Nigeria similar bodies may not exist in the same way.

## Tourism and UK BAME-LED Social Enterprises

When viewed from a Black and Minority Ethnicity (BAME) standpoint, ethnicity and race are not necessarily the same. Race is a social construct primarily based on natural characteristics or qualities that are similar in appearance while ethnicity is more encompassing and includes issues relating to ancestry, religious inclinations, language, or country of origin (Law Society, 2023). Mostly viewed as a better way to categorize individuals, ethnicity is frequently preferred when describing people because of the negative undertones race often evokes.

There is a significant BAME population in the United Kingdom. They make up 14% of the UK population, and at 3%, Black Britons are the largest of the ethnic minorities (O'Neill, 2023). Of these, 1.3 million people are black African, and they make up 1.8% of the combined English and Welsh population (BBC, 2020). The largest African cultural group by country of origin is Nigeria with 271,000 people (ONS Census data, 2022). That number also includes people of a South-East Nigerian heritage.

There are a reasonable number of BAME-led social enterprises in the UK and they seem to be growing at a faster rate than ordinary businesses and are more impactful (NHS Health Careers, 2015).

BAME-led organizations are mainly funded through government grants. An average of 81% of BAME social entrepreneur respondents identified government grants as their main source of funding (Harries and Miller, 2021). A 2021 study found that grants received by UK BAME social enterprises are comparatively the same as others (Sepulveda and Rabbevag, 2021). However, the investment amounts BAME-led organizations receive are often less, perhaps because they are thought to be smaller and newer. This highlights the need to further understand such funding discrepancies, and whether this is because of their smaller size as opposed to unacceptable bias (Sepulveda and Rabbevag, 2021).

## Creating Value through Social Innovation

Social tourism encompasses a range of initiatives aimed at promoting the access of disadvantaged individuals or groups to leisure and tourism activities (Lee and Baek, 2021). Innovation means generating new knowledge and ideas and making use of them. To do this, it is vital to know what already exists and understand how they were applied in the past. Social Innovation can be defined as addressing challenges by creating social value from novel concepts, new ways of doing things, or applying creative modifications to things that already exist (Young, 2006).

The role of social entrepreneurship involves transacting business as a means of tackling societal issues, embarking on community development, improving the quality of people's lives, and uplifting the environment (NHS Health Careers,

2015). Social entrepreneurship has two missions: the business mission i.e., profit-making; and the social mission i.e., uplifting the society. Social innovation exists within a social enterprise and takes place within a complex social eco-system (Bloom and Dees, 2007), focusing on different but interconnected issues (Oham and Macdonald, 2016).

Social Innovation can be evaluated and implemented using Action Research (AR). It informs community change by observing and reflecting on repeated cycles of actions and the resultant effects (Ekwugha, 2022). Action Research is a methodology for intentionally devising necessary changes and implementing those alterations (Pracht et al., 2022) (Figure 1.2).

In AR, change is successful when individuals reflect and gain insights into their whole situation (Ekwugha, 2022). Furthermore, reflection draws conclusions from specific situations and critically investigates the deeper factors involved in that situation (Table 1.1).

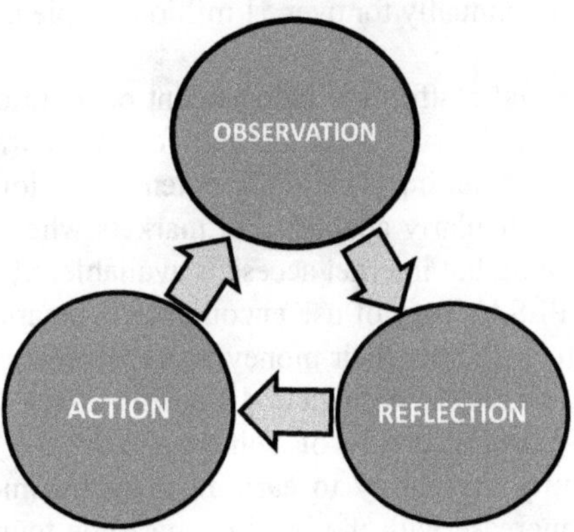


*Figure 1.2* Action Research (AR) cycle.

*Table 1.1* Action Research Implementation

| *Phases* | *Action* |
|---|---|
| Action | Situation<br>Objective<br>Planned improvement<br>Action |
| Observation | Semi-structured interviews<br>Questionnaires<br>Observation, notes<br>Recordings, evaluation of options |
| Reflection | Meetings with stakeholders<br>Determining available resources<br>Determining timescale<br>Monitor progress |

AR can be used to introduce social innovation to develop the potential of the South-East economy and support human capital development. In 2022, about 735 million or 9.2% of the world faced chronic hunger, while 2.4 billion people do not have adequate food (United Nations, 2023). One way poverty in rural communities can be reduced is by the less developed nations embracing digital technologies and having improved access to connectivity (Reiter, 2023). In 2022 there were 5.3 billion people online globally and 2.7 billion offline including 60% of Africans (Timuray, 2023). To overcome the challenges confronting the indigenous population particularly in financial matters, a viable solution could be the adoption of a financial technology such as M-PESA (Vodafone, 2023).

M-PESA, first implemented in Kenya in 2007, literally means mobile cash. 'Pesa' means cash in Swahili. It is Africa's largest and most advanced financial technology platform with two and a half times more transactions than American Express. It offers safety, security, and convenience, and handles over $364 billion in financial transactions annually for over 51 million people in seven African countries (Vodafone, 2023).

The M-PESA financial platform is independent of traditional financial institutions and governments and serves as an alternative to the current payment system, enabling transactions without intervention or potential exploitation from these entities. This makes it particularly adaptable in markets where traditional financial infrastructures are lacking but internet access is available. Much like the appeal of electronic credit, M-PESA's ease of use encourages a situation where individuals rarely feel the need to withdraw their money, thereby reducing the importance of physical cash (Vodafone, 2023; William and Tavneet, 2011).

The commission-based structure of being an M-PESA agent offers people in the local community the ability to earn an extra income from transactions. This income would increase with the level of maritime tourism giving the local community an incentive to further develop the area (William and Tavneet, 2011). Moreover, M-PESA also gives local merchants access to a larger market, making trade easier and attracting more customers who might be unable to use conventional means of transacting business or withdrawing cash (Vodafone, 2023) (Figure 1.3).

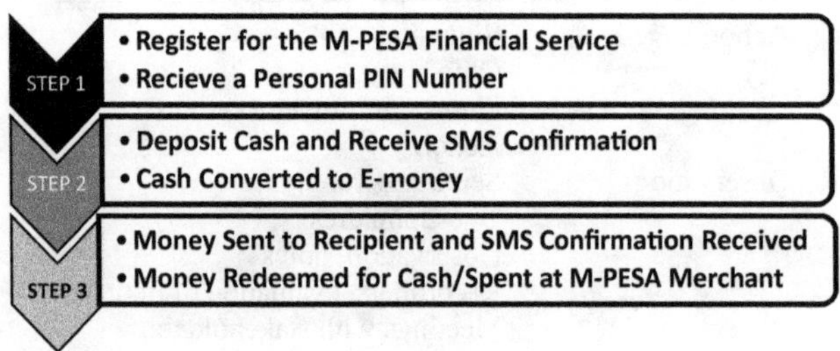


*Figure 1.3* M-PESA transaction in simple steps.

## Solutions and Recommendations

The logic model is made up of long-term goals, intermediate outcomes, enablers, activities, and inputs (see Figure 1.4). Here, we adapt the logic model as a conceptual framework for developing maritime tourism in South-Eastern Nigeria. The long-term goal involves developing a maritime tourism hub to benefit the rural population through the establishment of viable seaports in the region. The intermediate outcome involves attracting BAME-led UK social enterprises to buy into the South-East Nigerian tourism industry. This will be enabled by a finance-friendly environment where the safety and security of transactions is

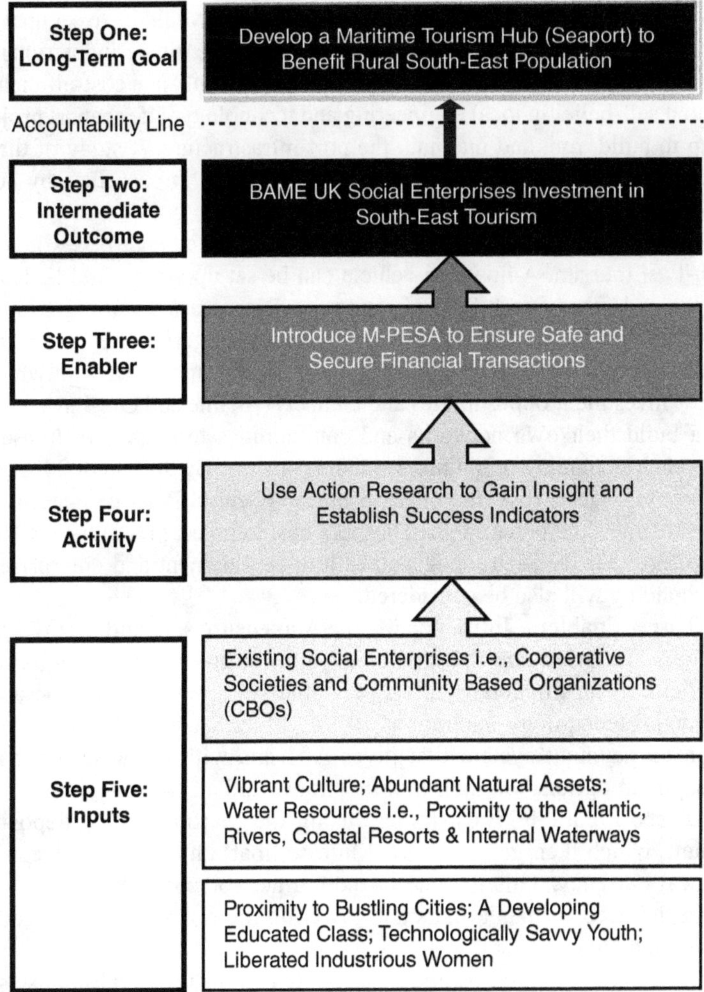


*Figure 1.4* Conceptual framework for developing maritime tourism in South-East Nigeria.

guaranteed using a digital platform tried and tested in the African market such as M-PESA. Action research tools will be used to conduct real-time investigations on ground to gain insight and establish success indicators. And the inputs are the various advantages replete in the area including the existing social enterprises, i.e., cooperative societies and community-based organizations, its vibrant culture, the abundant natural assets, and the numerous rivers and internal waterways that constitute its abundant water resources in addition to the region's proximity to the Atlantic Ocean. The inputs will also include proximity to bustling cities, a developing educated class, technologically savvy youth, and liberated industrious women.

Step One: Long Term Goal – Develop a maritime tourism hub (seaport) to benefit the rural South-East population. The long-term goal would be to set up a locally built and privately funded seaport that would utilize the innate industrious nature and innovativeness of the indigenous people. This will enhance cost-efficiency and bring about self-belief in local engineering and technology. The seaport project can be set up to build, run, and maintain the port infrastructure. A study of the entire spectrum of the maritime value chain will create social impact, thereby boosting youth employment and business start-ups in the region.

Step Two: Intermediate Outcomes – BAME UK social enterprises investment in South-East tourism. A financial vehicle can be set up where BAME-led social enterprises and other South-East Nigerians in Diaspora are willing to invest in this project. The Diasporas contribute large sums to Nigeria. They are more willing to invest than others because of social and kinship networks with which they can access investment opportunities and contacts (Akanle and Ola-Lawson, 2021). Diaspora build their own networks and communities to engage in transactional activities such as sending remittances and investments. By the end of 2022, foreign remittances to Nigeria rose to almost $5 billion (Central Bank of Nigeria, 2023). This demonstrates the potential the Diaspora has, to invest in South-East Nigerian social projects. Funds for training youths into employment and enterprise in the marine economy will also be considered.

Step Three: Enabler – Introduce M-PESA to ensure safe and secure financial transactions. To overcome the challenges confronting the indigenous population particularly in financial matters, a viable solution could be the adoption of a financial technology such as M-PESA. M-PESA is a cheap, easy-to-use phone-based payment system that gives Africans with no access to banking, online payment options via their mobile phones and a 2G network in three simple steps (see Figure 1.3). Customers using the platform make deposits into an account, which then gets converted into 'e-float' units which are equivalent to the local currency. This is done through a network of agents via SMS thus removing the need for bank visits and lowering the risk of robbery (Vodafone, 2023; Yeoman, 2014).

Step Four: Activity – Use Action Research to gain insight and establish success indicators. AR observes situations with a view to evaluating, measuring, and reflecting on the performed activities and their outcomes. This involves having a deep insight into the needed transformation and understanding that progress depends on a proper situational analysis, accurately pin-pointing all possible solutions, and determining and utilizing the best solution for that situation (Bennett, 2004). AR can be a useful instrument in the recruitment process and in the evaluation of different projects and reporting progress to ease the workload, delegate responsibilities and create ownership.

Step Five: Inputs – Make use of existing social enterprises, the vibrant culture of South-Easterners, the abundant natural assets and water resources, proximity to bustling cities, a developing educated class, technologically savvy youth, and industrious women. South-East Nigeria is blessed with many rivers like the Niger River that discharges into the Atlantic Ocean, and others such as the Imo River, Cross River, Anambra River, Ebonyi River and so on (Enyidi, 2017). The region also boasts many thriving commercial cities. The indigenes are also renowned for their commerce and international trading activities across Africa and the world. However, the absence of a viable seaport to drive tourism and facilitate economic development is a major setback.

Aspects of the conceptual framework including the involvement of BAME-led UK social enterprises, the M-PESA digital platform, the Action Research tool, and the various inputs, have been discussed or highlighted earlier. The long-term goal of establishing a sustainable maritime industry spearheaded by building viable seaports is now addressed.

### *Developing Maritime Tourism in South-East Nigeria*

South-East Nigeria, West Africa is blessed with several major rivers that empty themselves into the Atlantic Ocean. Compared to other ethnic peoples in Africa, they possess arguably the highest number of commercial cities for its small landmass including Aba, Nnewi, Onitsha, and Port Harcourt. However, they lack considerable ports that would facilitate economic development and the growth of tourism in the region.

Establishing seaports in this region must also be considered from an entrepreneurship perspective as Schumpeter (1954) describes entrepreneurs as lynchpins of economic development. Entrepreneurship captures the development of the old into the new as businesses and products evolve and emerge through the intricacies of the change process (Anderson, 2015). Such a project can be a major catalyst to the industrialization of the region if links are established with other sectors of the economy e.g., agriculture, manufacturing, ship maintenance, and decommissioning, etc. Anderson further points out that change is challenging for others but an opportunity for entrepreneurs as it becomes a Schumpeterian opportunity for capturing new markets with products, services, and processes. This is in line with Opportunity Recognition theory where Kirzner (1973) and Stevenson and Jarrillo (1990) note that market intelligence through the appreciation of societal needs and market failures can be exploited by an entrepreneur with the development of strategies to exploit such opportunities. An established seaport will be in a privileged position to do this and can attract the necessary material and moral support. Reynolds et al. (2002) found that exploiting opportunities and reallocating resources to meet the new business objectives underscores what entrepreneurs do. Any entrepreneurs embarking on a seaport project must consider how to reallocate resources, e.g., human capital, rivers, lakes, waterways, land assets, environment, etc. Capital or material limitation fears must be jettisoned as de Lange (2016) found that entrepreneurs facing investment constraints who partnered with similar entrepreneurs formed successful alliances creating a dynamic cluster of investible firms leading to de-risking. Therefore, apart from opportunity recognition and creating change, entrepreneurs have a professional responsibility to stimulate investor confidence through a process of de-risking. This must be a critical objective of any investor in the seaport project. There is scope to work with local and international partners and global institutions such as the Alaigbo Development Foundation led by Professor Uzodinma Nwala.

The social enterprise landscape in South-East Nigeria is rooted in rural ideology. The enterprises often operate without any form of government recognition, support, or legal structure, and most times their identities are blurred even as they serve as agents of rural development. Such businesses need reforming, reengineering, and repositioning as seen in other climes.

Those involved in social entrepreneurship with connections to the Global South through bonding social capital need to get involved in social enterprise development and capacity building. This will build capacity and increase social impact in places because social enterprises are a force for good (Roy et al., 2014; Royce, 2007; Steiner and Teasdale, 2017; World Economic Forum, 2021). Those without ties to the developing world should become allies by nudging and supporting their colleagues to invest their time and resources. Training, mentoring, and providing resources for just one person in the Global South can significantly impact a community and add value to the UN Sustainable Development Goals (Wongtschowski, 2015).

To develop maritime tourism in South-East Nigeria, a few suggestions are hereby made. To consider the tourism and culture potential in the value chain of the maritime industry, ports previously used in colonial times for the export of palm produce, which was South-East Nigeria's main export and the engine of its industrial revolution in the past, should be identified. Also, slavery, which is a crime against humanity, had an impact on the region. For example, the Arochukwu, Onitsha and Oguta ports may have been used as major ports for exporting slaves in those days. Locating the slave port through which Olaudah Equiano (c.1745–1797) was transported will bring much visibility to the region. To boost tourism, applications for UNESCO World Heritage status for such ports can be made.

The proposed seaport project could adopt Ndibe sand beach, which has not been given government approval, as a potential seaport site. Cross River would be a great seaport site possessing a very natural estuary and inland seaport. In order to reduce the reliance on government and to avoid partisanship and poor management, an option could be the establishment of the first cooperative seaport in Africa, owned and managed by the people. This would be of immense importance since the people have a global outlook and need a seaport they can manage on their own.

A mapping exercise of universities and colleges should be conducted to ensure the provision of a trained workforce for the envisioned maritime industry, and to develop action plans that will mitigate any gaps in training while empowering the people. For example, experienced people from within the maritime economy can be recruited to train and develop skippers, captains, deck crews, ports managers, etc.

Resources should be developed to teach local maritime history to the youth for future generations to know that the South-East had engaged with the outside world several hundred years ago. The development of a maritime museum should be considered.

Exploring the feasibility of a shipbuilding and maintenance industry in the South-East will also involve the development of blacksmithing and iron craft industries.

De-risking is an entrepreneurial attribute among a range of attributes critical to fulfilling business objectives. We must de-risk the Nigerians of South-Eastern origin in Diaspora by investing in seaports and the required infrastructures.

## Conclusion

The chapter explored the contribution of social enterprises to the tourism potential in South-East Nigeria and how developing its latent maritime tourism potential in the long-term by embarking on a viable seaport project, will fulfil the objective of reversing the woeful economic fortunes of the region. Using a conceptual framework derived from the Logic Model, it suggested that the involvement of BAME-led social enterprises as an investment vehicle supported by the M-PESA financial platform in harnessing the many natural assets in South-East Nigeria including its vast water resources and other advantages, can uplift the social condition of the local people and bring about change. It investigated untapped potential and suggested innovative ways of moving forward by combining the use of Action Research with Theory of Change principles to measure change and ensure success. It concludes that broader strategic plans and policies are needed to grow and develop social enterprises in South-East Nigeria (Oham and Okeke, 2022) and that expected change can happen when social entrepreneurial educators arise in tackling governmental, market, and societal failure (Baggot, 2013; Beugre, 2017; Oham and Macdonald, 2016).

**Case Study: Clays Advice and Training Centre (CLATC)**

***Introduction***

Set up in the UK in early 2000 by a group of Christians who were planting a church, Clays Advice and Training Centre (CLATC) was situated near a large housing and cooperative estate. The social enterprise is volunteer-led and does not have any paid staff due to its meagre income. Nevertheless, the impact of its projects in Africa is very high as it utilizes different types of capital to achieve its objectives – social, economic, cultural, symbolic, environmental, spiritual, and human capital (Oham and Macdonald, 2016). It demonstrates the role of UK BAME-led social enterprises in creating social value in the global south. CLATC have run or supported projects in the UK, Mombasa Kenya, Accra Ghana, and Orlu Nigeria.

In this case study, we discuss a current project run by CLATC called 'Meputa Oruaka', a phrase for handiwork in Igbo the local parlance, in Orlu. CLATC currently runs its projects from an uncompleted building site which it has been developing. It plans to complete this project and set up another project in a river and fishing community in the South-East.

By video recordings sent through WhatsApp, the project is inspiring people in the UK to volunteer and contribute to the project's objectives. The videos also serve as virtual tourism, inspiring people on community development in a local African village. There are plans to turn the videos and pictures into a short film or documentary on community development.

Monitoring and Evaluation: CLATC uses Action Research to review its projects and develop future response to need. Discussions and reflections on problems and solutions are discussed on a regular basis at its meetings with members. This gives the organization the potential to continue with the project.

***Background of the Project***

The Meputa Oruaka (Handiwork) concept was conceived in 2016 and piloted in 2022 at Umuagbada, a village in the Orlu Local Government Area of Imo state, Nigeria. Umuagbada became the pilot site for this project for five main reasons. First, there was a high level of poverty within the community due to lack of jobs and skills, and a decline in interest for primary and secondary education; second, there was the sense of general moral decay among the old and young; third, the activities of the Nigerian army against the youths perceived as terrorists within the area led them to resort to idolatry i.e., the seeking of protection from otherworldly spirits; fourth, the absence of any advisory and rehabilitation centre for trauma healing, skill identification and acquisition, and purposeful leadership in the area was glaring; And finally, because of Diaspora connections to the community. The vision and mission statements are as below:

*Vision and Mission Statement*

Vision: To commit to massive community development and poverty reduction and have the gospel of Christ preached in all the host communities.

Mission: To identify people with skills that can be developed/enhanced into an economic advantage; to enable people acquire skills for their economic advancement; to promote the Asset-Based Community Development (ABCD) model to help reduce poverty; to organize the indigenes into a co-operative society for proper administration and promotion of mutual benefit in micro credit facility distribution; and, to use the tools mentioned above as a means to share the love of Jesus Christ.

***Ongoing Projects and Work Progress***

Five seminars were held within the host and at the project centre at various times (see Table 1.2).

Various workshops were also held where training was given on catering and baking, alternative energy (solar energy installation), and Christian discipleship (see Table 1.3).

A cooperative society was set up to organize the locals into an organization for proper administration and promotion of mutual benefits in a micro credit facilities distribution centre (see Table 1.4).

Seed business loans were disbursed as interest free micro-credit facilities given to the members of the cooperative to enable micro-production in each area of strength for economic growth (see Tables 1.5–1.7).

*Table 1.2* Seminars

| *Seminar* | *Topic/Text* | *Attendance* |
|---|---|---|
| Seminar 1 | Necessities of spiritual investment | 15 |
| Seminar 2 | Seed business | 12 |
| Seminar 3 | The joy of starting small | 15 |
| Seminar 4 | Staring and planning your business | 10 |
| Seminar 5 | Assets and community development | 10 |

*Table 1.3* Workshops

CAKE CRAFT WORKSHOP:
FONDANT WORKSHOP
ALTERNATIVE ENERGY WORKSHOP – Solar Energy Installation

*Cake Craft Training*
Students were exposed and trained on:

1 English cake craft ranging from wedding cakes, birthday cakes and party cake.
2 Traditional cake craft majoring on traditional marriage in Igbo culture with special attention to; kola nut, garden egg, traditional beads, and bangles and
3 Palm wine.

(*Continued*)

*Table 1.3* (Continued)

Total number trained were 8
Total contact made were 4

*Catering and Baking*

Attention was given to the following: fish roll, bread, pepper roll, meat pie, burns, egg roll, dough nuts, cakes, puff puff, chin chin.
Total number persons trained were 10
Total contacts made were 5

*Training on Solar Energy and Installations*

A free training on solar energy and free installations on five households was completed. The theoretical training completed at the centre and trained persons were used to carry out the installations as a proof of full understanding of what was taught.
Total number of students: 10
Instructors: 2

*Discipleship Class*

A special discipleship class commenced every Mondays, so far only one class has been held in June and 14 persons attended.

*Table 1.4* Co-operative Society

| **Co-operative Society** |
|---|
| Members formed and registered UMUAGBADA YOUTHS FOR CHRIST MULTI PURPOSE COOPERATIVE SOCIETY LIMITED under the Nigeria co-operative society Act 2004 with REGISTRATION number: IMC 0877, SRC:00986DC. |
| The total registered members: 10 |

*Table 1.5* Seed Business Loan (First Disbursement)

| *S/no.* | *Name* | *Amount (N)* | *Return (N)* |
|---|---|---|---|
| 1 | Beneficiary 1 | 300,000 | 300,000 |
| 2 | Beneficiary 2 | 100,000 | 100,000 |
| 3 | Beneficiary 3 | 50,000 | 50,000 |
| 4 | Beneficiary 4 | 280,000 | 280,000 |
| | **Total** | **730,000** | **730,000** |

*Second Disbursement*

| *S/no.* | *Name* | *Amount (N)* | *Return (N)* |
|---|---|---|---|
| 1 | Beneficiary 5 | 170,000 | NIL |
| 2 | Beneficiary 6 | 280,000 | NIL |
| | **Total** | **450,000** | |

Total amount disbursed – N730,000
Total amount returned – N280,000
Outstanding – N450,000

*Table 1.6* Seed Business Loan (Second Disbursement)

| *Date* | *Description* | *Debit* | *Credit* | *Balance* |
|---|---|---|---|---|
| | Balance from last account | | 32,702 | |
| | Poultry loan repayment by at various installments | | 200,00 | |
| | Loan to for car repair and uber registration | 170,000 | | |
| | Loan to Mr. innocent for purchase of two used tyres | 30,000 | | |
| | ***Summary of the 2nd disbursement*** | ***200,000*** | ***232,702*** | ***32,702*** |

*Table 1.7* Seed Business Loan (Third Disbursement)

| *Date* | *Description* | *Debit* | *Credit* | *Balance* |
|---|---|---|---|---|
| | Credit balance from 2nd disbursement | | 32,702 | |
| 7/12/2022 | Electronic funds transferred from Diaspora at two separate periods (N452,500 and N384,300) through WISL GLOBAL LIMITED | | 836,800 | |
| | Instruction on disbursements | | | |
| | Cake training/craft (materials, aprons, head covers) | 100,000 | | |
| | Assistance to CAPRO for SMS NNOKWA ANAMBRA STATE | 100,000 | | |
| | 5 nos. of local oven @20,000 | 100,000 | | |
| | 3 nos. of iron doors @20,000 | 60,000 | | |
| | Trainer's fee | 100,000 | | |
| | 20 days consultancy @15,000 to | 300,000 | | |
| | Travel documentation | 80,000 | | |
| | **Summary of the 3rd disbursement** | **840,000** | **869,502** | **29,502** |

A popular local edible nut known as bitter kola (Garcinia Kola) was identified for cultivation. Bitter kola is in high demand in the country and overseas due to its medicinal properties. The cost per kilogram is on the rise and the consumption rate is high. A lead person has been appointed to oversee the project. An Asset-Based Community Development (ABCD) framework was adopted to achieve the vision. The members believe that there is potential within the community to identify assets, and when discovered can become an economic advantage leading to the development of the community with less external assistance. To discover the assets within the community, a framework for a 3 × 5 model also known as '3 LIQUD', was designed. By this, the members meant they would focus on three main areas and five approaches. The three main areas of focus are: agriculture, talents, and gifting, and human capital. The five approaches to the framework are: Lead person – getting a connector to the resources; identify the area/location and map them out; quantify the resource using available market scales; build understanding with individuals and those concerned; and develop the assets to meet economic value.

*Table 1.8* 2023 Budget

| *S/No.* | *Description* | | | *Amount (N)* |
|---|---|---|---|---|
| 1 | Materials for catering for the 2nd batch students | | | 65,000 |
| 2 | Cake crafts and tools | | | 60,000 |
| 3 | Soap making and toiletries/chemicals and tools | | | 120,000 |
| 4 | Paint productions | | | Xxxx |
| 5 | Agric seminar and free seedlings | | | 120,000 |
| 6 | Grants to 1st batch @20,000 for five persons | | | 100,000 |
| 7 | Grants to 2nd batch @20,000 for 15 persons (no can be reduced) | | | 300,000 |
| 8 | Graduation ceremony cost | | | 150,000 |
| 9 | **Seed loan applications** | | | 900,000 |
| | 1. | V O | 50,000 | |
| | 2. | K A | 50,000 | |
| | 3. | U D | 150,000 | |
| | 4. | N C | 100,000 | |
| | 5. | O A | 300,000 | |
| | 6. | A C | 250,000 | |
| | Total | | 900,000 | |
| 10 | **Running cost** | | | XXXX |
| | **Total budget** | | | **1,915,000+** |

***Future Training and Project Proposals***

To further progress the project the following additional training courses are advised: soap and toiletries production; paint production; and solar energy installation (with inverters). These training options were requested for by some locals to ensure inclusiveness, as our current training workshops are mainly targeted at women. A proposal was made by the team leader to partner with another non-governmental organization (NGO) to engage in agricultural and farming skills capacity training for the locals in the use of new high yielding seedlings to enhance local production. Such training would include setting up farms and seed beds for maximum production; correct application of seedlings; and maximizing the use of KUCH 99 seedlings. There is also a proposal that a certain amount be set aside for the graduating students as start-up or business production grants in the area of skill acquired. A proposal to provide free medical outreaches and free glasses as a means of reaching out to the people with the gospel message has also been tabled. A well-crafted sign on the wall of the centre is needed for public awareness.

***Challenges and Recommendations***

The execution of the project has met with the following challenges. There is a high level of insecurity within the environment including kidnapping and similar vices. The imposition of sit-at home orders by separatist groups has led to inconsistency in training times and dates. Some of the locals have an attitude of indifference and so far, have not taken advantage of the services on offer. More finance is required to properly execute the project, make the centre conducive for learning, increase the training portfolio and ensure consistency in the disbursement of micro-credit facilities. The tendency towards religious embellishments could be a distraction from the primary objective of disseminating the gospel of God's kingdom to every household with simplicity.

The leadership team therefore, recommends that additional training be adopted to promote inclusiveness; funding be assessed and released to undertake trainings and the implementation of the framework to make the centre conducive for learning; and that evangelism and outreaches be conducted with or without medical outreaches within the host community.

**Case Questions**

1 Review the case study stating the benefits and challenges of running social enterprises in Africa.
2 How can maritime tourism potential be developed through BAME social enterprises?
3 Give suggestions on how CLATC can address funding challenges and scale up its projects.

**Key Terms and Definitions**

**Action Research**: A method of investigating a scenario, observing what needs to change, how to go about making that change, evaluating the results, and reflecting upon the process and its effects.

**BAME**: An umbrella term describing people of Black, Asian and Minority Ethnicities.

**Diaspora**: People domiciled in foreign countries who retain filial attachment to their countries of origin.

**Logic Model**: Graphic models that help structure projects into clearly identifiable outcomes, inputs, and activities.

**Maritime Industry**: The movement of men, materials, goods and services across the seas, rivers, and numerous waterways of the world.

**Social Enterprises**: Businesses set up to address social problems where the profits are ploughed back into the community and used for their benefit.

**Theory of Change**: A concept describing the sequence of changes tracking the causal relationships in a project from the desired long-term outcomes to the initial designs and challenges that undergird it.

**Tourism**: Travelling for business or pleasure and the industry that provides the goods and services driving it.

## References

Akanle, O., & Ola-Lawson, D.O. (2021). Diaspora networks and investments in Nigeria. *Journal of Asian and African Studies*, 002190962110529. https://doi.org/10.1177/00219096211052970.

AliAmri, M. (2019). Social and solidarity economy: A step forward toward ending poverty in Tunisia. Retrieved from https://blogs.worldbank.org/arabvoices/social-and-solidarity-economy-step-forward-toward-ending-poverty-tunisia.

Anderson, A.R. (2015). Conceptualising entrepreneurship as economic 'explanation' and the consequent loss of 'understanding'. *International Journal of Business and Globalisation.* https://doi.org/10.1504/IJBG.2015.067432.

Ashley, C., & Poultney, C. (2005). *Pro-poor Tourism Pilots in Southern Africa.* Overseas Development Institute. Retrieved from https://doi.org/en/about/our-work/pro-poor-tourism-pilots-in-southern-africa/.

Baggot, R. (2013). *Partnership for Public Health and Well-Being, Policy and Practice.* London: Palgrave Macmillan.

Belcher, B., & Halliwell, J. (2021). Conceptualizing the elements of research impact: Towards semantics standards. *Humanities and Social Sciences Communications*, (8), 1–6. https://doi.org/10.1057/s41599-021-00854-2. S2CID 236461259.

Belcher, B., & Palenberg, M. (2018). Outcomes and impacts of development interventions: Toward conceptual clarity. *American Journal of Evaluation*, 39(4), 478–495. https://doi.org/10.1177/1098214018765698. S2CID 149485744.

Bennett, M. (2004). A review of the literature on the benefits and drawbacks of participatory action research. *First Peoples Child and Family Review*, 1(1), 19–32. https://doi.org/10.7202/1069582ar.

Beugre, C. (2017). *Social Entrepreneurship, Managing the Creation of Social Value*. Oxon: Routledge.

Bloom, P.N., & Dees, J.G. (2007). Cultivate your ecosystem. *Stanford Social Innovation Review*, 6(1), 47–53. https://doi.org/10.48558/QWAW-VP62.

Borzaga, C., Salvatori, G., & Bodini, R. (2019). Social and solidarity economy and the future of work. This paper draws on a work that was previously published by the ILO and is available at: https://www.ilo.org/wcmpsp5/groups/public/-ed_emp/-emp_ent/-coop/documents/publication/wcms_573160.pdf, https://doi.org/10.1177/14744740211034479.

Brest, P. (2010) *The Power of Theories of Change. Stanford Social Innovation Review.* Spring 2010; Stanford Graduate School of Business; Leland Stanford Jr. University, Stanford CA94305-5015. https://www.sc4ccm.jsi.com

British Broadcasting Corporation (2020). BAME we're not the same: Black African - Exploring culture, identity and heritage. Retrieved from https://www.bbc.com.

Burns, P. (2011). *Entrepreneurship and Small Business, Start-Up, Growth and Maturity.* Basingstoke: Palgrave Macmillan.

Central Bank of Nigeria (2023). Nigeria remittances - 2023 data - 2024 forecast - 2008–2022 historical - Chart - News: Trading economics. Retrieved from https://www.tradingeconomics.com.

Clark, H., & Anderson, A.A. (2004). Theories of change and logic models: Telling them apart (PowerPoint Slides). Retrieved from https://www.theoryofchange.org/wp-content/uploads/toco library/pdf/TOCs and Logic Models forAEA.pdf.

Clark, H., & Taplin, D. (2012). *Theory of Change Basics: A Primer on Theory of Change.* New York: Actknowledge. Retrieved from https://www.actknowledge.org/resources/documents/ToC-Tech-Papers.pdf.

Claus, R., & Belcher, B. (2020). Theory of change. *Swiss Academies of Arts and Sciences.* https://doi.org/10.5281/zenodo.3717451.

Cooperatives Europe (2015). *Building Inclusive Enterprises in Africa: Cooperative Case Studies*. Cooperatives Europe. Retrieved from https://coopseurope.coop.

Cunha, L. (2012) The definition and scope of tourism: A necessary inquiry. *COGITUR Journal of Tourism Studies*, 5, 91–114. http://hdl.handle.net/10437/5239

Dave, M. (2021). Retracted: Resilient to Crisis: How cooperatives are Adapting Sustainability to overcome COVID-19-induced Challenges. *International Journal of Rural Management (Sage Journal)*. https://doi.org/10.1177/097300522.1991624

De Lange, D.E. (2016). A social capital paradox: Entrepreneurial dynamism in a small world clean technology cluster. *Journal of Cleaner Production*, 139, 576–585.

Develtere, D., & Papoutsi, G. (2021). Rebuilding and realizing a resilient goal society through cooperatives. Paper for the Export Group Meeting on the Role of Cooperatives in Economic and Social Development: Recover Better from the COVID-19 Pandemic.

Egorov, V., & Inshakov, A. (2021). Cooperation as an integral part of the social and solidarity economy (SSE). *SHSWeb of Conference*, 94(5), 01009. Retrieved from https://www.shs-conferences.org.

Ekekwe, N. (2021). A Nigerian model for stakeholder capitalism. *Harvard Business Review*. Retrieved from https://hbr.org/2021/05/a-nigerian-model-for-stakeholder-capitalism.

Ekwugha, M. (2022). Adaptive leadership in micro social enterprise teams: Exploring innovative healthcare partnerships. In C.A. Oham (Ed.), *Cases on Survival and Sustainability Strategies of Social Entrepreneurs* (pp. 22–47). Hershey, PA: IGI Publishers. https://doi.org/10.4018/978179987724-0.ch002.

Enyidi, U. (2017). Potable water and national water policy in Nigeria: A historical synthesis, pitfalls and the way forward; *Journal of Agricultural Economics and Rural Development*, 3, 105–111.

Funnel, S., & Rogers, P. (2011). *Purposeful Program Theory: Effective Use of Theories of Change and Logic Models*. San Francisco, CA: Jossey Bass.

Harries, E., Hodgson, L., & Noble, J. (2014). Creating your own theory of change. Retrieved from https://www.thinknpc.org.

Harries, R., & Miller, S. (2021). Community business: The power on your doorstep. Power to Change 2020 Impact Report. Retrieved from https://www.powertochange.org.uk.

Kirzner, I.M. (1973). *Competition and Entrepreneurship.* Chicago, IL: University of Chicago Press.

Lee, J., & Baek, J. (2021). Sustainable growth of social tourism: A growth mixture modelling approach using heterogenous travel frequency trajectories. *International Journal of Environmental Research and Public Health*, 18(10), 5241.

Leiper, N. (1979). The framework of tourism: Towards a definition of tourism, tourist, and the tourist industry. *Annals of Tourism Research*, 6(4), 390–407.

Netto, A. (2009). What is tourism? Definitions, theoretical phases and principles. In J. Tribe (Ed.), *Philosophical Issues in Tourism* (pp. 43–61). Bristol: Channel View Publications.

NHS Health Careers (2015). Social enterprises – Health careers. Retrieved from https://www.healthcareers.nhs.uk.

Nwankwo, C.V. (2021). Examining the role of age grade associations in financing real estate development in Ohafia, Nigeria. *JOSR Journal of Economics and Finance (JOSR-JEF)*, 12(1), Ser. Vi.

Office for National Statistics (2022). Census data. Retrieved from https://www.ons.gov.uk.

Ogoko, C. (2022). Empowering youths through social enterprise: Light recreated assembly – Enriching lives through social entrepreneurship. Retrieved from https://lightrecreated.info/.

Oham, C., & Macdonald, D. (2016). *Leading and Managing a Social Enterprise in Health and Social Care*. London: Community Training Partners.

Oham, C.A., & Okeke, O.J. (2022). Strategic formulation and implementation of social entrepreneurs. In C.A. Oham (Ed.), *Cases of Survival and Sustainability Strategies of Social Entrepreneurs*. IGI Publishers. https://doi.org/10.4018/978-1-7998-7724-0.ch001

Ojiagu, N.C., & Usman, A.U. (2023). Value addition to sustainable economic growth of Sub-Sahara Africa: The path of social and solidarity economy. *Cross Current International Journal of Economics, Management and Media Studies*, *5*(3), 49–53.

O'Neill, A. (2023) United Kingdom – Ethnicity. Retrieved from https://www.statista.com.

Osiri, J.K. (2020). Igbo management philosophy: A key for success in Africa. *Journal of Management History*, 26(3), 295–314. https://doi.org/10.1108/JMH-10-2019-0067.

Overseas Development Institute (ODI) (2023). Tourism in poor places: Who gets what? Retrieved from https://www.odi.org.

Pracht, D., Toelle, A., & Broaddus, B. (2022). Action research: A methodology for organizational change. 2022(1) UF/IFAS EXTENSION 4-H. *Youth Development Program*. https://doi.org/10.32473/edis-4H424-2022.

Reiter, J. (2023). UNGA: How digital technologies can accelerate progress towards SDGs. Paper presented at the ITU, United Nations Development Programme for SDG Digital at the UN Headquarters, ECOSOC Chamber, New York.

Reuters (2020). Pandemic costs Africa travel, tourism almost $55billion. Retrieved from https://www.reuters.com.

Reynolds, P.D., Bygrave, W.D., Autio, E., Hay, M., & Marion Kauffman Foundation (2002). *Global Entrepreneurship Monitor 2002 Executive Report*. New York: GEM.

Roy, M.J., Donaldson, C., Baker, R., & Kerr, S. (2014). The potential of social enterprise to enhance health and well-being: A model and systematic review. *Social Science & Medicine*, 123, 182–193. https://doi.org/10.1016/j.socscimed.2014.07.031. Epub 2014 Jul 12. PMID: 25037852.

Royce, M. (2007). Using human resource management tools to support social enterprise: Emerging themes from the sector. *Social Enterprise Journal*, 3(1), 10–19. https://doi.org/10.1108/17508610780000718.

Sharpley, R. (2022). Tourism and development theory: Which way now? *Tourism Planning and Development*, 19(1). https://www.tandfonline.com/doi.org/10.1080/21568316.2021.2021475.

Schumpeter, J. (1954). *History of Economic Analysis*. New York: Oxford University Press.

Sepulveda, L., & Rabbevag, S. (2021). *Minoritised Ethnic Community and Social Enterprises*. Power to Change. Retrieved from https://eprints.icstudies.org.uk/id/eprint/243.

Steiner, A., & Teasdale, S. (2017). Unlocking the potential of rural social enterprise. *Journal of Rural Studies*, 70, 144–154. https://doi.org/10.1016/j.jrurstud.2017.12.021.

Stevenson, H.H., & Jarrillo, J.C. (1990). A paradigm of entrepreneurship: Entrepreneurial management. *Strategic Management Journal*, 11, 17–27. https://www.jstor.org/stable/2486667.

The Law Society (2023). A guide to race and ethnicity terminology and language, June 27, 2023. Retrieved from https://www.lawsociety.org.uk.

Thirumoorthi, T., & Wong, K.M. (2015). Tourism Chapter 24. In A. Idris, S. Moghavvemi, & G. Musa (Eds.), *Selected Theories in Social Science Research* (pp. 300–325). Kuala Lumpur: UM Press.

Timuray, S. (2023). Ensuring inclusion for all through improving access to connectivity. Retrieved from https://www.vodaphone.com.

United Nations (2023). The Sustainable Development Goals report 2023: Special edition: Towards a rescue plan for people and planet. Retrieved from https://unstats.un.org/sdgs.

United Nations Industrial Development Organization (UNIDO) (2017). The role of the social and solidarity economy in reducing social exclusion. *Budapest Conference Report*, June 1–2, 2017.

United Nations World Tourism Organisation (UNWTO) (2020). Secretary General's policy brief on tourism and COVID-19. Retrieved from https://unwto.org.

Weiss, C.H. (2011). Nothing as practical as good theory: Exploring theory-based evaluation for comprehensive community initiatives for children and families. Retrieved from https://www.semanticscholar.org.

William, J., & Tavneet, S. (2011). *Mobile Money: The Economics of M-PESA*. National Bureau of Economic Research. Retrieved from https://www.nber.org/papers/w16721.

Wongtschowski, A. (2015). Social enterprise capabilities and human development. DPU Working Paper No.174, Special Issue on Capability Approach in Development Planning and Urban Design.

World Bank Group (2018). *Tourism Theory of Change*. Tourism for Development Knowledge Series. Washington, DC. Retrieved from https://www.worldbank.org; https://hdl.handle.net/10986/35459.

World Economic Forum (2021). *6 Ways Social Entrepreneurs Are Saving Lives during India's Covid-19 Crisis: The Role of Social Entrepreneurs*. World Economic Forum Retrieved from https://weforum.org.

Vodafone (2023). *What Is M-Pesa*? Retrieved from https://www.vodafone.com/about-vodafone/what-we-do/consumer-products-and-services/m-pesa.

Yeoman, K. (2014). *M-PESA Helps World's Poorest Go to the Bank Using Mobile Phones*. Christian Science Monitor. Retrieved from https://www.csmonitor.com/World/Making-a-difference/Change-Agent/2014/0106/M-PESA-helps-world-s-poorest-go-to-the-bank-using-mobile-phones.

Young, R. (2006). For what it's worth: Social value and the future of social entrepreneurship. In A. Nicholls (Ed.), *Social Entrepreneurship: New Models of Sustainable Social Change* (pp. 56–73). Oxford: Oxford University Press.

# 2 Exploring Destination Managers' Approach to Shock Advertising

## The Case of Southeast Asian Countries and Turkey

*Augusta Evans*

### Introduction

Studies on the effectiveness of shock tactics in tourism are few. This chapter examines the perception of destination managers (DMs) in the utilisation of shock advertising to tackle bad tourist behaviour. It aims to investigate the effect that culture and religion have on DMs' reaction to shock advertising. Recommendations for practice and future research directions are discussed.

### Background

The efficacy and distinctiveness in utilising shock tactics in grabbing customer's interest and actions are well documented (Zafran and Masud 2023). As far back as the 1980s, advertisements like these were created to discourage smoking, to promote healthy eating, to encourage the wearing of seatbelts and contributions to charities.

Shock advertising could be defined both by the effect it has on an audience and its distinctive components. In the first instance, shock advertising awakens a perception of disconformity with established community norms and conducts. Shock advertising are useful strategies that stand out of the norm and useful for targeting social problems (Mukattash et al. 2021). With respect to the components of shock advertising, authors point out how shock appeals such as sex (Sawang 2010), fear appeals (Mukherjee and Dubé 2012) and violent appeals (Lewis et al. 2007) are the distinctive traits of shock advertising.

Most of the studies, for instance, Yukako et al. (2021) researched its impact on road safety, climate change (Jin and Brown-Devlin 2022), consumer brand products (Lee et al. 2020), and Mayer et al. (2019) researched the outcome of utilising shock tactics in the medical sector. Studies on the efficacy of shock ads, however, produce conflicting findings. For example, sex ads are being employed from the 1960s and have become more prevalent in developed economies as advertisers employ more inventive and imaginative techniques to maintain their competitive advantage and boost sales (LaTour and Henthorne 1994), however, there are conflicting results as pertaining their effectiveness. Nevertheless, despite the

DOI: 10.4324/9781003369967-4

attractiveness of employing sex in advertising, there are substantiation that it can be detrimental to marketers that are in need for a quick surge in sales and brand awareness (Samson 2018).

Dangerous tourist behaviours such as spree drinking, prostitution, illegal drug consumption and more are common in numerous tourist destinations. DMs can prevent undesirable effects and outcomes of misbehaviours by providing prevention information and taking legal action. There has been very few research in employing shock appeals in the travel sector despite the benefits of shock advertising (Evans et al. 2019). Albeit some locations lately introduced shock strategies campaigns to reduce and stop misbehaviours.

Asia in general and the South and Middle East of Asia specifically count as attractive tourist destinations both for the internal flows as well as for the external markets such as Europe, North America and Australia. However, most of them experience tourist's misbehaviour: ignoring local laws and roles, culture and habits, as well as ethical norms and values create potential damages that affect the tourist and destination (Yu et al. 2022). Hence, this chapter investigates DMs viewpoint in the utilisation of shock tactics, the role of culture and how shock advertising affect destinations image and prevent tourist's misbehaviour.

The background study addresses tourist misbehaviours and its impact on destination management. Following, it defines shock advertising from a cultural perspective and in terms of its main components in using this style of advertising, perceived impact on destination image. Then the issues are introduced and finally recommendations and future research directions are presented.

## Shock Advertising

Shock advertising has been defined and analysed following two main perspectives: the first one is linked to the audience's reactions and cultural contexts of shock advertising, while the second deals with its components. According to the first perspective. According to Venkat and Abi-Hanna (1995) shock advertising that purposely upsets its viewers while Pflaumbaum (2011) defines it as advertising that surprises the target audience and may elicit an undesirable response and pushes various taboos. Skorupa (2014, p. 8) defined shock advertising "as advertising that is intended to cause dread, upset and queries societal and communal norms"; It's components are addressed in the following.

## The Components of Shock Advertising

The second perspective, relating to shock advertising components, as it is typically considered to deliberately scare and offend its audience rather than doing so accidentally (Gustaeson and Yssel 1994). Furthermore, Andersson and Pettersson (2004) outlines shock advertising as consisting of three primary elements which are: (a) distinctiveness, (b) ambiguity, (c) transgression of norms.

### Distinctiveness

An original and compelling advertisement is required that have not been over flogged so that it can be recalled by the consumers and leave a lasting impression. Any attempt at replication will lessen the impact of the advertisement's surprise element and reduce its capacity for provocativeness. Pope et al. (2004) discovered that unique stimuli enhance brand evaluation by increasing interest to the advertisement, retention, and recovery. The degree of congruence between an ad and the product it is meant to promote may have an impact on how offensive it is perceived by an audience (Christy and Haley 2007). One study discovered that shocking advertisement and the display of graphic images, prohibited behaviours, terrifying pictures and misbehaviours are better tolerated in non-profit marketing more than in for-profit advertising (Van Putten and Jones 2008), making the communication and its medium crucial.

Further research has been done to explore distinctiveness, in relation to the content of the message. It became evident that incongruence of message as it pertains to an audience's schema was essential (Vézina and Olivia 1997). Heckler and Childers (1992) propose that the processing of data that is inconsistent has a huge impact on how well the communication's content is remembered. It indicates that consumers probably react favourably to the incongruity as they stop, ponder, recall and reflect on the brand.

### Ambiguity

Ambiguity of an advert arises when its meanings can vary depending on the consumer's interpretation and in turn creating a shocking effect to the consumers. Ambiguity can arise in the perception of not just the ad message but the marketers' objectives. Ambiguity is when a consumer starts to question the advertising's relevance to the goods or services, and what message is being conveyed.

Howard and Sheth (1969, p. 158) suggest that stimulus ambiguity causes stimulation and ultimately generates some experimental actions. The research purports that stimuli ambiguity "is deficit of clearness of the stimulus exhibit in communicating the graphic and evaluation characteristics of the brand, merchandise and the disposition of intentions". Yet, further research reveals that for persuasion to occur, the consumer must have some comprehension of the advertisement Vézina and Olivia (1997). There is still a need for more research to determine the appropriate degree of ambiguity, how it influences viewers' grasp of the ads, and how this ultimately influences their behaviour.

### Transgression of Norms and Taboos

Advertising that breaks taboos or represents something that is not typically accepted as the norm is said to be more effective. This is one feature that can distinguish and draw attention to an advertisement. In some cases, people believe that advertising is lowering the bar for decency and communal norms. Sex appeal is seen as a taboo at times, as it occasionally causes a lot of backlash as it is perceived as immoral and prejudicial, but it can be a successful method of informing the end users. Even though defying convention can make customers feel bad. According to Virvilaitė and Matulevičienė (2013), individuals are at times influenced to purchase goods or service after feeling negative emotions to counteract their effects. Although it ought to be said that shock advertising does not always provoke undesirable emotions, it can also elicit positive emotions like pleasure, which is why Sabri (2012) claims that shock advertising can have an optimistic or adverse impact on consumers' sympathy or hostility for such advertisements. In the following the link between shock advertising and culture is presented.

## Cultural Dimension of Shock Advertising

Advertisers are increasing cognisant of diversity in numerous communities because of the expansion of social media and the internet. Recognising consumer behaviour in numerous cultural contexts has been the subject of extensive research to help us better appreciate diversity. According to Hofstede (1983) cultural factors can influence how an audience reacts to shock advertising. Virvilaitė and Matulevičienė (2013) in their research found that collective societies respond significantly adversely than countries with individualistic cultures to advertisements that contain sexual content. Sawang (2010) investigated how sexual content is perceived by consumers in Asian and American cultures. They discovered that Americans viewed sexual and nudist-themed advertising more favourably than Asians. In contrast to audience from collective cultures like Malaysia, Taiwan, and China. Fam and Waller (2003) discovered that customers from individualistic cultures such as New Zealand would be slightly upset by provocative goods. Additionally, according to An and Kim (2006), sexually suggestive goods and advertisements were less acceptable in Korea (a more communal society) than in America (a more individualistic society), as these products were perceived as being against customs, principles and views.

However, Anabila et al. (2015) discovered that sex ads can be utilised as a shock tactic in a communal culture like Ghana because viewers won't be deterred from buying the goods, however he recommended that advertisers should endeavour to ensure a good correlation between the merchandise and the advertisement by using a sensible degree of sex languages. Marketers need to be careful of moral standards and avoid upsetting customers, harming children and reaching audiences that weren't their intended target.

Hall (1977) distinguished between two kinds of cultures: high- and low-context cultures, which were categorised based on their messages. Geographic regions like the Middle East, Africa and South America are instances of high-context societies. Their religions and cultures place a strong emphasis on social interactions, communal ties, and the coding and sharing of messages among community members. However, low-context cultures, which include countries in the Northern America and Western Europe, are more individualistic and rational in their interpretation. They also overtly incorporate mass information in their communication. The method of interaction is crucial in establishing if shock advertising is received constructively or adversely. Lui et al. (2009) states that although high cultures are collectivistic and encourage unfavourable consumer buying habits, individualistic cultures encourage good conduct. An and Kim (2006) discovered that Korean consumers find sex ads and addictive goods to be unpleasant compared to American consumers. It may be since these goods are severely controlled in high-context societies, where they are viewed as ethically repugnant and immoral. Phau and Prendergast (2001) discovered that while customers in Singapore with tertiary degree find an ad off putting, the effect may not spill to other products of the organisation regarding their buying intention. The media used in shock ads was also analysed by the writers, who discovered that this influenced consumers' tolerance levels. Specifically, consumers are less tolerant of shock ads when aired in male's or female's glossy magazine than when they appear in newspapers, billboards, or direct mail. Individuals' high- or low-context cultural aspects are significant, however other influential factors exist such as the medium used in its execution, the recipient's context and their educational background.

An important component of cultural orientation is religiosity. Religious commitment is said to be "the extent whereby an individual follows their religious orientations, content, and rituals and in their day-to-day routines" (Worthington et al. 2003, p. 85). It can be further defined as having two parts, internal religiosity, which deals with individual religious experiences and external, this indicates the degree to which a person follows practices of a spiritual association to attain communal approval and individual gain (Donahue 1985; Gorsuch and McPherson 1989; Putrevu and Swimberghek 2013). Religion is essential in the moral viewpoint of people, as the more religious an individual is, the more he follows absolute moral laws and perceives ethnic problems better than minority religious individuals (Hunt and Vitell 2006). For example, ads that deal with sensitive topics such as abortion, child abuse, can create an adverse reaction and dissonance with the end user and the business's brand.

According to Fam et al. (2004, p. 537), "variations in religious associations oftens affect individual's lifestyle and settlement, their preferences, food preferences, and interpersonal relationships". Religion and more traditional, conservative viewpoints are strongly correlated. According to Boddewyn and Kunz (1991, p. 14), Muslim nations are more hostile to the "penetration of western advertising, ideas, and methods", although Christian nations slightly traditional (Parry et al. 2013). In essence we have significant distinctions in how people in Muslim and Christian nations respond to commercials for alcohol and those with a sexual theme. In contrast to Christian countries, nations that are predominantly Muslim are strongly against such advertisements. Religion has a significant impact on our attitudes towards a specific shock advertisement and our social interactions (Fam et al. 2004). Sex ads or depictions are prohibited in Islamic nations but are permitted in Christian nations (Boddewyn and Kunz 1991). Fam et al. (2004), a person's likelihood of being offended by the advertising of a contentious product increases with their level of religiosity. This might be the case because ardent Muslims, Christians and Hindus are more conventional in respect to others that are easier going.

Gender is an important factor to consider when examining culture. However, there appears to be contradictory evidence despite the abundance of studies on the impact of shock advertising on variations in sexual category sensitivity. For instance, Chu et al. (2006) discovered that females, regardless of cultural background, exhibit greater affectionate empathy for victims in advertisements than do men. Although some people occasionally exhibit more masculine traits than feminine, or at times a combination of both traits (Yoon and Yeuseung 2014). Men typically find violent humour hilarious than females (Kazarian and Martin 2006). As stated by these authors, there were no appreciable variations observed in the degree of response from males and females to humorous violent behaviour in commercials; and be attributed to disposition traits and inclinations (Brown et al. 2010). However, since some women may possess more masculine characteristics than others, problems like mixed gender identity may also come up. But according to Urwin and Venter (2014), enterprises should look for more creative ways to stand out from the competition rather than employing shock tactics as a marketing tactic, regardless of a person's gender, personal identity, religion or values.

Numerous studies indicate that females have more tendency to be upset by shock advertising (Chu et al. 2006). Dahl et al. (2003) in their study discovered that females tend to have an adverse response to sexual shock tactics than males and that organisations should ensure that the right tactics are employed at the target audience to achieve a favourable result.

According to Christy (2006), shock advertising offends women more than it does men, because typically women uphold values that are associated with justice, integrity and family. Additionally, they contend that even when an advertisement is not specifically aimed at women, advertisers must pay close attention to their sensitivities as women are assumed the media gatekeepers and have sole and secondary responsibility in making purchasing choices (Christy 2006) (Figure 2.1).

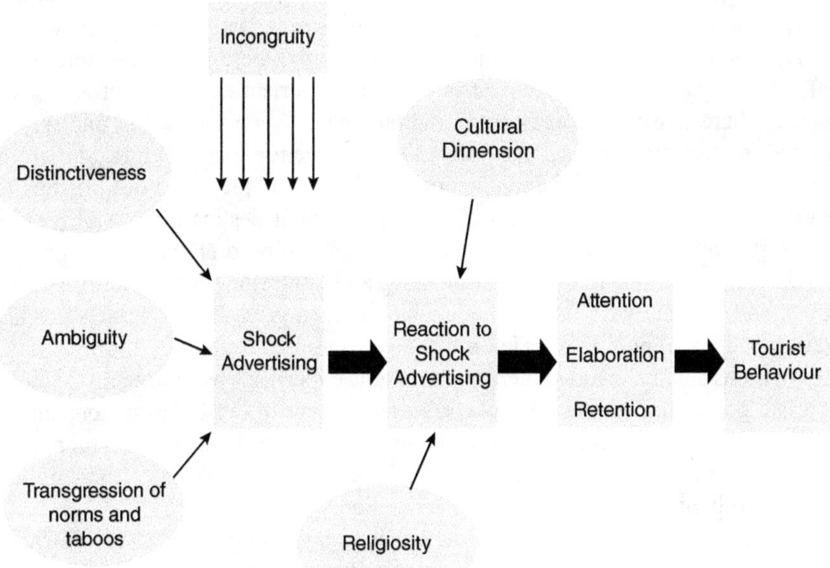


*Figure 2.1* Conceptual framework.

## Issues for Further Discussion

### *Tourist Misbehaviour, Destination Management, and Destination Image*

An important area in marketing and tourism is the study of tourist behaviour (Cohen et al. 2014). Recognising the conduct and feelings of travellers is crucial because it boosts the economy overall. Tourism has a noteworthy impact on the society, lifestyle, patterns, social life, traditions, beliefs, and values (García et al. 2015) so it's critical to comprehend how visitors behave.

According to Pearce (2005), comprehending tourist actions is beneficial for everyone interested in the travel industry. Investors use the knowledge gathered on visitor actions for a variety of purposes, including planning vacations, developing business plans, marketing goods and controlling challenging tourist behaviour. Loi and Pearce (2012) grouped tourist misbehaviour into the following categories: (a) behaviours involving others, e. g, smoking in the public area, dropping litter and spitting in public, using swear language, (b) solo individual actions and (c) slightly illicit or dodgy behaviours. Dangerous sexual behaviour is another example of tourist misbehaviour; for example, Berdychevsky and Gibson (2015) highlight a rise in dangerous sexual encounters among youths, and Hughes et al. (2011) follow along in a similar vein that young adults are responsible for most tourist misbehaviours, including incidents involving alcohol that raise the risk of violence. Blackden (2004) catalogued some shocking tourist behaviours, such as violent behaviour, nudity, stealing and scamming.

To reduce the issues that arise from tourist's misbehaviour to the destinations, other visitors and stakeholders, it is essential that these annoying behaviours are managed. In the current environment, bodies that have been given the responsibility of managing a tourist destination are known as "Destination management organisations" (DMOs) (Durašević 2015). Recently, there have been intensified efforts on destination management in respect to tourism development and sustenance. There is a lot of pressure on nations that rely on tourism in one way or another to create a contemporary organisation structure that will integrate all the advancements made, maintain them and consistently improve the destinations for tourists and stakeholders (Durašević 2015). A tourist destination's improvement and sustainability are greatly dependent on its capacity to effectively manage its assets, engage in the work of coordinating and controlling activities in its locations, and connect with nearby societies (Plzáková et al. 2014). DMOs are also very proactive in putting policies and projects into place that will improve tourist destinations and make travel there a memorable experience by producing top-notch tourism goods and benefits. Successful management of these destinations aids in boosting their negotiating position with the rest of the world, generating more prospects for marketing. Pike (2008) suggested that one of the DMOs' responsibilities is to actively market the destination's reputation.

Destination image can be viewed either as an individual's own knowledge about the location (raw materials, infrastructure, society, community and economy value), or their respective perception regarding the location (Li et al. 2015). Echtner and Ritchie (2003, p. 41) also defined destination image as "beliefs that tourists have about a location". Destination image has three constructs which are cognitive (beliefs and factual knowledge that an individual has of the characteristics and characteristics of the location), affective dimension (personal attitude and emotions that they have about the location) and finally the conative image (which implies the individual's future behaviour or intentions) (Guzman-Parra et al. 2016). Cognitive image affects beliefs, which individuals formed from their cultural background and affects their attitude formation and behaviour (Hofstede 2001). Conative dimension is "analogous to behaviour as they are the action component of the image" (Kaplanidou 2009, p. 252). The perceived destination image encompasses the expectation that the tourists have about the destination and how they are met during their visit which will in turn result in their satisfaction or dissatisfaction (Law and Cheung 2010).

The cognitive and affective construct of destination image affect tourist behavioural intentions (Enright and Newton 2005; Hallmann et al. 2015; Ritchie and Crouch 2003). The model proposed by Hallmann et al. (2015) indicates that behavioural intention is the outcome of tourist's perception of the destination. "Destination image has a great influence on tourists' behaviour" (Pechlaner et al. 2013, p. 125).

In recent times there have been many studies on the impact of destination image on behaviour intent and revisit intention and it was found that a positive destination image tend to lead to favourable repurchase decision and intended behavioural change (Leung et al. 2011). However, destination image is not constant and affected by internal and external factors (Pechlaner et al. 2013; Tasci and Gartner 2007).

To prevent harmful links with socioeconomic problems like political unrest, antisocial behaviour, cruelty, homicides and risky sexual conducts. It is crucial that the organisations responsible for destination management collaborate effectively to develop a tourist product and location that perfectly complements the idealised image of the destination they are trying to project. Next, we'll talk about some of the difficulties DMs have when using shock advertising.

### *DMs Understanding of Shock Advertising and Its Components*

Many DMs had used or had some exposure to scary ads and were able to understand what they found shocking. Dahl et al. (2003) proposed that marketers increasingly utilise unusual advertising to capture consumers concentration and enhance awareness, exposure, and brand recognition. Dahl et al. (2003) described shocking images as those "suggestions of gore, excreta, fumes, mortality, nauseating signs, foul language, lewd actions, and racial epitaphs".

Although, shock depicts the setting in which an advertisement is broadcast as well as its subject matter. Latour et al. (1990) discovered that the advertisement depicts offense not just through its subject content but also through its setting. Given that shock tactics and the illustration of violence, criminal activity, scary depictions and misbehaviour are bearable in non-profit marketing compared to for-profit marketing, the content and context of an advertisement are extremely important (Van Putten and Jones 2008). It was suggested by Vézina and Olivia (1997) propose that shock ads work well to increase perception and draw interest.

However, the drawbacks of employing shock ads were described using phrases like “harmful brand image”, “trigger outrage”, “distress”, and “discrimination”. Shock commercials have many advantages, but they can also have the opposite effect of intended results and incite offense or fear in viewers, particularly if they foster the use of, repetitive actions (Henley and Donovan 1999). According to Machová et al. (2015, p. 111) “shock ads can be a successful weapon if applied efficiently” because there is a fine line between shocking someone and making them feel uncomfortable.

### *Barriers to the Effective Application of Shock Advertising*

DMs want to use shock advertising, but they run into so many obstacles and resistance that they either don’t use it at all or employ the strategy sparingly. Senior manager’s absence of enthusiasm, belief, ethnical norms, fright of legal implications, public backlash and an outright ban on advertisements were a few of the obstacles. Governmental and regulatory bodies also influence the kind of adverts that are aired as most of the tourism boards are public parastatals.

### *The Role of Culture in Shock Advertising*

Culture has been a central theme as tourists behave in a certain way as there are from different cultural background and religion. Tourists are generally known to let themselves go when they are on holiday and hence don’t behave the way they normally do at home. Also, there is the issue of cultural shock when they visit a destination and not used to what is the norm there. Tourists are not aware of cultural restrictions and sometimes people expose themselves indecently in public places which is unacceptable in some destinations. Finally, some shock ads are restricted by the regulatory bodies which highlights those destinations with a advent religious standing, some of the advertisements may not be displayed or broadcast for instance sexual appeals (Boddewyn and Kunz 1991).

### *Management of Tourists Misbehaviour and Impact on Destination Image*

Tourist misbehaviour is detrimental to the destination and should be controlled and supervised. Everyone who is in some capacity involved in the tourism industry should work together to achieve this. The management of local tourist locations and infrastructures, however, appears to be lacking (Pike 2008). According to Adeyinka-Ojo et al. (2014), national, regional and local businesses typically make up tourism organisations. Majority of the participants responded that these businesses cannot control the destinations by themselves rather visitors and residents should be key players. Pearce and Schänzel (2013) suggested that for successful destination management all interested parties must participate in the process. However, many studies do not include tourists, even though they are important stakeholders who can affect whether any scheme or plan put in place for effective resource management is successful or unsuccessful.

DMs use visitors' perceptions of their destination to determine whether they would return or suggest it to friends and family. Since tourists' perceptions influence how they organise their holidays, image is "more important than reality" (El Kadhi 2008). Assaker et al. (2011) claim that when a customer has a favourable impression of a place, their intention to return grows over time. The perception of the destination is important because the goal of DMs is to draw visitors and promote their location. The capacity to manage a distinctive and pleasant location image is a critical component of any organisation's branding and placement development (Ekinci 2003).

DMs believed that shock commercials would have an adverse impact on their destination because visitors might associate the ads' content with local events and steer clear of them.

### *Solutions and Recommendations*

The primary goals of this study were to examine how DMs view and utilise shock advertising in the context of tourism, as well as the effects on the destination and the significance of culture. Since shock advertising can be seen as a socially and culturally taboo, DMs were less likely to use it, supporting Sawang's (2010) findings that North Americans where likely probable to employ sexual appeals than Asians. The efficacy of such strategies for drink-driving, social campaigns and emphasising healthiness, however, discourages the use of shock advertising in many locations.

It is important to explain to the DMs the advantages of using shock tactics so they can make better decisions about their strategy. Additionally, DMs encountered challenges when attempting to use shock advertisements, including red tape in governmental regulations, monitoring bodies, visitors, who can decide to avoid their destination. DMs can involve senior managers in the proposal procedure and authorisation stages to get around some of the barriers.

DMs should employ a mild shock strategy which would be consistent with their ethnic upbringing if their goal is to raise awareness of specific tourist misbehaviour that be an issue in their location. As moderate incongruity improves memory, increase attention, and generates fulfilment when an outcome is achieved, this won't turn off both domestic and foreign tourists (Mandler 1982). However, given that they anticipate receiving worldwide visitors, DMs should consider the significance of adopting a multicultural approach (Engelbart et al. 2017). DMs can use shock advertising as a potent tactic to influence people to alter their attitudes and behaviours (Mehta 2000). According to Banyte et al. (2014) shock marketing can be used to address social problems and alter behaviour and as the current study highlighted, this may be useful in identifying and addressing intrusive tourist behaviours.

### *Future Research Suggestions*

This research focused on the viewpoint of DMs and did not consider tourist's reaction to shock advertising. Further research should aim to combine the viewpoints of DMs' and visitors, which will help in understanding both the marketers' and that of tourists. This would be beneficial to DMs as it would assist them in designing tools to manage and curtail tourist misbehaviour as it is crucial in sustainability of their destinations. It is also imperative for DMs to assess the effect of these ads in respect to how it affects their destination image and if tourist would recommend, revisit their destinations after viewing these ads. Given that shock advertising is the focus of this research, it will be helpful to examine how tourists interpret these kinds of appeals to better understand and control their behaviour.

Future studies should consider using interviewing DMs and respondents from the hospitality industry, like lodging facilities, resorts, and tourist destinations. The DMs should have a similar or congruence cultural background as their perception of the shock ad campaign may differ due to cultural factors. Concentrating on a single location might be more beneficial and thorough or destinations with cultural similarities and religion.

## Conclusion

The pros and cons of tourism have been the subject of countless discussions and debates. Just like how behaviour is controlled in other corporate parastatals, airports, colleges and places of worship, so also should tourist behaviour be controlled. With the rise in tourism, there seem to be ongoing issues with controlling visitor behaviour in hotels, resorts, and other major tourist destinations. Shock ads are a successful tool for mainstream and social marketing strategies. It might help control obnoxious tourist behaviour and improve the quality of life for both locals and visitors and boost the economy.

### Case Study: Managing Tourist Misbehaviour

***China National Tourism Administration***

Tourists are known to act in various manners before, during and after visiting a destination. This concept is known widely as tourist behaviour (Soliman and Abdelmoaty 2021). Tourist behaviour is an important element that plays a vital role in determining the successful expansion of tourism services; hence an awareness of tourist behaviour or misbehaviour is a prerequisite in the advancement of tourism. An in-depth understanding of tourists' behaviour is beneficial for all tourism stakeholders (Pearce 2005).

According to Tsaur et al. (2019) tourist misbehaviour can be defined as acts displayed by tourist purposely or unconsciously that disregards the destination customs, rules, policies and laws. According to Volgger and Huang (2019) tourist irresponsible behaviour range from moderately defiant to customs and norms to corrupt conduct and illicit crimes. Deviant tourists' behaviour includes incidents of vandalism, violence, binge drinking, irresponsible sexual conduct, recreational drug taking and damaging the environment. Deviant tourists' behaviours are on the rise and with the emergence of social media these behaviours are wildly seen, and additionally, they harm other stakeholders and the host communities. To lessen the effects of adverse behaviour and ensure the continued growth of this industry, it is crucial that it be managed. This harmful behaviour can be changed with the aid of social marketing.

China is one of the most populous countries in the world, in 2021 its population was estimated at over 1.4 billion (World Bank 2022). International tourists' arrival from China in 2021 was estimated 32 million compared to 145.3 million in 2019 (Statista 2023). Chinese tourists are also known to have a diverse behaviour in relation to their host because of their cultural diversity, differences in demographics and increased purchasing power in the global travel market (Ng 2021). According to Melubo and Kisasembe (2022) Chinese tourists act in ways that are different to other tourists such as not making eye contact when speaking, travelling in large groups as this is an integral part of their culture, they also tend to spend and demand more luxurious items, better infrastructures like fast speed internet and many have religious and superstitious beliefs for instance lucky numbers. For destinations to remain sustainable understanding tourist behaviour is a key aspect and Chinese tourist behaviour needs to be fully explored since they represent a large sector of the tourism market.

Due to the numbers of Chinese tourists a year, there have been numerous research (Ng 2021, Wen and Meng 2021; Ying et al. 2019) in attempting to understand why and how Chinese outbound tourists misbehave in various context. For instance, Wen and Meng (2021) in their research proposed that there is an increase of the use of illicit drugs among Chinese tourists, and that although the use of such drugs in the western world is rampant, the use of any leisure drugs is greatly frowned upon and seen as deviant behaviour. Furthermore, Ng (2021) in attempting to understand the reason why Chinese tourists may speak loudly in public in a country like Japan attribute this to the confidence that they possess. This confidence they ascertain may have resulted from the rapid growth of their economy and their attempt to promote the image of their country assuming superiority over their host countries.

According to Wang et al. (2023) Chinese outbound tourists has raised fears that deviant behaviour does not only subject destinations to numerous adverse impacts but also destructs the national image of the country. Many concerns have been raised all over the world due to the many incidents of Chinese tourists' misbehaviours in the press and literature such as a teenager an example of this was when a Chinese teenager defaced a 3,500-year-old Egyptian temple. Other examples according to the Deputy

Prime Minister of China includes talking too loud in public areas, carving names and symbols on tourists' attractions, not taking cognisant of host traffic rules, spitting in public places and many other uncivilised behaviours. He insisted that this damaged the national image of the Chinese people and cannot be ignored (Guardian 2023).

Tourists' misbehaviour impact destinations in numerous ways: socio-culturally, economically, and environmentally, however few governments have taken the necessary actions to curb such adverse impacts. Stakeholders fear that any regulatory action may hamper tourists visit intentions and have a simultaneous effect on tourism revenue. The Chinese government is known globally to exercise a strong influence on its citizens (Ng 2021). In this respect, in other to curb some of the annoying behaviour of its populace in tourist destinations worldwide, the government created a list that blacklisted any individuals found to misbehave, this list was published, and it affected their credit ratings and free travel movement (Gong et al. 2018). More recently, the Chinese government in Beijing made more attempts to promote its image and the Guide to Civilized Behaviours of Outbound Chinese Tourists was issued (China National Tourism Administration 2006). It devised a law and regulation to manage and minimise tourists' misconduct and deviancy that was published in the "The Regulations of Deviant Tourist Behaviour Management". Penalties for visitors' bad behaviour were registered, and they comprised things like "damaging the environment, destroying infrastructures, or i violating the norms of locations and registered on visitors' credit files" (Li and Chen 2017, p. 4). Twenty tourists have breached the manifesto's guidelines since it was issued, and they were all penalised and defamed publicly. Other governmental authorities such as Thai tourism have also issued some etiquette manuals to Chinese tourists to help educate them on their culture and pro environmental behaviour (Gong et al. 2018). Although the Chinese government have taken steps to address this problem by naming and shaming the offending tourists, there is still more that can be done by destinations to manage and control tourists' misbehaviour.

**Case Study Questions**

- What would you consider to be tourist misbehaviour? Do you think that this is a global issue? Who should manage this?
- Did the Chinese government get it right? Is this an effective way to manage tourists' misbehaviour?
- Can you suggest some other management tools that could be used by the Chinese government in curtailing tourists' misbehaviours?

**Key Terms and Definitions**

**Destination Image**: Hallmann et al. (2015, p. 95) defines it "as visitors' and retailers understanding of the characteristics or appeals that are accessible in a location and contributes to the portrayal, advertising, unification and conveyance of the locations' product circulation". Many tourism destinations often use advertising and integrated marketing communication strategies to affect the way that tourists view their destinations in essence projecting an image. Destinations sway image formation indirectly through numerous means such as through news, journalism, that are facilitated by external intermediaries or directly by their own marketing activities (Govers et al. 2007). "Destination image has a strong impact and impact on tourists' conduct" (Pechlaner et al. 2013, p. 125).

**Destination Management**: This refers to a collective structure and procedure that brings together all stakeholders, integrating their activities to ensure an articulate destination image and enjoyable tourists experience (Fragidis and Kotzaivazoglou 2022).

**Destination Management Organisations (DMOs)**: The development of a specific tourist destination is actively aided by DMOs, which include participants from the public and corporate sectors. They do this by providing details about tourist attractions and activities, fostering a positive reputation, preventing adverse environmental effects, and advertising appropriate behaviour between visitors and destinations (Plzáková et al. 2014).

**Incongruity in Advertising**: Advertising incongruity describes a disparity amid a component of advertising stimulus for example this could a picture, words, video, music, actor/actress with a current schema that a person have about that stimulus (Cao et al. 2020). For instance, in an ad for Evian water a video was shown of a baby roller skating in a diaper.

**Tourists Misbehaviour**: Tourist misbehaviour on the other hand is defined as acts that are breaking norms, customs, regulations and guidelines (Tsaur et al. 2019). Fullerton and Punj (2004, p. 1239) defined it as "actions by customers, which disrupt the usually acknowledged standards of behaviour in buying pattern" thereby "characterise the ugly, adverse characteristics of the customer".

**Schema**: According to Lee et al. (2020) advertising is processed and assessed by an individual's norms and schemas. An individual schema is their interpretation of perception which helps to process the message and use their understanding and interpretation in comparable contexts.

**Shock Advertising**: This is advertising that violates an individual's norms and may trigger a positive or negative reaction (Lee et al. 2020).

**Disclosure Statement**: The author report there are no competing interests to declare.

## References

Adeyinka-Ojo, S. F., Khoo-Lattimore, C., and Nair, V. (2014). "A Framework for Rural Tourism Destination Management and Marketing Organisations". *Procedia - Social and Behavioural Sciences*, 1(144), 151–163.

An, D., and Kim, S. (2006). "Attitudes toward Offensive Advertising: A Cross-Cultural Comparison between Korea and the United States", paper presented at the *Annual Conference of the American Academy of Advertising*, Reno, Nevada, March 30–April 2.

Anabila, P., Tagoe, C., and Asare, S. (2015). "Consumer Perception of Sex Appeal Advertising: A High Context Cultural Perspective". *The IUP Journal of Marketing Management*, 14(4), 34–55.

Andersson, S., and Pettersson, A. (2004). "Provocative Advertising. The Swedish Youth's Response". Master Thesis, Marketing and E-commerce, Lulea University of Technology, Sweden.

Assaker, G., Vinzi, V. E., and O'Connor, P. (2011). "Examining the Effect of Novelty Seeking, Satisfaction, and Destination Image on Tourists' Return Pattern: A Two Factor, Non-Linear Latent Growth Model". *Tourism Management*, 32(4), 890–901.

Banyte, J., Paskeviciute, K., and Rutelione, A. (2014). "Features of Shocking Advertizing Impact on Consumers in Commercial and Social Context". *Innovative Marketing*, 10(2), 35–46.

Berdychevsky, L., and Gibson, H. J. (2015). "Sex and Risk in Young Women's Tourist Experiences: Context, Likelihood, and Consequences". *Tourism Management*, 51, 78–90.

Blackden, P. (2004). *Holidaymakers from Hell: Shocking Behaviour by Tourists Abroad*. London: Virgin.

Boddewyn, J. J., and Kunz, H. (1991). "Sex and Decency Issues in Advertising: General and International Dimensions". *Business Horizons*, 34(5), 13–20.

Brown, M. R., Bhadury, R. K., and Pope, N. K. L. (2010). "The Impact of Comedic Violence on Viral Advertising Effectiveness". *Journal of Advertising*, 39(1), 49–65.

Cao, S. et al. (2020). "A Facilitatory Effect of Perceptual Incongruity on Target-Source Matching in Pictorial Metaphors of Chinese Advertising: Eeg Evidence". *Advances in Cognitive Psychology*, 16(1), 1–12.

China National Tourism Administration. (2006). Guide to Civilized Behaviours of Outbound Chinese Tourists. Available at http://www.china.com.cn/policy/txt/2006-10/02/content_7212276.htm (Accessed on 01/08/2023).

Christy, T. P. (2006). "Females' Perceptions of Offensive Advertising: The Importance of Values, Expectations, and Control". *Journal of Current Issues & Research in Advertising*, 28(2), 15–32.

Christy, T. P., and Haley, E. (2007). "The Influence of Context on College Students' Perceptions of Advertising Offensiveness". In American Academy of Advertising, Annual Conference.

Chu, T. Q., Seery, M.D., Ence, W.A., Holman, E. A., and Silver, R. (2006). "Ethnicity and Gender in the Face of a Terrorist Attack: A National Longitudinal Study of Immediate Responses and Outcomes Two Years after September 11". *Basic and Applied Social Psychology*, 28(4), 291–301.

Cohen, S. A., Prayag, G., and Moital, M. (2014). "Consumer Behaviour in Tourism: Concepts, Influences and Opportunities". *Current Issues in Tourism*, 17(10), 872–909.

Dahl, D. W., Frankenberger, K. D., and Manchandra, R. V. (2003). "'Does It Pay to Shock?'. Reactions to Shocking and Non-shocking Advertising Content among University Students". *Journal of Advertising Research*, 43(3), 268–280.

Donahue, M. J. (1985). "Intrinsic and Extrinsic Religiousness: Review and Meta-Analysis". *Journal of Personality and Social Psychology*, 48(2), 400–419.

Durašević, S. (2015). "Tourism in Montenegro: A Destination Management Perspective". *Tourism*, 63(1), 81–96.

Echtner, C. M., and Ritchie, B. J. R. (2003). "The Meaning and Measurement of Destination Image". *Journal of Tourism Studies*, 14(1), 37–48.

El Kadhi, W. (2008). *Cross-Cultural Destination Image Assessment: Cultural Segmentation versus the Global Tourist: An Exploratory Study of Arab-Islamic and Protestant European Youths' Pre-visitation Image on Berlin*. Hamburg: Druck Deplomica Verlag.

Ekinci, Y. (2003). "From Destination Image to Destination Branding: An Emerging Area of Research". *e-Review of Tourism Research*, 1(2), 21–24.

Engelbart, S. M., Jackson, D. A., and Smith, S. M. (2017). "Examining Asian and European Reactions within Shock Advertising". *Asian Journal of Business Research*, 7(2), 37–56.

Enright, M. J., and Newton, J. (2005). "Determinants of Tourism Destination Competitiveness in Asia Pacific: Comprehensiveness and Universality". *Journal of Travel Research*, 43(4), 339–350.

Evans, A. I., Adamo, G. E., and Czarnecka, B. (2019). "European Destination Managers' Ambivalence toward the Use of Shocking Advertising". In E. Bigne, and S. Rosengren (Eds.), *EAA Advances in Advertising Research*. Springer Gabler, Wiesbaden.

Fam, K. S., and Waller, D. S. (2003). "Advertising Controversial Products in the Asia Pacific: What Makes Them Offensive?". *Journal of Business Ethics*, 48(3), 237–250.

Fam, K. S., Waller, D. S., and Erdogan, Z. B. (2004). "The Influence of Religion on Attitudes towards the Advertising of Controversial Products". *European Journal of Marketing*, 38(5/6), 537–555.

Fragidis, G., and Kotzaivazoglou, I. (2022). "Goal Modelling for Strategic Dependency Analysis in Destination Management". *Journal of Tourism, Heritage & Services Marketing*, 8(2), 3–15.

Fullerton, R. A., and Punj, G. (2004). "Repercussions of Promoting an Ideology of Consumption: Consumer Misbehavior". *Journal of Business Research*, 57(11), 1239–1249.

García, F. A., Vázquez, A. B., and Macías, C. R. (2015). "Resident's Attitudes towards the Impacts of Tourism". *Tourism Management Perspectives*, 1(13), 33–40.

Gong, J., Detchkhajornjaroensri, P., and Knight, D. W. (2018). "Responsible Tourism in Bangkok, Thailand: Resident Perceptions of Chinese Tourist Behaviour". *International Journal of Tourism Research*, 21(10), 221–233.

Gorsuch, R. L., and McPherson, S. E. (1989). "Intrinsic/Extrinsic Measurement: I/E-Revised and Single-Item Scales". *Journal for the Scientific Study of Religion*, 28(3), 340–348.

Govers, R., Go, F. M., and Kumar, K. (2007). "Promoting Tourism Destination Image". *Journal of Travel Research*, 1(46), 15–23.

Guardian. (2023). Chinese Tourists Warned over Bad Behaviour Overseas. Available at https://www.theguardian.com/world/2013/may/17/chinese-tourists-warned-behaving-badly-wang-yang (Accessed on 11/07/2023).

Gustaeson, B., and Yssel, J. (1994). "Are Advertisers Practicing Safe Sex?". *Marketing News*.

Guzman-Parra, V. F., Vila-Oblitas, J. R., and Maqueda-Lafuente, F. J. (2016). "Exploring the Effects of Cognitive Destination Image Attributes on Tourist Satisfaction and Destination Loyalty: A Case Study of Málaga, Spain". *Tourism & Management Studies*, 12(1), 67–73.

Hall, E. T. (1977). *Beyond Culture*. Garden City, NY: Anchor.

Hallmann, K., Zehrer, A., and Mülle, S. (2015). "Perceived Destination Image: An Image Model for a Winter Sports Destination and Its Effect on Intention to Revisit". *Journal of Travel Research*, 54(1), 94–106.

Heckler, S. E., and Childers, T. L. (1992). "The Role of Expectancy and Relevancy in Memory for Verbal and Visual Information: What Is Incongruency?". *Journal of Consumer Research*, 18(4), 475–492.

Henley, N., and Donovan, R. J. (1999). "Threat Appeals in Social Marketing: Death as a 'Special Case'". *International Journal of Non-profit and Voluntary Sector Marketing*, 4(4), 300–319.

Hofstede, G. H. (1983). "National Culture in Four Dimensions". *International Studies of Management and Organization*, 13(2), 46–74.

Hofstede, G. H. (2001). *Culture's Consequences: Comparing Values, Behaviours, Institutions, and Organizations across Nations*, 2nd edn. London: Sage.

Howard, J. A., and Sheth, J. N. (1969). *The Theory of Buyer Behavior*. New York: Wiley.

Hughes, K., Bellis, M. A., Calafat, A., Blay, N., Kokkevi, A., Boyiadji, G., Mendes, M., and Bajcàrova, L. (2011). "Substance Use, Violence, and Unintentional Injury in Young Holidaymakers Visiting Mediterranean Destinations". *Journal of Travel Medicine*, 18(2), 80–89.

Hunt, S. D., and Vitell, S. J. (2006). "The General Theory of Marketing Ethics: A Revision and Three Questions". *Journal of Macro Marketing*, 26(2), 143–153.

Jin, E., and Brown-Devlin, N. (2022). "When Others Are Here: The Combinative Effects of Social Presence and Threat Appeals in Climate Change Message Effectiveness". *Mass Communication and Society*, 25(1), 25–50.

Kaplanidou, K. (2009). "Relationships among Behavioral Intentions, Cognitive Event and Destination Images among Different Geographic Regions of Olympic Games Spectators". *Journal of Sport & Tourism*, 14(4), 249–272.

Kazarian, S. S., and Martin, R. A. (2006). "Humor Styles, Culture-Related Personality, Well-Being, and Family Adjustment among Armenians in Lebanon". *Humor*, 19(4), 405–423.

Latour, M. S., Pitts, R. E., and Snook-Luther, D. C. (1990). "Female Nudity, Arousal, and Ad Response: An Experimental Investigation". *Journal of Advertising*, 19(4), 51–62.

LaTour, M. S., and Henthorne, T. L. (1994). "Female Nudity in Advertisements, Arousal and Response: A Parsimonious Extension". *Psychological Reports*, 75(3), 1683–1690.

Law, R., and Cheung, S. (2010). "The Perceived Destination Image of Hong Kong as Revealed in the Travel Blogs of Mainland Chinese Tourists". *International Journal of Hospitality & Tourism Administration*, 11(4), 303–327.

Lee, M. S. W., Septianto, F., Frethey-Bentham, C., and Gao, E. (2020). "Condoms and Bananas: Shock Advertising Explained through Congruence Theory". *Journal of Retailing and Consumer Services*, 57, 1–11.

Leung, D., Law, R., and Lee, H.-A. (2011). "The Perceived Destination Image of Hong Kong on Ctrip.com". *Internal Journal of Tourism Research*, 13, 124–140.

Lewis, I., Watson, B., and Tay, R. (2007). "Examining the Effectiveness of Physical Threats in Road Safety Advertising: The Role of the Third-Person Effect, Gender, and Age". *Transportation Research Part F: Traffic Psychology and Behaviour*, 10(1), 48–60.

Li, T., and Chen, Y. (2017) "The Destructive Power of Money and Vanity in Deviant Tourist Behaviour". *Tourism Management*, 61(4), 152–169.

Li, J., Faizan, A., and Woo Gon, K. (2015). "Re-examination of the Role of Destination Image in Tourism: An Updated Literature Review". *E-Review of Tourism Research* (eRTR), 12(3/4), 191–209.

Loi, K. I., and Pearce, P. L. (2012). "Annoying Tourist Behaviors: Perspectives of Hosts and Tourists in Macao". *Journal of China Tourism Research*, 8(4), 395–416.

Lui, F., Hong Cheng, H., and Li, J. (2009). "Consumer Responses to Sex Appeal Advertising: A Cross-Cultural Study". *International Marketing Review*, 26(4/5), 501–520.

Machová, R., Seres, H., and Zsuzsanna, T. (2015). "The Role of Shockvertising in the Context of Various Generations". *Problems and Perspectives in Management*, 13(1), 104–112.

Mandler, G. P. (1982). "The Structure of Value: Accounting for Taste'. Affect and Cognition". *The 17th Annual Carnegie Symposium on Cognition*, Lawrence Erlbaum, Hillsdale, NJ, 30–36.

Mayer, J., Zainuddin, N., Russell-Bennett, R., and Mulcahy, R. F. (2019). "Scaring the Bras Off Women: The Role of Threat Appeal, Brand Congruence, and Social Support in Health Service Recruitment Coping Strategies". *Journal of Service Theory and Practice*, 29(3), 233–257.

Mehta, A. (2000). "Advertising Attitudes and Advertising Effectiveness". *Journal of Advertising Research*, 40(3), 67–72.

Melubo, K., and Kisasembe, R. (2022). "We Need Chinese Tourists, But Are We Ready? Insights from the Tanzanian Safari Industry". *Journal of China Tourism Research*, 18(1), 185–202.

Mukattash, I. L., Dandis, A. O., Thomas, R., Nusair, M. B., and Mukattash, T. L. (2021). "Social Marketing, Shock Advertising and Risky Consumption Behavior". *International Journal of Emerging Markets*, (1), 1–28.

Mukherjee, A., and Dubé, L. (2012). "Mixing Emotions: The Use of Humor in Fear Advertising". *Journal of Consumer Behaviour*, 11(2), 147–161.

Ng, S. L. (2021). "Would You Speak Softly in Public? An Investigation of Pro-environmental Behavior of Chinese Outbound Tourists in Hong Kong". *Current Issues In Tourism*, 24(22), 3239–3255.

Parry, S., Jones, R., Stern, P., and Robinson, M. (2013). "'Shockvertising': An Exploratory Investigation into Attitudinal Variations and Emotional Reactions to Shock Advertising". *Journal of Consumer Behaviour*, 12(2), 112–121.

Pearce, P. L. (2005). *Tourist Behaviour: Themes and Conceptual Schemes*. Clevedon and Buffalo, NY: Channel View Publications.

Pearce, D. G., and Schänzel, H. A. (2013). "Destination Management: The Tourists' Perspective". *Journal of Destination Marketing & Management*, 2(3), 137–145.

Pechlaner, H., Dal Bò, G., and Pichler, S. (2013). "Differences in Perceived Destination Image and Event Satisfaction among Cultural Visitors: The Case of the European Biennial of Contemporary Art 'Manifesta 7'". *Event Management*, 17(2), 123–133.

Pflaumbaum, C. (2011). "Shock Advertising – How Does the Acceptance of Shock Advertising by the Consumer Influence the Advertiser's Designs?". Master Thesis, Curtin University.

Phau, I., and Prendergast, G. (2001). "Offensive Advertising". *Journal of Promotion Management*, 7(1–2), 71–90.

Pike, S. (2008). *Destination Marketing: An Integrated Marketing Communication Approach*. Oxford: Elsevier Butterworth-Heinemann.

Plzáková, L., Studnička, P., and Vlček, J. (2014). "Individualization of Demand of Tourism Industry and Activities of Destination Management Organizations". *Czech Hospitality and Tourism Papers*, 1535–1801.

Pope, N. K. L., Voges, K. E., and Brown, M. R. (2004). "The Effect of Provocation in the Form of Mild Erotica on Attitude to the Ad and Corporate Image: Differences between Cause-Related and Product-Based Advertising". *Journal of Advertising*, 33(1), 69–82.

Putrevu, S., and Swimberghek, K. (2013). "The Influence of Religiosity on Consumer Ethical Judgments and Responses toward Sexual Appeals". *Journal of Business Ethics*, 115(2), 351–365.

Ritchie, B. J. R., and Crouch, G. I. (2003). *The Competitive Destination: A Sustainable Tourism Perspective*. Oxon: CABI.

Sabri, O. (2012). "Taboo Advertising: Can Humor Help to Attract Attention and Enhance Recall?". *Journal of Marketing Theory and Practice*, 20(4), 407–422.

Samson, L. (2018). "The Effectiveness of Using Sexual Appeals in Advertising". *Journal of Media Psychology*, 30(4), 184–195.

Sawang, S. (2010). "Sex Appeal in Advertising: What Consumers Think". *Journal of Promotion Management*, 16(1–2), 167–187.

Skorupa, P. (2014). "Shocking Contents in Social and Commercial Advertising". *Creativity Studies*, 7(2), 69–81.

Soliman, S., and Abdelmoaty, G. (2021). "Domestic Tourism Challenges: Tourist Misbehavior". *Journal of Association of Arab Universities for Tourism and Hospitality*, 20(1), 195–219.

Statista. (2023). Overseas Visitor Arrivals in China from 2010 to 2020 with an Estimate for 2021. Available at https://www.statista.com/statistics/234785/international-tourists-arrivals-in-china/ (Accessed on 11/07/2023).

Tasci, A. D. A., and Gartner, W. C. (2007). "Destination Image and Its Functional Relationships". *Journal of Travel Research*, 45(4), 413–425.

Tsaur, S.-H., Cheng, T.-M., and Hong, C.-Y. (2019). "Exploring Tour Member Misbehavior in Group Package Tours". *Tourism Management*, 71, 34–43.

Urwin, B., and Venter, M. (2014). "Shock Advertising: Not So Shocking Anymore. An Investigation among Generation Y". *Mediterranean Journal of Social Sciences*, 5(21), 203–214.

Van Putten, K., and Jones, S. (2008). "It Depends on the Context: Community Views on the Use of Shock and Fear in Commercial and Social Marketing, Partnerships, Proof and Practice". *International Non-profit and Social Marketing Conference*, (7), 15–16.

Venkat, R., and Abi-Hanna, N. (1995). "Effectiveness of Visually Shocking Advertisements: Is It Context Dependent?". In Administrative Science Association of Canada Proceedings. *Journal of Advertising Research*, 43(3), 268–280.

Vézina, R., and Olivia, P. (1997). "Provocation in Advertising: A Conceptualization and an Empirical Assessment". *International Journal of Research in Marketing*, 14(2), 177–192.

Virvilaitė, R., and Matulevičienė, M. (2013). "The Impact of Shocking Advertising to Consumer Buying Behavior". *Economics and Management*, 18(1), 134–141.

Volgger, M., and Huang, S. (2019). "Scoping Irresponsible Behaviour in Hospitality and Tourism: Widening the Perspective of CSR". *International Journal of Contemporary Hospitality Management*, 31(6), 2526–2543.

Wang, T., Xiuli, Z., Yu, W., Xue, L., and Yueyue, G. (2023). "A Broader Social Identity Comes with Stronger Face Consciousness: The Effect of Identity Breadth on Deviant Tourist Behavior among Chinese Outbound Tourists". *Tourism Management*, 94, 104629.

Wen, J., and Meng, F. (2021). "Research Design in Socially Deviant Tourist Behavior Studies: A Mixed-Method Approach". *Tourism Analysis*, 26, 83–88.

World Bank. (2022). Population, Total – China. Available at https://data.worldbank.org/indicator/SP.POP.TOTL?locations=CN (Accessed on 11/07/2023).

Worthington, E. L., Nathaniel, J. R., Wade, G., Hight, T. L.,Ripley, J. S., McCullough, M. E., Berry, J. W., Schmitt, M. M., Berry, J. T., Bursley, K. H., and O'Connor, L. (2003). "The Religious Commitment Inventory--10: Development, Refinement, and Validation of a Brief Scale for Research and Counseling". *Journal of Counselling Psychology*, 50(1), 84–96.

Ying, T., Wen, J., and Shan, H. (2019). "Is Cannabis Tourism Deviant? A Theoretical Perspective". *Tourism Review International*, 23, 71–77.

Yoon, H. J., and Yeuseung, K. (2014). "The Moderating Role of Gender Identity in Responses to Comedic Violence Advertising". *Journal of Advertising*, 43(4), 382–396.

Yu, I., Yang, M., Fan, D., and Zeng, K. (2022). "Can Travelling Abroad Experiences Trigger Tourist Misbehaviours? The Role of Moral Relativism". *Current Issues in Tourism*, 2(1), 1–9.

Yukako, N., Kazuko, O., Ritsu, K., and Goro, F. (2021). "The Effectiveness of Group Discussion in a Threat Appeal - Based Road Safety Education for Young Cyclists". *Japanese Journal of Traffic Psychology*, 37(1), 16–29.

Zafran, M., and Masud, S. (2023). "Consumer's Response to Fear Appeals and Their Effectiveness in Advertising: Cross-Culture Comparison of Finnish and Pakistani Consumer's Attitude towards Threat Appeals". *Economy and Market Communication Review*, 25(1), 95–112.

# 3 Post-Brexit Tourism

## Customer Perspectives

*Isabel Dolores Jimenez Jimenez*

### Introduction

After 47 years of alliance Union with the European, the United Kingdom voted to leave the European Union in a referendum in 2016. Since the United Kingdom's departure from the European Union, the limitations of the present commercial connections between the United Kingdom and the European Union have created new barriers for United Kingdom's travel businesses.

### Background

After years of intensive negotiation, postponements, and re-negotiations, the final withdrawal agreement was agreed in late 2019 (BBC, December 17, 2019). On December 30, 2020, the Trade and Cooperation Agreement was signed (The EU-UK Trade and Cooperation Agreement, April 30, 2021). The historical, legal, political, and economic ties that existed between the United Kingdom and the European Union before and after they joined brought a complex and long procedure that took more than four years of negotiations. The United Kingdom and the European Union agreed and signed the termination contract. In line with Article 50 of the Treaty on European Union, it established the terms for the United Kingdom's orderly exit from the European Union.

An essential agreement that regulates the relationships between the European Union and the United Kingdom is the Withdrawal Agreement. This includes but is not limited to, civil rights in the United Kingdom and the European Union, as well as the transitional period that ended on December 31, 2020.

To implement the termination contract, action was needed at both the European Union and Member State levels, as well as action in the United Kingdom. The United Kingdom formally left the European Union with an economic agreement at the end of 2020.

In the summer of 2021, the Financial Times reported that "the picture on trade and employment has been drowned by the economic impact of the Coveid-19 pandemic", highlighting the difficulty in differentiating between the effects of COVID-19 and Brexit on the nation and on specific firms.

DOI: 10.4324/9781003369967-5

Ramifications of the Brexit process still dominate the political agenda as tensions with the European Union remain high. The Government enacted the Brexit Freedoms Act, sometimes referred to as the European Union Retained Laws Act, in September 2022. By the end of 2023, it intends to eliminate the dominance of European Union law in the United Kingdom and grant the governing body greater authority to change, replace, or abolish European Union legislation that has been incorporated into domestic United Kingdom law (by the European Parliament).

Despite the initial severe outcome that Brexit had on this business and the effects of the present worldwide financial crisis brought on by the conflict in Russia and Ukraine, where the yearly inflation rate is at a record high, the tourism sector in Britain continues to be rebounding since COVID-2019.

One of the most significant economic sectors in the United Kingdom is travel and tourism. With nearly £130 billion in economic impact and three million employees, it ranks as the fourth largest industry in the nation. The strong travel and tourism flows between the United Kingdom and the European Union are also important to take into account, as the United Kingdom has the sixth-largest tourism sector in the world (House of Commons, 2018). For instance, according to ABTA and Deloitte (2016), is the European Union both the primary source market and the primary destination for tourists from the United Kingdom? The free flow of people, goods, and capital within the European Union has made it easier to achieve this interdependence. Therefore, Brexit may jeopardize this freedom of movement and affect how smoothly travel and tourism operate (ABTA & Deloitte, 2016).

Leaving the European Union affected the entire industry. Public concerns and perceptions of the risks associated with such events have a far-reaching impact on travellers' travel and decision-making, and consequently on the travel industry as a whole (Ma et al., 2020).

The full impact of Brexit on the global travel industry has not yet been felt, according to a World Travel Market Industry Report study from November 2021. Throughout the report, more than 700 senior professionals from all over the world were questioned about whether their companies had experienced any particular pressures because of Brexit. The majority of respondents who acknowledged a Brexit effect were unfavourable, and nearly half (45%) of respondents said they hadn't noticed any difference as a result of Brexit. Only 8% of respondents indicated a positive impact, compared to 24% who indicated a negative impact and 23% who were unsure or unsure of the impact's magnitude.

The trade agreement between the United Kingdom and the European Union resulted in changes to regulations, these changes affected both incoming and outgoing journeys between the United Kingdom and other members of the European Union. Several of the biggest mobile phone providers in the United Kingdom announced that the surcharge-free roam for tourists that was required by law while the United Kingdom was a member of the European Union ended. Costs increased due to this change (Lau, 2022).

Along with the COVID-19-related problems, there were also concerns with passport expiration dates, driving licenses, insurance, resort personnel levels, immigration lineups at airports, and other things. Additionally, companies and customers were impacted by the combined effects of Brexit and COVID-19. While hiring new employees was different, cross-border tax, refunds, fulfilment, and accounting remained difficult issues (ABTA, 2021).

According to the director of World Travel Market London, Simon Press, the sector managed to avoid the worst effects of Brexit after the COVID-19 pandemic ensured and led what would have been the first truly successful post-Brexit holiday season. However, as COVID restrictions began to ease, concerns about visas, financial protection, security, tax exemptions, health insurance, and other issues related to Brexit emerged (Team & Team, 2021).

In the latest study, published in the November 2022 London World Travel Market study, a rising global cost of living was the biggest threat to the travel industry in 2023. "Which of the following is most likely to affect your business in 2023?" according to a survey of travel industry experts. More than 50% said, "The cost of living in general." 13.4% of respondents raised the price of gasoline, while 9.6% said they were concerned about energy prices (Team & Team, 2022).

More than two-thirds of those who took part in the survey (67.8%) attributed their participation in the cost of living, the price of gas, and the price of electricity. The report confirmed warnings from the World Travel and Tourism Council that economic headwinds could threaten industry recovery worldwide.

In September, the Group of 20 (G20) tourism ministers were briefed on the issues by the World Travel & Tourism Council (WTTC), stating that it was more important than ever for the public and commercial sectors to work together. Julia Simpson, President and CEO of the World Travel & Tourism Council hammered home the point by stating that factors such as escalating energy costs, rising living expenses, a workforce shortage, constrained airspace, and climate change "all threaten the full recovery of our sector."

The United Kingdom Government promised to curb energy costs for individuals and companies throughout the winter, but worries about hard times in the economy persisted as recessions loomed across Europe due to the intervention of Russia into Ukraine and its constraint on natural gas supply.

In compliance with the United Kingdom's small spending plan and package of tax cuts, the pound fell to a record low, escalating concerns about the cost of living and increasing the cost of travel for Britons. Rising costs on holidays has made leisure travel prohibitive for some, particularly at the low-cost end of the market.

This article, therefore, examines the implications of Brexit on the United Kingdom's travel and tourism sector (Figure 3.1).

## The Implications of the Brexit for the Travel Industry

Brexit's effects on employment in the United Kingdom are sector-specific; job losses occur in certain industries, like banking, while job growth occurs in others, like agriculture. Brexit has also affected the ability of the United Kingdom's to recruit qualified foreign employees, leading to a skills gap in several industries. Overall, Brexit has had wide-ranging repercussions, with some industries suffering more than others. While there is room for development in some industries,

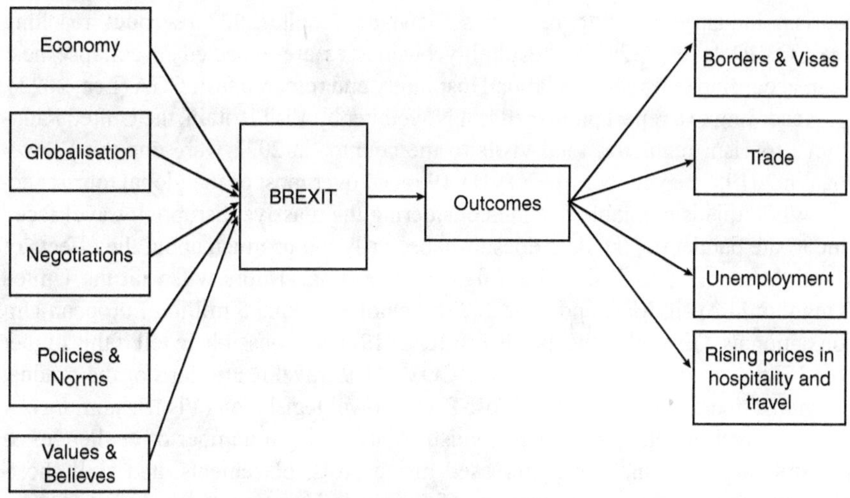


*Figure 3.1* Key factors affecting Brexit and their consequences in the travel industry.

*Table 3.1* Top Five European Union Nations and Markets for the United Kingdom in 2018

| *Number of Inbound Visits* | (In *Millions*) | *Number of Outbound Visits* | (In *Millions*) |
|---|---|---|---|
| 1 France | 3.69 | 1 Spain | 15.62 |
| 2 Germany | 3.26 | 2 France | 8.56 |
| 3 Ireland | 2.78 | 3 Italy | 4.33 |
| 4 Spain | 2.56 | 4 Ireland | 3.22 |
| 5 Netherlands | 1.95 | 5 Portugal | 2.82 |

*Source:* (Horsfield, 2020).

Brexit has severely impacted the United Kingdom's economy as a whole, resulting in less trade, decreased investment, and decreased economic growth (Statista).

The travel and tourism sectors of Europe and the United Kingdom are interconnected structurally and physically. Over 54 million people, or more than 75% of all UK visitors who travel overseas, visit other EU nations as their primary destination. In terms of numbers, the European Union is the country's major source of tourism, contributing to approximately 25 million or 65% of all international tourists. Each year, the United Kingdom economy receives more than £10 billion in visitors from the European Union, while the UK economy receives approximately £22.4 billion in tourism from the European Union (Hall, 2020). The top five European Union nations and markets for the United Kingdom in 2018 are as follows (Table 3.1):

As many as 40.9 million tourists arrived in the UK in 2019, so it has long been a popular tourist destination in Europe. The number of tourists fell significantly to 11.1 million due to these incidents. But as the health crisis subsided in the rest of the world and related restrictions began to relax in the following years, the British tourism industry began to recover somewhat. Despite 2022 revenues reaching pre-COVID levels, 32% of hospitality businesses are expected to collapse next year, according to trade association Hospitality and research firm CGA (Lee, 2023).

According to a report published last November by Visit Britain, the United Kingdom's tourism regulator, total visits to the country in 2022 were one-third lower than in 2019. The year before COVID-19 raced over most of the global tourist sector. While this is a notable decline considering the massive disruption travel faced due to the pandemic, the data doesn't necessarily tell us much about the effects of Brexit. According to Visit Britain figures, 8 million visitors will visit the United Kingdom in April, May, and June 2022, including around 5 million European Union nationals. Given that the data is just for 2019, it's impossible to tell if this higher trend is the result of the lifting of the COVID-19 travel restrictions or the gradual normalization and acceptance of post-Brexit travel legislation (VisitBritain.org).

The travel, hotel, and services industries are facing a number of challenges as a result of Brexit, including increased hiring costs, placements, and skill shortages due to a decline in the number of European Union workers. Although over fifty-three percent (53%) said Brexit had no effect on recruiting, 26% said it had an adverse impact (VisitBritain.org).

Employees travelling from European Union nations will be subjected to a points-based immigration process in the future and will be required to verify that they are qualified when applying for jobs. Employers that depend upon European Union citizens who do not meet the new employment visa criteria are going to encounter shortages of employees in the near future. Furthermore, global businesses, such as entertainment, tour operators, cruise lines, shipping, and aviation firms, frequently employ highly flexible foreign personnel and depend on the rapid deployment of workers across the world, even in the European Union. While Brexit increases the amount of paperwork for firms, it also opens up chances to train local workers and recruit people from other areas of the globe (Deloitte, James Yearsley).

In a similar way, visa limitations may have an influence on the entertainment business, making it more difficult and expensive for United Kingdom performers to travel to and to perform in Europe. Artists and their teams from the United Kingdom, including crew and goods vendors, will now be required to apply for visas, receive a sponsor's certificate, and submit extra paperwork such as merchandise tax forms and permits. To safeguard their tour-based profits, entertainment companies and artists will need to reassess their operating expenditures and agreements (Tourist of England).

Brexit has added certain obstacles to business travel and operating in the European Union for the international services industry. Where professional qualifications are required, such as auditing and legal services, consultation, etc., the professional services industry is likely to face obstacles. Such companies must determine if their service is subject to national restrictions in individual member nations and guarantee that all responsibilities are fulfilled. To avoid additional legal risk, travel management businesses will need to update corporate travel rules and conduct rigorous monitoring to ensure that customers and travellers understand the extent of the agreement. Individual business visitors must also deal with tougher customs restrictions and the need for visas, work permits, insurance, and other documents. Although Brexit-related complexities boost the value travel management services provide their clients, it will be costly due to the necessity for in-depth country-level awareness of norms and regulations (gov.uk).

### *Parameters Evaluated after Brexit*

Brexit has generated uncertainty, and impacted consumers, industry players, and policymakers. Among the parameters evaluated are the unemployment rate, trade, and the consequences on the travel industry:

#### *Unemployment*

The unemployment rate falls progressively through 2020 before peaking in 2021 and remaining for six months. Meanwhile, gross weekly wages climb gradually over time, and the Gross Domestic Product (GPD) index rises through 2020 nevertheless occurs throughout COVID-19. Particularly, trading declined the most after Brexit. Furthermore, the influence of these events differs across the United Kingdom and among sectors.

Northern Ireland and Wales have been identified as regions most severely impacted by Brexit, with earnings and labour numbers dropping in industries that include accommodation, building, and global commerce. Finance, science, and healthcare, on the other hand, have shown an increased contribution to the entire Gross Domestic Product (GDP) of Britain in the post-Brexit age, with favourable impacts. Commerce suffered the most severe repercussions of any variable analysed in the United Kingdom. At the beginning of 2021, the nation's economic condition is defined by an essential dynamic: demand for goods increases more rapidly than supply, leading to shortages of goods, congestion, and price increases (Office for National Statistics).

#### *Trade*

Brexit has caused an enormous impact on trade between the United Kingdom and the European Union. The United Kingdom's exit from the European Union caused it to renegotiate trade treaties with the union, resulting in greater obstacles to trade and prices. As a consequence, the volume of trade between the United Kingdom and the European Union has declined, with specific sectors, such as the automotive sector. In addition, Brexit has had an impact on investment in the United Kingdom, causing many firms to move operations to other European Union nations. This has resulted in employment losses and a slowing of the country's economic growth. However, the developing partnership with the European Union provides chances for investment in industries like tech and banking (Office for National Statistics).

*Hospitality and Tourism*

According to a study by the United Kingdom Parliament, tourism is the fastest-growing sector in Britain, employing over 3 million people and contributing more than £130 billion to the economy. The tourist sector in the United Kingdom is highly developed, ranking sixth in terms of value in the world. With over 20 million visitors in 2017, London, Europe's most visited city, has by far the biggest number of international tourists from the United Kingdom (ABTA Travel Association).

As stated by the Association of British Travel Agencies (ABTA), the United Kingdom's outbound tourism sector supports more than 840,000 jobs in the country and contributes more than £49 billion a year in Gross Value Added (GVA) to the economy. The European Union receives around 70 million visits from the United Kingdom each year, generating more than €40 billion in revenue for businesses in local communities across the European Union. According to surveys and interviews with over 100 industry Chief Executive Officers (CEO), the amount of British nationals working in European Union countries has decreased to 69% since 2017. The consequences of losing these employees far outweigh the impact on individuals, or higher costs for the business of United Kingdom companies, although both are important (ABTA).

For years, the industry has relied on the types of tasks performed by temporary and seasonal workers as a route to attracting talent.

***Statistics & Facts***

The United Kingdom's travel and tourism industry's both direct and indirect value added to Gross Domestic Product (GDP) in 2022 was 4.6% lower than in 2019 (the year before COVID-19). In total, these industries contributed around £237 billion to Gross Domestic Product (GDP) in 2022. The number of residents entering the United Kingdom in 2021 fell to just over 6 million, the lowest in 20 years, while the number of tourists entering the United Kingdom rose to just over 31 million in 2022.

However, international arrivals remained lower than before the pandemic. Similarly, United Kingdom outbound tourist arrivals increased by nearly four times in 2022, although the country reported about 20 million fewer outbound trips than in 2019. Whether it was before, during, or after the pandemic, holidaying continued to be the primary reason for United Kingdom outbound travel. Spain was the top destination for British holidaymakers in 2022, with France, Greece and Portugal coming in second and third respectively (Office for National Statistics).

***Data Protection***

The EU General Data Protection Regulation (GDPR) remains in the United Kingdom legislation and comply with the Data Protection Act of 2018. Although the fundamental principles, rights, and responsibilities stay unchanged, there are some consequences for the laws governing the transfer of personal data for which the United Kingdom is awaiting sufficiency determinations from the European

Commission. A draft judgement has just been published recommending that the United Kingdom seek an adequacy determination, but it is unclear whether that adequacy determination would be consistent with the current. If complete data adequacy is not granted in the UK, companies will need to review their data protection practices and contracts and place appropriate legal protections. Businesses that rely heavily on information flows, including: Travel and Transportation, Professional Business Services, Betting and Gambling, Entertainment, and Hospitality, and some industries may experience considerable changes.

#### *Rising Prices Could Deter Tourists*

Travel companies specializing in United Kingdom travel are been hit by rising prices, claiming that the price of hotel rooms and other services has increased since Brexit, prompting them to increase prices. According to researchers at the London School of Economics and Political Science, Brexit has raised the volume of documentation required to trade with the European Union, which has raised prices in the United Kingdom. The researchers found that this has also caused an increase in entertainment and tourism costs. Of course, there are other factors at play when it comes to cost increases, including rising energy costs. British inflation reached its highest level in 40 years at the end of last year.

### The Present and Future of the Travel Industry after Brexit

According to an article by Simon Phippard (*What Does Brexit Mean for the Travel Industry?*, 2021), the post-Brexit Withdrawal Agreement provides no guidance for the travel sector, and although the detailed provisions of the Trade and Cooperation Agreement (TCA) include provisions on aviation services and safety, they do not mention explicitly the travel industry.

Brexit marks a major turning point in Britain's current and future worldwide. The effects on travel and tourism destinations are significant. On the one hand, British citizens in Britain and around the globe will be regulated by a tighter travel mobility programme. On the contrary, the United Kingdom government's immigration laws will impact European nationals living and working in Britain and also seasonal workers from Europe working in the tourism, leisure and hospitality sectors. Regardless of your political or ideological views on Brexit, there's no question that Brexit signals a shift away from Visa-free travel. There are a number of significant obstacles for United Kingdom travel businesses wishing to offer holidays in Europe. Employee mobility is by far the most important of these obstacles, as it allows United Kingdom travel companies to employ British nationals to work within the European Union and to deliver trips to British tourists (Mark Tanzer, ABTA).

#### *Holiday Packages*

European Union legislation has had a significant impact on the sale of package holidays and other travel products in the UK for many years. Currently, many of these stem from the European Union's Package Travel Directive (PTD2) of 2015, which imposes obligations on anyone who organizes or sells package tours, or who sells or organizes related tours.

The Package Travel and Related Travel Arrangements Regulations 2018 introduced the Package Travel Directive (PTD2) to the UK market and continue to apply post-Brexit. The 2018 regulations provide for refunds and repatriation in the event of bankruptcy and impose additional obligations on package tour operators at the point of sale, making it easier for consumers to purchase package tours or associated travel agreements (LTAs). Package tour operators are required to provide information on whether the consumer has purchased a package or a linked travel agreement (LTA) and are responsible for the work of subcontractors (such as airlines and hoteliers).

### *Air Travel Organizers License (ATOL)*

The Air Travel Organisers' Licensing Scheme (ATOL) is a form of financial protection that applies to bookings of package trips that include a flight. The Air Travel Organizers (ATOL) is distinct from travel insurance.

The United Kingdom's Air Travel Organisers (ATOL) programme enforces the requirements of the Package Travel Directory on hotels (PTD2), and its predecessors, to protect against bankruptcy. Package holiday organizers and the Linked Travel Arrangements (LTA) must provide ATOL protection under the new Regulations 2018. The more package holidays you book, the more likely it is that ATOL protection will be required. ATOL protection isn't compulsory for long-term arrangements, but you'll need some form of protection (i.e. insurance or deposit). Sometimes ATOL is given as part of a marketing strategy to increase consumer confidence. If an organizer is insolvent while you're abroad, the ATOL fund (which is managed by your CAA) ensures that you'll be able to come home. If your business is insolvent before you travel overseas, you'll have the option of claiming a refund from your ATOL fund.

After Brexit, any UK organizer selling to European customers will no longer be recognized as an insolvency protector by the remaining EU Member States. ATOL will continue to cover bookings you've already made. However, if you're selling a package holiday to an EU country after 1st January 2021, you'll need to comply with local insolvency rules. The United Kingdom (through the Civil Aviation Authority) no longer acknowledges European Union-based bankruptcy protections for European Union-based organizers

Selling to British clients. The United Kingdom has enacted a Statutory Instrument that requires EU traders actively marketing package holidays to United Kingdom customers to follow United Kingdom insolvency regulations. United Kingdom consumers may be vulnerable to buying a package holiday from an EU-based trader and should request relevant insolvency protection before purchasing.

### *Examples of How Brexit Is Affecting the Tourism Industry*

#### *Rising Prices Might Discourage Visitors*

A number of European travel companies specializing in British holidays have warned of increasing hospitality prices in the United Kingdom. A German travel businessman told DW that Brexit has caused him to increase his prices. He has been organizing custom Scotland holidays since the mid-90s for affluent German, Austrian, and Swiss tourists. Four or five years ago a ten-day Scotland holiday for two people would have been priced at €6,000–€8,000, but now he must charge twice as much.

*Staff Shortages become Tougher in Hotels, Bars and Restaurants*

Brexit, according to experts at the London School of Economics and Political Science, has increased the number of paperwork required for conducting business with European Union nations, which is causing Britain's increasing inflation. The expense of doing business with the European Union is also at the heart of the rise in the cost of doing business in the hotel and tourism sectors. Other factors, such as rising energy costs, have also contributed to the rise in UK inflation (School of Economics).

During the Brexit negotiations, immigrant workers of European Union citizenship have been a major concern for UK companies. As a part of the European Union, the tourism sector benefits from a wider labour pool. In 2016, the majority of hotel and other lodging employees in the UK came from other European Union countries. Hotels and restaurants rely on European Union nationals more than any other sector to fill jobs that are difficult to fill. This is often done to fill labour-market skill gaps.

Businesses reliant on European Union nationals who do not fulfil the new work visa standards will undoubtedly suffer skilled labour shortages in the coming years. Multinational corporations, such as those in the entertainment, travel, cruise, shipping, and aviation industries, frequently employ highly mobile workers from all over the globe and rely on their capacity to put individuals anywhere in the world, including the European Union. Employers face certain administrative hassles as a result of Brexit, but it also opens them chances to hire international workers and train local personnel.

*Visitors to the UK*

The majority of international tourists to the United Kingdom come from the European Union. In 2017, 67% of visitors to the United Kingdom came from European Union nations, with France, Germany, Spain, and the Netherlands accounting for a large portion. Eight of the top 10 nations whose visitors visit the United Kingdom are from the European Union.

Joss Croft (CEO, UK inbound) of UK's tourism trade association says that "UK will continue to be a fantastic place to visit", but wants to see "employment as well as travel agreements" between both the United Kingdom and the European Union countries, similar to Australia's working holiday visa model, which allows under 30s to work while on holiday. Croft said that this may give "a significant temporary" opportunity for labour for the difficult tourism sector, as well as "cultural exchange". "We know that young people come here, and they come back, and they're more likely to spend money and do business with the UK," he added (UK Inbound).

According to UK Parliament research, tourism is the fastest-growing economic sector in the UK, employing more than 3 million people and contributing more than £130 billion to the UK economy last year. The tourist sector in the United Kingdom is one of the most developed in the world, ranking sixth in terms of value. London, Europe's most popular city, received about 20% of all international visitors to the UK in 2017 (second only to Edinburgh, which received 2 million).

## Solutions and Recommendations

ABTA Chief Executive Mark Tanzer said that as a consequence of withdrawal from the European Union, United Kingdom workers under the posted staff directive cannot now be employed in the EU travel industry. Instead, the fact that British nationals are not allowed to work for an EU period of up to 90 days every 180 days has proved to be a challenge for many businesses in the sector, affecting roles such as rep in resorts and tour guides. Mark Tanzer thinks expanding the youth mobility programme is a sensible and realistic approach. And even if it doesn't completely solve the problem, it would be a good bridging solution, allowing young people from the United Kingdom to work in the European Union and inversely, helping in the filling of some of the staffing – and also skills gap that exist in both the outbound and inbound sectors (ABTA). There are also a number of other Brexit-related issues, such as making sure that UK visitors are aware and prepared for the new EU entry and exit restrictions, which have been postponed until the year-end and will be implemented simultaneously. Some issues are more complex than others. For example, coach drivers, who spend over 180 days per year in Europe but cannot apply for a visa in any EU country, find the 90-day restriction particularly difficult. There is no quick fix for this problem, but it is essential to continue to raise it with the government so that it remains on the agenda. The fact that this will also impact goods drivers is a reminder of how important it is to work together with non-travellers (ABTA).

## Future Research Directions

Brexit has been a contentious political subject for a while, yet it still affects us in many different ways. It is crucial to examine the relationship between consumer faith in the UK and EU governments and consumer perceptions of uncertainty and consumer decisions, such as travel plans.

The biggest motive for leaving the nation is for holidays: In 2022, United Kingdom citizens spent around £60 billion overseas, up from £15.3 billion in 2021. The majority of the money was spent on holidays (70%), followed by visits to friends and family (20%) and work travels (8%). According to a poll, two-thirds (66%) of British people intend to spend at least some money on holidays overseas in 2023, whereas 27% do not expect to spend money on vacation abroad or just plan to holiday in the United Kingdom.

Other studies point to a forecast that each British tourist will spend an average of £6,684 on holidays between 2024 and 2025 and will travel three times a year. An indicator of this recovery is the data presented by Spain, a natural destination for the UK, which received some 7 million 755 thousand 540 tourists from the United Kingdom in the first half of 2023, which represented a 20% growth in relation to the same period of the year. 2022. It is estimated to reach the figure of more than 17 million by the end of this year.

According to an analysis carried out in 2022 by York Aviation on behalf of ABTA Travel Association, the United Kingdom's international travel sector is expected to rise 15% by 2027, compared to 2019 trends. This article suggests the importance of the quality and price ratio will be the main motivation for choosing the destination. The robustness and solidity of the United Kingdom economy and its tourism industry indicate a recovery in travellers that, according to experts, will have its peak by 2025 with approximately 24 million tourists will once again recorded.

Government and the Travel Industry must continue working together on the recovery of the tourism sector. Members of the Travel Industry should attend future conventions to address the policy goals for supporting United Kingdom travel firms and to emphasize the importance of the outbound travel sector to the United Kingdom's economy.

Future meetings with Travel Organisations and Members of the Parliament were to discuss statistics, which demonstrate that the United Kingdom's outbound travel sector supports more than 800,000 employees and contributes £49 billion to the United Kingdom economy. According to ABTA research, the industry will expand by 15% by 2027 compared to 2019 levels, outpacing predictions of a 10.3% rise in the UK economy generally during that time.

Finally, it is important to talk about the youth mobility plan to be expanded to include European Union nations so that United Kingdom travel companies may hire United Kingdom personnel in European Union locations, according to ABTA, which will also use the occasion to outline particular policy concerns the sector wants the government to solve.

## Conclusion

First and foremost, UK tourism continues to be optimistic about the future. A solution already exists, at least partially, to address some of the issues highlighted in this report: an extension to EU countries of the current Youth Mobility Plan (YMS). This can be achieved over time in line with the points-based immigration system introduced after Brexit. It is a simple policy solution, and it is a priority that the Government moves towards re-negotiation and negotiations with various European Union countries. It is important that policymakers across the European Union work together to rebuild the conditions that have been built over decades.

**Case Study**

Due to the Brexit immigration regulations, the travel industry is suffering the effects of staffing. Many Europeans employed in the travel industry have left the United Kingdom to find employment elsewhere or to return to their European Union country of origin. The travel industry has long relied on low-paid labour from its European Union nations, which has ceased to exist. Prior to Brexit, the free movement of labour allowed for easy recruitment of seasonal, skilled, and temporary workers.

According to University of Oxford's Migration Observatory, between June 2019 and June 2021, the number of European Union workers engaged in the tourism business in the United Kingdom fell by 25%.

The hospitality sector in the United Kingdom is still understaffed in summer 2023, and firms fear they will struggle to keep up. To alleviate post-Brexit shortages, the travel industry is appealing for European Union workers to be permitted into the United Kingdom.

Brexit resulted in a negative effect for many workers in Britain and has implications for future immigration policies. The majority of arguments in favour of inward migration for employment-related reasons have focused on the advantages for local economies and the improvement of cultural variety.

It has been determined that migrant labourers improve cultural variety. In several industries, migrant workers make up a sizeable share of the workforce. The fourth largest employment in the United Kingdom is the Travel and Tourism industry, consisting of restaurants, accommodation, Tour Operators, and Airlines, among other establishments. According to the Labour Force Survey 2014, 1–4, migrant workers made up 25% of the workforce overall, and 34% of the labour force in the "food preparation and leisure" sector.

The Trade Body for Hospitality in the United Kingdom "Hospitality" has said that there is a "serious crisis" in staffing, with vacancies 48% greater than before COVID-19. Employers in hotels, tour operators and airlines claim a need for additional European Union employees. European service staff are unable to work in the UK because Brexit has removed the EU's free movement of workers. A recent article in the New York Times newspaper reported that 11% of restaurants across London had to cut hours because of staff shortages. Businesses that used to hire Italian, Spanish or Greek cannot do that anymore due to the Immigration regulations (UKHospitality.org.uk).

According to Deloitte's market projection for 2023, human shortage-related disruptions would certainly intensify and remain until 2025. This prediction was supported by rising inflation. One example of the shortage of staff is the Tour Operator Travel Simply Tours based in central London. The tour operator has been suffering with a shortage of staff for almost two years, and has gone from having more than ten reservations agents working from the office to four travel agents, which has affected the volume of reservations that the travel agents can do, and the quality of the services they offer as the employees have been working for more than eight hours a day, affecting their performance in the reservations process and having a few disappointed clients.

Travel Simply Tours is worried that this problem might continue for longer, stopping it for growing the business. One of the main obstacles that the tour operator's human resources department is having in addition to the lack of personnel is that most of the potential candidates for the job are not interested in working in the office, they would rather work hybrid, or full time work from home.

**Case Questions (Three Questions)**

***Questions***

1 How can the Travel Simply Tours improve the performance of its employees?
2 What measures can the Tour Operator take to employ more travel agents?
3 Can the government and the travel industry work together on having a better understanding of tourism productivity to find solutions to these problems?

***Solution***

Travel Simply Tours may need to modify its hiring practices, make investments in programmes for the growth and development of current employees, and look into alternate labour markets to meet these issues. Some of the strategies employed to address this shortage of staff within the Tour Operation business can be:

***Diversifying Recruitment Sources***

To recruit domestic staff, offering training programmes and apprenticeships.

The Company can also employ students of the Travel industry who can work on a part time basis. Travel Simply tours can diversify its workforce by recruiting staff from non-EU countries.

### *Remote Work and Flexibility*

The business can also expand their offer to include remote workers who can work from their home country.

### *Join Trade Organizations*

Another solution can be to join trade organizations and lobby for better immigration laws for the tourist industry with other tour operators. Government policy changes may be influenced through cooperative efforts.

### *Training Programmes*

In order to close the skills gap, Travel Simply Tours might create training and development programmes for its current employees to assist them in obtaining the knowledge required for particular jobs. For instance, they could provide instruction in specialized directing methods or language courses.

### *Retention Methods*

Travel Simply Tours can employ retention techniques to keep current employees, such as offering competitive compensation, alluring perks, and chances for professional progression. Employee retention is higher when they are happy with their jobs.

### *Market Expansion*

Look at emerging areas and industry segments where a lack of qualified workers might not be as acute. Travel Simply Tours may be able to minimize its dependency on known markets and more effectively adapt to changing labour demographics by doing this.

### *Results*

Over time, Travel Simply Tours successfully handled their workforce problem by putting these strategies into practice. They expanded their number of potential employees, trained their current employees with new courses and certification, offered hybrid jobs to all of their employees, giving them the opportunity to work from home three days and week, and formed alliances with nearby educational institutions giving to students the opportunity to learn and have the work experience needed for their careers. Travel Simply Tours also improve the perks for the employees, increasing their salaries and bonus in 5% By making these steps, Travel Simply Tours not only lessened the employment deficit but also set itself up for long-term development and flexibility.

**Key Terms**

- **ABTA**: British Travel Agents Association. The largest travel association, ABTA, represents travel agencies and tour operators who offer holidays in the United Kingdom. ABTA was created in 1950, with more than 1,200 members and 900 tour operators. Its Chief Executive is Mark Tanzer.
- **ATOL**: Air Travel Organisers Licensing. The UK Financial Protection Scheme that protect consumers if their travel organizer should bail. It provides financial security when booking a package holidays with a United Kingdom Travel Company.
- **CAA**: The Civil Aviation Authority is in charge of aviation safety regulation in the United Kingdom. The Civil Aviation Authority is a statutory company in the United Kingdom that controls and regulates all elements of civil aviation.
- **GDPR**: The General Data Protection Regulation (GDPR) of the European Union. GDPR is a European Regulation on data privacy that applies to the European Union and the European Economic Area. The General Data Protection Regulation (GDPR), which went into effect on May 25, 2018, establishes a framework for the collection, processing, storage, and transfer of personal data.
- **PTD2**: A set of rules that must be followed by persons selling and booking package holidays and associated travel arrangements. With regard to package travel, Council Directive 90/314/EEC(3) outlines a number of significant consumer rights, particularly with regard to information requirements, traders' liability for package performance, and protection from the insolvency of an organizer or retailer.
- **WTM**: World Travel Market-The world's largest and most significant travel and tourism event in London that inspires, educates, sources, and benchmarks travel professionals while providing exhibitors with a platform to do business and market their products to a global media audience.
- **WTTC**: World Travel and Tourism Council – The world's foremost authority on the economic and social contributions of travel and tourism. Its members work with governments to raise public awareness of the travel and tourism industry.
- **Brexit**: Abbreviation for "British" and "Exit". The term was used prior to the referendum on June 23, 2016, in which a majority of the United Kingdom voters decided to exit the European Union.

## References

ABTA & Deloitte. (2016). What Brexit Might Mean for UK Travel. In *www.abta.com.* https://www.abta.com/industry-zone/reports-and-publications/what-brexit-might-mean-for-UK-travel

BBC News. (2019, December 17). *Brexit bill to rule out extension to transition period. BBC News.* https://www.bbc.co.uk/news/election-2019-50818134

Hall, D. (2020). *Brexit and tourism: Process, impacts and non-policy* (Vol. 86). Channel View Publications.

Horsfield, G. (2020, May 21). *Travel trends - Office for National Statistics.* https://www.ons.gov.uk/peoplepopulationandcommunity/leisureandtourism/articles/traveltrends/2019#:~:text=There%20were%2093.1%20million%20visits,of%207%25%20compared%20with%202018

Lau, P. L. (2022). *The Murky Waters of the Metaverse: Addressing Some Key Legal Concerns.*

Lee, D. (2023, February 24). *A third of hospitality at risk of going under, despite record sales - CGA.* CGA. https://cgastrategy.com/a-third-of-hospitality-at-risk-of-going-under-despite-record-sales/

*Local Government Association Briefing the Impact of Brexit on Tourism and Creative Industries.* (2018). In https://www.local.gov.uk. House of Commons. https://www.local.gov.uk/sites/default/files/documents/LGABriefing_BrexitandTourism_HoC_April18_v2.pdf

Ma, H., Chiu, Y., Tian, X., Zhang, J., & Guo, Q. (2020). *Safety or travel: Which is more important? The impact of disaster events on tourism. Sustainability*, 12(7), 3038. https://doi.org/10.3390/su12073038

*Mobile roaming in the EU after Brexit.* (2023, September). https://commonslibrary.parliament.uk/research-briefings/cbp-8649/.

On, H. (2023, April 26). *How does Brexit impact the tourism sector? Hospitality ON.* https://hospitality-on.com/en/niches-tourism/how-does-brexit-impact-tourism-sector

Team, W. and Team, W. (2021, November 1). *Industry report: Brexit's effect on travel still to come | WTM Global Hub. WTM Global Hub* [Preprint]. https://hub.wtm.com/press/wtm-london-press-releases/industry-report-brexits-effect-on-travel-still-to-come/.

Team, W., & Team, W. (2022, November 11). *Industry Report: Cost-of-living is biggest challenge for travel in 2023 by far, says WTM poll | WTM Global Hub.* WTM Global Hub. https://hub.wtm.com/press/wtm-london-press-releases/industry-report-cost-of-living-is-biggest-challenge-for-travel-in-2023-by-far-says-wtm-poll/

The October 2019 EU-UK Withdrawal Agreement. (n.d.). https://researchbriefings.files.parliament.uk/documents/CBP-8713/CBP-8713.pdf.

*The EU-UK Trade and Cooperation Agreement.* (2021). European Commission. https://commission.europa.eu/strategy-and-policy/relations-non-eu-countries/relations-united-kingdom/eu-uk-trade-and-cooperation-agreement_en

*Tourism: statistics and policy.* (2023, November 7). https://commonslibrary.parliament.uk/research-briefings/sn06022/. https://commonslibrary.parliament.uk/research-briefings/sn06022/

*Travel Law Today.* (2021, June). https://www.abta.com/sites/default/files/media/document/uploads/5023%20TLT%20Spring%2022_online.pdf.

Section 2

# Sustainable Practices and Environmental Considerations

# 4 Environmental Sustainability in the Tourism Sector

*Claudia Sevilla-Sevilla, Ainhoa Rodriguez-Oromendia and Julio Navio-Marco*

## Introduction

Environmental sustainability in the tourism sector is crucial to preserve natural resources and protect biodiversity for future generations. By adopting sustainable practices, responsible tourism reduces carbon footprints, mitigates environmental degradation, and fosters a positive impact on local communities, ensuring a sustainable and thriving industry.

## Background

Travel and tourism are among the fastest-growing sectors globally, with industry forecasts projecting 1.8 billion international arrivals by 2030 (WTTC, 2019). However, due to its growth and impact, environmental sustainability has become a critical issue receiving increasing attention in recent years. Tourism not only contributes significantly to global greenhouse gas emissions (WTTC and International Transport Forum, 2019) but also consumes natural resources, leading to local ecosystem degradation and loss of habitats. The sector's footprint will be key due to both negative and positive externalities it generates. Therefore, promoting sustainable responses to societal needs and customer demands is necessary.

Implementing policies and practices aimed at reducing the environmental footprint of tourism, such as energy and water conservation, waste reduction, and the use of renewable energy sources, is gaining prominence to preserve the main source of income for many tourist destinations. Additionally, the perception of a destination as sustainable can significantly impact tourists' experiences, purchasing behavior, and loyalty. Thus, there is a need for greater emphasis on sustainable destinations, driven by the tourism industry and public administrations, to promote policies that foster the "sustainable" component.

This chapter focuses on examining how tourism research has integrated sustainability as a key factor in the success of tourism destinations. It explores the correlation between sustainable tourism policies, effective destination management and overall destination success. Furthermore, the perception of destinations as sustainable can profoundly influence tourists' experiences, purchasing decisions, and loyalty to those destinations.

DOI: 10.4324/9781003369967-7

## The Concept of Sustainable Tourism

The word "sustainability" has become a common term in politics, economics, environmental management, and research, but much of the work done either does not contain explicit definitions of what we mean by sustainability, or the word is replaced by other terms such as "sustainable development," without specifying the approach from which that term is approached and the two are used practically as synonyms. Defining sustainability is still a lively debate and there are hundreds of related definitions. Sustainability is closely related to the search for positive approaches that focus on ecosystem conservation, ensuring adequate levels of biodiversity in the face of global change.

The World Tourism Organization (UNWTO) defines Sustainable Tourism as "Tourism that takes full account of its current and future economic, social, and environmental impacts, addressing the needs of visitors, the industry, the environment, and host communities." The concepts of "sustainable development" and "sustainability," as understood in the first two decades of the 21st century, stem from the Brundtland Report (1987). It suggests that the truly sustainable progress needs to address simultaneously interconnected issues of the economy, environment, and social well-being (Johnston, Everard, Santillo, & Robert, 2007). Sustainability has become a new paradigm of development applied to any economic activity, including tourism (Eusebio, Kastenholz, & Breda, 2014; Kastenholz, Eusebio, & Carneiro, 2018).

Viñals and Teruel (2021) point out that the basic ideas defining the concept of environmental sustainability relate to responsible human interaction with the environment. This definition is framed within the paradigm of "sustainable development" described in the Brundtland Report (1987), which advocated for the rational use of resources and globally contrasted the economic development model based on environmental sustainability. Decades later, environmental sustainability remains an ongoing and unresolved issue. Hence, it forms the basis of the global framework for international cooperation: the 2030 Agenda for Sustainable Development and its Sustainable Development Goals (SDGs). The tourism sector can play an important role in achieving the SDGs by addressing interlinked social, economic, and environmental aspects (Johnston et al., 2007).

Tourism is a global phenomenon that relies on the utilization of territory and its resources, which, in turn, constitute its greatest business asset (Otero, 2007). The degradation or overexploitation of resources not only diminishes the environmental wealth of a destination but also its social (Büscher et al., 2017; Fletcher, 2019) and economic value due to the loss or degradation of resources and the resulting decline in the quality of the tourism offering, impacting its competitiveness (Viñals & Teruel, 2021).

Today it is necessary to go beyond the mere communication of sustainable initiatives. Sustainable tourism policies in promoting responsible practices and collaboration for long-term development are essential.

## Sustainability in Tourism Destinations: Beyond the Communication of Initiatives

Sustainable tourism should be understood as a form of management that should be applied to any type of product or tourist destination. According to the definition of the World Tourism Organization (UNWTO), a tourism destination is a physical space, with or without administrative or other boundaries, where a visitor can stay overnight. It is an aggregation, in one place, in the tourism value chain, and a basic unit of sectoral analysis. A destination can incorporate different stakeholders and can expand networks to form larger destinations. It is also intangible, with a corporate identity that can influence its competitiveness in the market.

Most tourist destinations nowadays aim for sustainable development as a development strategy (Choi & Sirakaya, 2006; Eusebio, Kastenholz, & Breda, 2014; Kastenholz, Eusebio, & Carneiro, 2018; Morgan, 2012). Achieving necessary visibility in search engines requires a series of techniques and strategies to improve the online positioning of a website (Moran & Hunt, 2005; Zhang & Dimitroff, 2005; Xiang, Pan, Law, & Fesenmaier, 2010). Destination Management Organizations (DMOs), in order to successfully promote their products and services to potential visitors, must ensure that relevant information is visible and accessible (Buhalis, 2000; O'Connor & Frew, 2002; Xiang et al., 2010).

Online marketing communication processes, as Buhalis and Law (2008) pointed out, have become a mainstream interaction stream where customers are dynamic targets to direct promotional messages toward. The internet drives the reengineering of the entire process of tourism product production, communication, and delivery, as well as drives interactivity among partners who can design specialized products and promotion to maximize the added value provided to individual consumers (Navio-Marco, Ruiz-Gómez, & Sevilla-Sevilla, 2018). Communicating sustainability performance can be a powerful communication tool with various stakeholder groups of an organization (Font, Walmsley, Cogotti, McCombes, & Häusler, 2012). Communication with various internal and external stakeholder groups of the organization is a necessary requirement to promote sustainability (Chan & Hawkins, 2012; Delmas, 2001). Information related to sustainability and sustainable development can positively affect the overall image of the company and purchasing behavior (Chan, Hon, Chan, & Okumus, 2014; Chen, 2015; Choi, Parsa, Sigala, & Putrevu, 2009; Lee, Hsu, Han, & Kim, 2010; Teng, Wu, & Liu, 2013), and it is an important issue for hospitality and tourism (Chen, 2015). It is becoming increasingly common for organizations that incorporate sustainability into their strategy to highlight their commitments using social media, online and offline promotional materials, and even by integrating environmental responsibility into the company's mission and vision (de Souza Cavalcani & Teixeira, 2015). Academic research on destination management and marketing conducted between 2005 and 2016 indicates growth patterns related especially to the development, the management, competitiveness, and sustainability of destinations (Avila-Robinson & Wakabayashi, 2018).

It has been criticized in the academic literature that the existence of initiatives to promote sustainability is merely a facade designed to attract consumers with a growing awareness of sustainable initiatives (Jones, Hillier, & Comfort, 2014). The communication of organizations focuses on making known the different initiatives that are implemented, raising questions about a serious commitment to sustainability, as well as whether certain environmental practices are merely a disguise for cost-saving policies by the organization (López-Gamero, Molina-Azorín, & Claver-Cortes, 2011). There is also no consensus in the academic debate on management tools to promote sustainability (such as ISO 14001 or EMAS, for example), which are primarily used to promote the company's image (Boiral, 2017; Font et al., 2012). On the other hand, recent studies indicate that an approach to communicate sustainability, as often used by marketers, is unlikely to be as persuasive as an experiential or hedonistic approach (Grant, 2007; Hanna, Font, Searles, Weeden, & Harrison, 2018; Malone, McCabe, & Smith, 2014; Rex & Baumann, 2007; Villarino & Font, 2015).

Today, there is a risk that research in the tourism sector remains at such an abstract level that it hinders its practical application in the sector in general and in tourist destinations in particular (Alonso-Muñoz, Torrejon-Ramos, Medina-Salgado, & Gonzalez-Sanchez, 2023). Probably due to the need to make sustainability more tangible, the sector is immersed in a considerable proliferation of certificates (Spenceley, 2019). This reality, far from helping, contributes to generating even more confusion than guidance or clear guidelines between companies and tourists. Being able to establish agreed indicators recognized by the tourism sector would allow knowing the level of sustainable performance achieved by the company in comparison with the sector. With regard to concrete actions and practices related to environmental sustainability, strategic planning of energy consumption and reduction of emissions is necessary. Both companies and key stakeholders should consider options for alternatives related to current energy use and associated benefits (Kornilaki & Font, 2019). For this, the cooperative role of public administrations and companies is crucial. Compliance with the 2030 Agenda can be a differentiating element for tourist destinations. It is necessary to incorporate a specialized orientation in the inclusion of the SDGs in the strategic objectives of companies and institutions. The consulting firms seem to have taken on this task and, curiously, the academic world is not incorporating the 2030 Agenda among its main approaches (Alonso-Muñoz et al., 2023).

## Sustainable Tourism Policies: Promoting Responsible Practices and Collaboration for Long-Term Development

Sustainable tourism should reduce tensions provoked by the complicated interactions among the tourism companies, visitors, the environment, and the communities hosting tourists (Bramwell & Lane, 1993), within a regulatory and political framework that facilitates and supports the drive toward sustainability.

Sustainable tourism promotes the sustainable long-term development of tourism (Ahmed, Bin Mokhtar, Lim, Hooi, & Lee, 2021; Khan, Lim, Ahmed, Tan, & Bin Mokhtar, 2021). To prevent the deterioration of natural resources, offer a better tourist experience, and improve the well-being of host communities, more ambitious sustainable tourism policies must be implemented and destination management and activity monitoring has to be improved (Goffi & Cucculelli, 2018).

The roles of Destination Management Organizations (DMO) are inevitable in tourism management and development (Khan, Khan, Lim, Tan, & Ahmed, 2021). Planning and management are key factors in tourism development. To achieve sustainability tourist destinations must apply suitable policies for sustainable development (Byrd, Bosley, & Dronberger, 2009; Yuksel, Bramwell, & Yuksel, 1999). In this line, environmental and cultural resources are crucial factors for the development of a tourist destination; therefore, these resources must be developed responsibly through sustainable tourism policies and destination management (STPDM) (Cucculelli & Goffi, 2016; Goffi & Cucculelli, 2018; Su, Huang, & Huang, 2018). It has been established that socially responsible practices in tourist destinations promote sustainable development (Su & Swanson, 2017).

### *Examples of Good Practices in Sustainable Tourism Policies*

Capacity management: Establishing limits on tourist carrying capacity is essential to avoid overloading and preserve the quality of life of local communities. For example, visitor control systems such as issuing permits or limiting access to sensitive areas can be implemented.

Promotion of environmental conservation: Sustainable tourism policies should promote the protection and preservation of the natural environment. This can include the creation of protected areas, promoting low-impact practices and environmental education for visitors and the local community.

Support for the local economy: Sustainable tourism policies should promote local economic development by supporting local businesses and enterprises. For example, promoting local products and services, creating markets and sales spaces for local producers, and providing business skills training to enhance competitiveness.

Social and cultural inclusion: Sustainable tourism policies should ensure the participation and equitable benefit of local communities, including minority groups and vulnerable populations. This involves promoting community participation in decision-making, preserving cultural identity, and respecting local traditions.

Education and awareness: Sustainable tourism policies should foster education and awareness among both tourists and the local community. This can be achieved through information campaigns, educational programs in schools and tourist centers, and promoting responsible practices among visitors.

Public-private collaboration: Collaboration between the public and private sectors is essential for effectively implementing sustainable tourism policies. This involves cooperation in planning, project implementation, and resource allocation to achieve common objectives.

### *How to Promote Sustainability at Travel Destinations*

Everything indicates that sustainability is, and should be, a priority for tourist destinations and it will be the challenge that destination management currently faces. However, if the sustainability of tourism destinations is to be managed, it must first be measured (Aguirre, Zayas, Gomez-Carmona, & López Sánchez, 2022). Despite the existence of international indicators, such as the European Tourism Indicator System for Sustainability (ETIS) or the sustainability indicators proposed by the UNWTO (UNWTO, 2005), these tools are not fully implemented due to the scarcity of available information and the need to adapt existing indicators to current information requirements (Aguirre et al., 2022; Gasparini & Mariotti, 2023; Tudorache, Simon, Frenţ, & Musteaţă-Pavel, 2017).

The work of Font, Torres-Delgado, Crabolu, Palomo Martinez, Kantenbacher and Miller (2023) highlights the mounting pressure faced by Destination Management Organizations (DMOs) to incorporate sustainability into their policies and the need to properly plan and manage future tourism and its impacts (McLoughlin & Hanrahan, 2019). Sustainable tourism indicators can also be useful (Castellani & Sala, 2010; Valentin & Spangenberg, 2000): by monitoring sectoral development to facilitate policy and practice evaluation, measuring the progress of the sector and developing suitable strategies for a desirable future and communicate knowledge by providing quantitative data that obtain a more holistic understanding of tourism phenomena in their spatial context.

According to Font et al. (2021), the World Tourism Organization has promoted indicators as a key management tool for tourism planners, as they provide information on issues and areas of concern (impacts, product quality, threats, etc.), facilitate the assessment of the performance of tourism plans and provide evidence for evaluating the planning and policy framework. The linear use of indicators in policy formulation follows the principles of evidence-based policy. Indicators are subject to biases and inherent limitations. They can be influenced by data availability, measurement decisions, and conceptual frameworks used. If indicators are not properly selected, measured, and interpreted, biased or incomplete results can be obtained. It is necessary to supplement their use with a broader and holistic approach, considering the complexity of the issues, understanding causal relationships and context, acknowledging the limitations of indicators, and promoting participation and diversity of perspectives in the decision-making process.

External factors such as COVID 19, excessive tourism, pollution, conflicts among key players, gentrification among others can significantly impact travel destinations. To ensure high-quality travel experiences for tourists and promote the well-being of relevant stakeholders at these destinations, it is crucial to prioritize sustainable destination management (Ahn, 2022).

### *Benchmarking*

Benchmarking is a strategic management tool used by organizations to compare their performance with industry best practices or competitors. It provides insights into improvement, setting goals, and enhancing overall competitiveness. Four examples of good practices in tourism management have been selected and are shown below.

## Segittur

In Spain, the State-Owned Company for the Management of Innovation and Technological Tourism (SEGITTUR) developed the methodology called "Smart Tourist Destinations" (DTI) with the aim of promoting sustainable tourism development and ensuring tourist satisfaction, as well as the quality of life for the local population. The areas covered include sustainability management, environmental conservation, preservation of culture, and socio-economic development of the local community (SEGITTUR, 2020). The diagram provides a brief description of each area and the specific components that each one addresses to manage sustainability at the destination (Figure 4.1).

### *Formentera*

Formentera is a small island located in the Balearic Islands, Spain. It has taken significant measures to preserve and protect its natural environment and promote

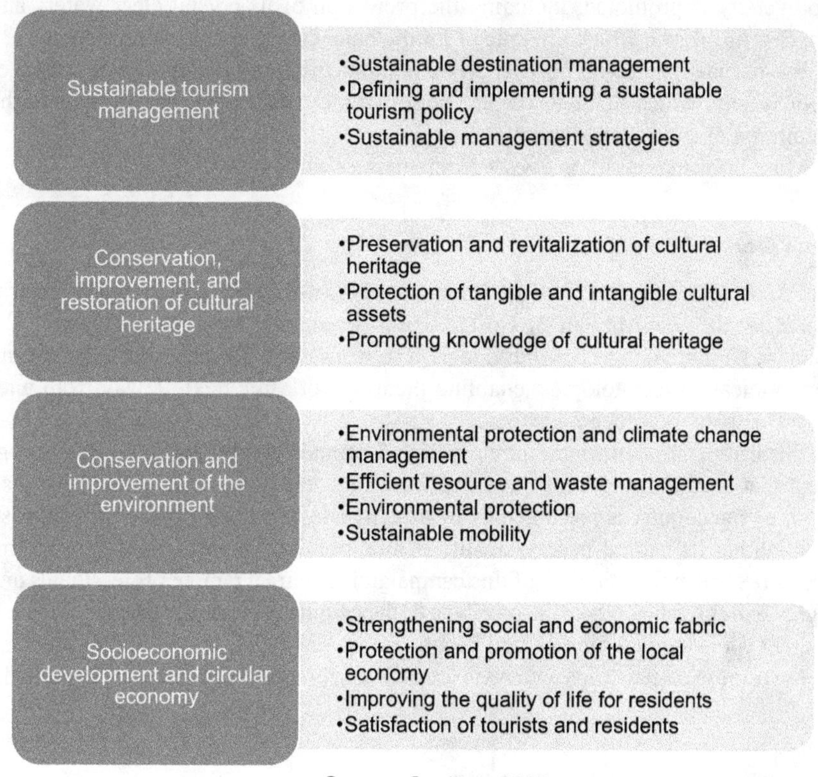


*Figure 4.1* Smart Tourist Destinations methodology – brief description of each area and the specific components that each one addresses to manage sustainability at the destination.

*Source:* Segittur (2022).

sustainability in various areas. Some of the sustainable practices implemented on the island include:

The Insular Territorial Plan of Formentera (PTI) establishes measures to limit the construction of new tourist establishments and control urban growth. These measures include the prohibition of large hotel complexes and the promotion of more sustainable alternatives such as rural tourism and vacation rentals. Additionally, a limit has been set on the island's tourist accommodation capacity to prevent saturation and preserve the quality of life for local residents. This means that only a certain number of hotel rooms and tourist accommodations are allowed on the island, and compliance with these limitations is strictly monitored.

Renewable energy: The island's investments in renewable energy sources, such as solar and wind energy, reduce dependence on fossil fuels and minimize greenhouse gas emissions.

Environmental protection: Formentera has implemented measures to protect its natural ecosystems. Protected areas have been established, and the conservation of biodiversity is promoted, including the protection of its crystal-clear waters and seagrass meadows, which are crucial for the balance of the marine ecosystem.

Waste management: The island has implemented an efficient waste management system, which includes the promotion of recycling, waste reduction, and the promotion of a circular economy.

More information: https://www.formentera.es/.

### *Costa Rica*

The "Costa Rica Digital Nomads Campaign," created in 2021, is an initiative promoted by the Government of Costa Rica to encourage long-stay tourism in the country. Digital nomads are individuals who work remotely using information and communication technologies, enabling them to work in a flexible way from anywhere with an internet connection.

The campaign aims to attract these digital nomads by offering them an environment that encourages them to live and work in Costa Rica for an extended period of time. The country is promoted as an attractive destination for this type of tourist, highlighting its natural beauty, quality of life, political stability, and digital infrastructure. The main objective of this campaign is to attract remote professionals and digital entrepreneurs who can contribute to the country's economic and social development while enjoying the natural beauty and quality of life that Costa Rica offers.

More information: https://www.visitcostarica.com/en/costa-rica/digital-nomads.

### *Sydney*

The Sydney "Single-use Items" campaign focuses on reducing the use of single-use products in businesses throughout the city. Single-use items are designed for one-time use and then discarded; such items include plastic packaging, straws, disposable cups and plastic bags, among others. Tourists are an important target for this campaign, as their participation and choices can have a significant impact on reducing the consumption of single-use items. The campaign promotes educating and raising awareness among tourists about sustainable alternatives, such as carrying reusable water bottles, using reusable shopping bags, or choosing restaurants and establishments that offer reusable or compostable packaging and utensils.

The regulation avoids the supply of lightweight plastic bags from 1 June 2022. The supply of other items has been prohibited as from 1 November 2022.

More information: https://www.sustainabledestinationpartnership.com.au/projects/single-use-items.

## The Role of Sustainability and Green Marketing in Enhancing Corporate Image and Customer Loyalty

Incorporating sustainability into the business practices is a goal for companies seeking to enhance their legitimacy in their stakeholders' eyes and also helps to improve perception, reputation, and customer loyalty (Sevilla-Sevilla, Mondejar-Jimenez, & Reina-Paz, 2019). The perception of a company's overall image significantly influences consumer buying behavior While consumers value sustainability, they are still price-sensitive. However, some studies have shown that consumers are willing to pay more for products or services from firms that prioritize sustainability in their business strategies (Sevilla-Sevilla, Mondejar-Jimenez, & Reina-Paz, 2019).

Most of the literature that addresses Corporate Social Responsibility (CSR) or sustainability defines it as a multidimensional construct encompassing three pillars: the economy, society, and the environment (Martinez & Rodriguez Del Bosque, 2013). However, measuring, and perceiving, sustainability from the perspective of consumers is still an open issue, despite the available research and the variety of methods for assessing consumer perceptions. Almost all studies face numerous limitations (Turker, 2009), and a variety of measurement tools have been utilized (Brown & Dacin, 1997; Carrigan & Attalla, 2001; Marin & Ruiz, 2007; Matute-Vallejo, Bravo, & Pina, 2011; Sen & Bhattacharya, 2001). As a result, there are instances where the findings are not consistent, primarily because of differences in how conceptual aspects like the perception of sustainability or corporate social responsibility (CSR) itself are defined (Peloza & Shang, 2011).

There is limited research available on consumers' understanding of sustainability and their preferences regarding the specific behavior of companies, which they value or wish to see improved (Jones, Hillier, & Comfort, 2014). Optimistic analyses confirm consumers' increased sensitivity toward environmental matters (Chan & Hawkins, 2012).

Several case studies have explored the significant positive impact of environmental values and awareness on tourists' intentions to engage in environmental behavior (Rodriguez-Oromendia, Reina-Paz, & Sevilla-Sevilla, 2013). Environmental concerns dominate the literature on sustainability, and greater emphasis should be placed on effectively combining economic and social goals with environmental goals (Chasin, 2014). Sustainability terms are an interconnected system encompassing environmental protection, economic performance, and social welfare (Glavic & Lukman, 2007).

The concept of sustainability and sustainable development has frequently been undermined by a subtle redefinition that panders to various interests and prioritizes traditional economic development as the driving force (Johnston et al., 2007; Sevilla-Sevilla, Mondejar-Jimenez, & Reina-Paz, 2019). The discussion involves different stakeholder perspectives and interests (Souza, Rosenhead, Salhofer, Valle, & Lins, 2015).

In the highly competitive tourism sector, the challenge lies not in acquiring information but in effectively interpreting that information to reduce risks in decision-making and simplify marketing strategies focused on one of the main stakeholders: customers. This challenge has become more pronounced today in the era of big data and artificial intelligence. Responsible use of natural resources, minimizing environmental impact, reducing waste production, addressing pressures on water and land, preserving biodiversity, using sustainable energy sources, safeguarding heritage, and maintaining the integrity of natural aspects and culture of destinations are key aspects of sustainability. The incorporation of sustainable practices in tourism investments can influence customer loyalty and attract new tourist segments (Bernini, Urbinati, & Vici, 2015).

Due to the environmental issues that afflict the world (global warming, environmental degradation, air pollution…), customers are increasingly considering green options to incorporate into their behavior that can benefit the planet (Hsieh, 2012; Jones, Hillier, & Comfort, 2014; Martin Martin, Jimenez Aguilera, & Molina Moreno, 2014).

### *The Concept "Green Marketing"*

Green marketing is the responsible management process that aims to identify, predict, and fulfill the needs of customers and society while also ensuring profitability and sustainability (Yadav, Kumar, & Swaroop, 2016). A manifestation of this compromise is through environmental certifications programs.

Environmental certifications improve environmental management because these schemes can help firms to work and focus on environmental performance (Qi, Zeng, Li, & Tam, 2012). While the significance of environmental certifications in the hospitality industry has been acknowledged, it is essential to gather more evidence on how these certifications can generate positive responses from consumers. Martinez, Herrero and Gomez-Lopez (2019) propose the need for further research in this area and recommending: First, strategies should be devised to increase the perception of the green features of environmentally certified hotels. Second, the hotel industry must effectively communicate the objectives of environmental certifications so that customers recognize the importance and value of operating environmentally certified hotels to reduce ecological degradation and trigger favorable customer responses. Third, a green positioning strategy should be developed and based on cognitive and affective aspects of corporate image.

The adoption of green marketing benefits the sector primarily for two reasons: First, the potential benefits of resource efficiency and related savings in consumption (e.g., water and energy consumption). Second, the market opportunities that arise from growing consumer interest in sustainability (Chen, Chen, Zhang, Xu, & long, 2018).

Green marketing in particular and tourism management in general leverages new technology by using innovative tools to measure environmental impact, analyze data, improve practices, etc. Technological advancements also enable new travel solutions in the tourism industry for sustainable development.

## New Technologies for a Sustainable Approach in Tourism: Reflections from Academia

The perspectives on global macroeconomic development point toward the necessary formulation of policies in key areas such as combating climate change, sustainable financing, and sustainable production and consumption (UN, 2019). The present situation compels us to develop new approaches to current systems of consumption and production, and to cultivate critical thinking within academia in response to this era of rapid changes, complexity, and growing conflicts, enabling us to rigorously examine the challenges posed by the immediate future (Boluk, Cavaliere, & Higgins-Desbiolles, 2017).

The remarkable advancements in technology now enable professionals in the tourism industry to generate highly personalized and relevant information for consumers in real-time (Buhalis & Foerste, 2015). It can be expected that in the next decade, ongoing progress in technology will foster greater communication and collaboration between tourists and stakeholders, satisfying the interests of both parties (Jovicic, 2019). The prospects for international tourism growth are contingent upon the necessary transformation of energy technical systems toward decarbonized sources and changes in consumption habits that ensure sustainability.

The meta-challenges identified by professionals in the tourism sector for global tourism today are (Table 4.1):

*Table 4.1* Meta-Challenges in the Tourism Sector for Global Tourism

| | |
|---|---|
| Globalization | This challenge encompasses the development of "new" source markets |
| Effects | To reduce negative effects and maximize positive effects |
| Balance | Tourism requires resources (nature, heritage, culture…), but there is a risk of destroying these values |
| Adaptation | Tourism is a system within a dynamic framework. More effective businesses need to be generated |
| Self-control | The need for strategic planning and professional management based on a set of shared objectives and values |

*Source:* Adapted from von Bergner and Lohmann (2014).

It is advisable for academia, businesses, and society as a whole to review, delve into, and reflect upon the identified meta-challenges, as the tourism industry cannot afford to ignore the challenges it faces. Tourism will either be sustainable or it will not be.

***The Advent of the Onlife-World***

Technology is enabling a significant shift: the coming together of the physical and digital worlds. To comprehend the transformations brought about by the explosion of new digital information and communication technologies, the European Commission promoted a document of philosophical nature: "The Onlife Manifesto: Being Human in a Hyperconnected Era" (Floridi, 2015). The term "onlife" is adopted to refer to the new reality in which the following transformations are observed (Table 4.2):

*Table 4.2* Onlife World

| | |
|---|---|
| Onlife World | Blurring of boundaries between the real and the virtual |
| | Blurring of boundaries between humans, machines, and nature |
| | Shift from scarcity to information abundance |
| | Transition from the primacy of things to the primacy of interaction |

*Source:* Adapted from Floridi (2015).

This new reality has a profound impact on society. The dependency on hyperconnectivity affects social, economic, cultural, and other aspects as a whole. The value of the onlife world is not created by the information itself, but by creating an environment capable of attracting the attention of the largest number of people for the longest possible time. This strategy underlies the business model of most successful Internet companies, with Google being a clear example. The crucial differentiator is not the production and distribution of content, but the filtering, contextualization, and organization of information (Floridi, 2015). The Manifesto raises two issues that need to be further addressed, also in the current tourism landscape: the role of technology and the relationship between power and knowledge.

The tourism sector is characterized by its dynamism and relies on intricate, non-linear interconnections, necessitating a significant presence in the digital realm. The spread of tourism should be seen as a constantly evolving process of interaction occurring within networks of activity systems, rather than a simple economic transaction (Jørgensen, 2017). Presently, the approach of enhancing market standing through the vertical integration of channels by tour operators and the provision of information significantly influences consumer decision-making (Macedo & da Silva, 2016). We must not forget that the online sphere is privatized. Today, technology is the soul of the tourism and travel industry (Drosos, Chalikias, Skordoulis, Kalantonis, & Papagrigoriou, 2017). Manipulation of online content is contributing to a general decrease in onlife freedom, along with an increase in mobile internet service disruptions and physical and technical attacks (House, 2017). Addressing the meta-challenges of global tourism requires the industry to effectively use technology to achieve common goals. Technological development per se is not the solution unless we change paradigms, and for that, academia must make an effort to address research that contributes to achieving the global challenges faced by the industry, which involve our way of thinking and acting. New opportunities are opening up in the tourism sector, but to seize them, it is necessary to deeply understand and develop a perspective on social and environmental issues to shape the future of the sector (WTTC, 2019).

### *The Contribution New Technologies Makes to Sustainable Tourism*

Technology, and ICTs in particular, could certainly make significant contributions to sustainability, but this potential remains theoretical (Gössling, 2020). "Hope is not a strategy," we are reminded by the World Economic Forum (2020). Gössling (2020) reminds us that the recent pandemic has highlighted the complexity and interconnection between the economy, society, and the environment.

In recent years, we have witnessed the development and introduction of the Internet of Things (IoT) applied to tourism, with sensors collecting data from the interactions between tourists and the environment. Short-range wireless communication technologies like Near Field Communication (NFC) and Radio Frequency Identification (RFID) enhance ubiquitous connectivity to facilitate new experiences for tourists interacting with their surroundings (Navio-Marco, Ruiz-Gomez, & Sevilla-Sevilla, 2019). Additionally, the use of beacons placed on traditional physical signage turns them into true digital tourist offices when connected to tourists' devices via Bluetooth or Wi-Fi.

There is a growing interest in aspects related to data analysis of satellite-provided information, such as geographical or geological studies linked to tourism. This includes landscape cartography, coastlines and landscape change, coastal and geological impact, tourism-related uses, livelihoods and lifestyles, climate and climate prediction in the tourism context, and ecology and nature conservation related to tourism, such as penguin observation, ecotourism, or deforestation. Special attention in this field is given to the conservation and maintenance of coral reefs, which are starting to generate significant interest due to their impact on tourism (Sevilla-Sevilla, Navio-Marco, & Ruiz-Gomez, 2020).

Drones have also been introduced for the observation and diagnosis of natural tourist spaces affected by natural phenomena that need to be analyzed to ensure their conservation (Sevilla-Sevilla, Mendieta-Aragon, & Ruiz-Gomez, 2023). Thus, phenomena such as erosion, which can be prevented by retaining vegetation (e.g., in the Canary Islands), sediment deposits for beaches and the conservation of these areas in the Tayrona National Natural Park in Colombia, beach conservation in Lesbos (Greece) and Australia, and the protection of geothermal vegetation in New Zealand. In Turkey, drones are used for coastal preservation in Karasu and geomorphological studies of sedimentation in Mount Akdag.

Robots will become common devices for many companies by the 2030s, as a response in part by labor shortages, the need to communicate with an increasing number of international travelers, and the requirement to create memorable customer experiences (Bowen & Morosan, 2018). The arrival of Artificial Intelligence (AI) in the tourism ecosystem, which also represents an area where sustainability is a topic to be further developed.

**Key Terms**

- **Sustainable Tourism**: Tourism that takes full account of its current and future economic, social, and environmental impacts, addressing the needs of visitors, the companies, the environment, and host communities (UNWTO, 2022).
- **Environmental sustainability**: responsible interaction of humans with the environment (Viñals & Teruel, 2021).
- **Destination management / marketing organization (DMO)**: A Destination Management/Marketing Organization (DMO) serves as the leading institutional entity responsible for encompassing various authorities, stakeholders, and professionals, thereby facilitating collaborative partnerships within the tourism sector, all working in unison toward a shared vision for a destination. The governance structures of DMOs exhibit a spectrum, ranging from a single public authority model to a public-private partnership framework. Their primary function lies in the initiation, coordination, and oversight of specific activities, including the implementation of tourism policies, the formulation of strategic plans, the development of tourism products, promotional and marketing endeavors, and the management of convention bureau activities. The scope and extent of DMO functions can span from the national to regional and local levels, contingent upon the prevailing and potential requirements, as well as the degree of decentralization within the public administration. It's important to note that not all tourism destinations have an established DMO in place, as highlighted by the United Nations World Tourism Organization (UNWTO).
- **Green marketing**: represents the management process responsible for identifying, anticipating, and satisfying the needs of customers and society in a profitable and sustainable way (Yadav, Kumar, & Swaroop, 2016).
- **Corporate Social Responsibility (CSR)**: multidimensional construct encompassing three pillars: the economy, society, and the environment (Martinez & Rodriguez Del Bosque, 2013).
- **Onlife**: integration of digital technologies and online interactions into our everyday lives (Floridi, 2015).

## Future Research Directions

Both academia and industry need to focus research on identifying practices and policies that promote sustainability and balance economic benefits with environmental, social, and cultural preservation.

Climate change poses significant challenges for tourism, as climate patterns affect the availability of natural resources and the stability of tourist destinations. Research should focus on understanding and mitigating impacts by developing adaptation strategies to ensure the resilience of tourist destinations and promoting responsible environmental policies and behaviors.

Regarding technology and digital transformation, the rapid evolution of technology is changing how tourists seek information, make travel decisions, and engage with tourist destinations. Research should explore how to leverage new technologies such as artificial intelligence, virtual and augmented reality, or the Internet of Things to enhance the tourist experience, destination management, and tourism promotion from a more sustainable perspective.

The growing amount of data generated in tourism presents challenges in terms of managing, analyzing, and utilizing information. Research should explore how to harness the potential of big data and analytics technologies to gain valuable insights about destinations, tourists, their behaviors and preferences, and use it to improve responsible decision-making in tourism planning and management.

## Conclusion

Regulatory environment and appropriate public policies positively contribute to tourism sustainability (Aktürk, 2022; Amoah et al., 2023; Bezvesilnaya, Shadskaja, Kozlova, Shelygow, & Alekseenko, 2020). Long-term thinking, the core of sustainability, and a holistic approach that recognizes the complexity and interconnectedness of society, economy, and the environment are necessary. Equity and justice should be ensured, promoting equal opportunities, reducing inequalities, and respecting human rights. Conservation of resources, avoiding overexploitation, and seeking ecosystem preservation and regeneration are essential. Collaboration and participation among stakeholders are also crucial.

In the context of tourist destinations, although progress has been made in specialized sustainability consulting, the academia has not fully incorporated the 2030 Agenda into its core approaches in the tourism sector.

We have seen innovative experiences and new practices that promote sustainability in tourism, from public policies, destinations, and the industry to consumers themselves, who can foster sustainable behaviors and choices. It is a two-way path: unsustainable practices lead to the deterioration of the tourism environment, ultimately diminishing the tourist experience. Reversing this spiral requires collective effort from all stakeholders.

The tourism industry is facing global challenges that demand a sustainable approach. The rapid advancement of technology provides opportunities for personalized information, enhanced communication, and collaboration between tourists and stakeholders. The convergence of the physical and digital worlds has profound societal implications. Embracing new technologies, along with a shift in mindset and actions, is crucial for the industry to achieve its common goals and shape its sustainable future.

**Case Study**

Reefs, underwater structures formed by corals and other organisms, are not only a stunning sight to behold but also play a crucial role in marine ecosystems. They provide a multitude of benefits, from supporting diverse marine species to protecting coastlines from erosion and wave impact. Moreover, these intricate ecosystems contribute to biodiversity, help in carbon capture and storage, and significantly contribute to climate change mitigation. In this context, Coastruction, a Netherlands-based company, has emerged as a trailblazer in the realm of reef conservation and restoration. Specializing in the creation of 3D-printed reefs, Coastruction's innovative approach employs sustainable practices and biomimicry to ensure the resilience and effectiveness of their artificial structures.

***The Importance of Reefs in Marine Ecosystems***

Reefs are like bustling cities underwater, teeming with life and activity. They provide crucial habitats and breeding grounds for an astonishing array of marine species, ranging from colorful corals to diverse fish, crustaceans, and other invertebrates. These ecosystems foster interconnected relationships, where each organism plays a pivotal role in maintaining balance and supporting the overall health of the ecosystem. With their intricate structures and symbiotic partnerships, reefs are vital to marine biodiversity.

Beyond supporting marine life, reefs serve as natural barriers that protect coastlines from the ravages of erosion and wave impact. They act as a buffer, reducing the force of waves and preventing coastal erosion during storms, protecting both human settlements and valuable coastal ecosystems.

***The Role of Reefs in Climate Change Mitigation***

One of the lesser-known contributions of reefs to the global ecosystem is their role in mitigating climate change. Coral reefs, for instance, are adept at capturing and storing carbon dioxide, making them significant players in carbon sequestration. This process helps reduce the amount of $CO_2$ in the atmosphere, mitigating the impacts of greenhouse gases and climate change.

***The Economic Importance of Reefs***

Aside from their ecological significance, reefs also hold economic importance, particularly for coastal communities. Many regions around the world rely on reef-based tourism, attracting visitors from far and wide to explore the breathtaking marine biodiversity. Snorkeling, diving, and underwater photography are popular activities that generate income for local economies.

In addition to tourism, fishing is a primary source of livelihood for many coastal communities. Reefs provide essential nurseries and habitats for various fish species, supporting fish populations that eventually spill over into surrounding areas, benefiting both commercial and subsistence fishing.

***Coastruction's Vision for Sustainable Reef Conservation***

Understanding the ecological, economic, and cultural importance of reefs, Coastruction has embraced a pioneering vision of sustainable reef conservation and restoration. They have harnessed the potential of 3D printing technology to create artificial reefs tailored to local species and environmental conditions.

By collaborating with environmental and area experts, Coastruction ensures that their 3D-printed reefs are precisely adapted to mimic the natural structures found in the region. The company's commitment to sustainability is evident in their choice of materials – opting for natural and locally sourced resources to minimize their carbon footprint and reduce environmental impact.

***Adopting a Circular Printing Process***

An essential aspect of Coastruction's sustainability approach is their circular printing process, which eliminates waste. This means that every part of the 3D-printed reefs that is not used is recycled or repurposed for future projects. By avoiding unnecessary waste, Coastruction ensures that their reef creation process remains environmentally friendly and resource-efficient.

***Biomimicry and Organic Design***

Coastruction draws inspiration from nature through biomimicry, an approach that imitates natural forms, structures, and processes to optimize the design of their artificial reefs. The geometrically complex organic shapes they create not only enhance material efficiency but also increase the effectiveness of the structures in providing habitats and supporting marine life.

***Seamless Integration with the Environment***

A key design principle employed by Coastruction is ensuring that their 3D-printed reefs seamlessly integrate with the surrounding environment. By mimicking the natural forms of reefs found in the region, the artificial structures appear less obtrusive and are readily adopted by key species. When marine life embraces their new habitat, the likelihood of the success of Coastruction's conservation efforts increases significantly.

Coastruction's innovative approach to reef conservation and restoration stands as a beacon of hope for the future of marine ecosystems. By utilizing 3D printing technology, sustainable materials, and biomimicry, the company successfully creates artificial reefs that not only support marine biodiversity but also protect coastlines and contribute to climate change mitigation. Through their visionary efforts, Coastruction inspires sustainable relationships with marine ecosystems and engages individuals and communities in the vital cause of reef conservation. As the world faces increasing environmental challenges, Coastruction's commitment to sustainability and the preservation of our oceans serves as a shining example of how innovation and conscious stewardship can pave the way for a brighter, more sustainable future.

**Case Questions**

1 What are the environmental and economic benefits of artificial reefs in marine ecosystems?
2 What challenges may arise during the installation and maintenance of artificial reefs?
3 How can the success of reef restoration projects and the effectiveness of artificial reefs in coastal protection be measured?

## References

Aguirre, A., Zayas, A., Gomez-Carmona, D., & López Sánchez, J.A. (2022). Smart tourism destinations really make sustainable cities: Benidorm as a case study. *International Journal of Tourism Cities*, *9*(1), 51–69.

Ahmed, M.F., Bin Mokhtar, M., Lim, C.K., Hooi, A.W.K., & Lee, K.E. (2021). Leadership roles for sustainable development: The case of a Malaysian green hotel. *Sustainability*, *13*(18), 10260.

Ahn, Y. (2022). City branding and sustainable destination management. *Sustainability*, *14*(1), 9.

Aktürk, G. (2022). A systematic overview of the barriers to building climate adaptation of cultural and natural heritage sites in polar regions. *Environmental Science & Policy*, *136*, 19–32.

Alonso-Muñoz, S., Torrejon-Ramos, M., Medina-Salgado, M., & Gonzalez-Sanchez, R. (2023). Sustainability as a building block for tourism - Future research: Tourism agenda 2030. *Tourism Review*, *78*(2), 461–474.

Amoah, J., Bankuoru Egala, S., Keelson, S., Bruce, E., Dziwornu, R., & Agyemang Duah, F. (2023). Driving factors to competitive sustainability of SMEs in the tourism sector: An introspective analysis. *Cogent Business & Management*, *10*(1), 2163796.

Avila-Robinson, A., & Wakabayashi, N. (2018). Changes in the structures and directions of destination management and marketing research: A bibliometric mapping study, 2005–2016. *Journal of Destination Marketing & Management*, *10*, 101–111.

Bernini, C., Urbinati, E., & Vici, L. (2015). Visitor expectations and perceptions of sustainability in a mass tourism destination. *3rd International Scientific Conference Tosee - Tourism in Southern and Eastern Europe 2015: Sustainable Tourism, Economic Development and Quality of Life, 3*, 1–17.

Bezvesilnaya, A.A., Shadskaja, I.G.E., Kozlova, N.A., Shelygov, A.V., & Alekseenko, E.V. (2020). Digital technology development in tourism and hospitality industry. *Eurasian Journal of Biosciences, 14*(2), 5561–5565.

Boiral, O. (2017). Corporate greening through ISO14001: A rational myth? *Organization Science, 18*(1), 127–146.

Boluk, K., Cavaliere, C.T., & Higgins-Desbiolles, F. (2017). Critical thinking to realize sustainability in tourism systems: Reflecting on the 2030 sustainable development goals: Guest editors. *Journal of Sustainable Tourism, 25*(9), 1201–1204.

Bowen, J., & Morosan, C. (2018). Beware hospitality industry: The robots are coming. *Worldwide Hospitality and Tourism Themes, 10*(6), 726–733.

Bramwell, B., & Lane, B. (1993). Sustainable tourism: An evolving global approach. *Journal of Sustainble Tourism, 1*, 1–5.

Brown, T.J., & Dacin, P.A. (1997). The company and the product: Corporate associations and consumer product responses. *Journal of Marketing, 61*(1), 68–84.

Brundtland, G.H. (1987). *Informe Brundtland*. Washington, DC: OMS.

Buhalis, D. (2000). Marketing the competitive destination of the future. *Tourism Management, 21*, 97–116.

Buhalis, D., & Foerste, M. (2015). SoCoMo marketing for travel and tourism: Empowering co-creation of value. *Journal of Destination Marketing & Management, 4*(3), 151–161.

Buhalis, D., & Law, R. (2008). Progress in information technology and tourism management: 20 years on and 10 years after the Internet—The state of eTourism research. *Tourism management, 29*(4), 609–623.

Büscher, B., Fletcher, R., Brockington, D., Sandbrook, C., Adams, W.M., Campbell, L., Corson, C., Dressler, W. Duffy, R., Gray, N., Holmes, G., Kellyn, A., Lunstum, E., Ramutsindela, M., & Shanker, K. (2017). Half-earth or whole earth? Radical ideas for conservation and their implications. *Oryx, 51*(3), 407–410.

Byrd, E.T., Bosley, H.E., & Dronberger, M.G. (2009). Comparisons of stakeholder perceptions of tourism impacts in rural eastern North Carolina. *Tourism Management, 30*, 693–703.

Carrigan, M., & Attalla, A. (2001). The myth of the ethical consumer – Do ethics matter in purchase behaviour? *The Journal of Consumer Marketing, 18*(7), 560–577.

Castellani, V., & Sala, S. (2010). Sustainable performance index for tourism policy development. *Tourism Management, 31*(6), 871–880.

Chan, E.S.W., & Hawkins, R. (2012). Application of EMSs in an hotel context: A case study. *International Journal of Hospitality Management, 31*(2), 405–418.

Chan, E.S.W., Hon, A.H.Y., Chan, W., & Okumus, F. (2014). What drives employees' intentions to implement green practices in hotels? The role of knowledge, awareness, concern and ecological behaviour. *International Journal of Hospitality Management, 40*, 20–28.

Chasin, F. (2014). Sustainability: Are we all talking about the same thing? State-of-the-art and proposals for an integrative definition of sustainability in information systems. In *Proceedings of the 2014 Conference ICT for Sustainability*, 342–351. Atlantis Press.

Chen, R.J.C. (2015). From sustainability to customer loyalty: A case of full service hotels' guests. *Journal of Retailing and Consumer Services, 22*, 261–265.

Chen, S., Chen, H.H., Zhang, K.Q., & Xu, X.L. (2018). A comprehensive theoretical framework for examining learning effects in green and conventionally managed hotels. *Journal of Cleaner Production, 174*, 1392–1399.

Choi, G., Parsa, H.G., Sigala, M., & Putrevu, S. (2009). Consumers' environmental concerns and behaviours in the lodging industry: A comparison between Greece and the United States. *Journal of Quality Assurance in Hospitality & Tourism*, *10*(2), 93–112.

Choi, H.C., & Sirakaya, E. (2006). Sustainability indicators for managing community tourism. *Tourism management*, *27*(6), 1274–1289.

Cucculelli, M., & Goffi, G. (2016). Does sustainability enhance tourism destination competitiveness? Evidence from Italian destinations of excellence. *Journal of Cleaner Production*, *111*, 370–382.

De Souza Cavalcanti, M.C., & Teixeira, R.M. (2015). Motivações e Ações Sustentáveis Implementadas por Empreendedores do Setor Hoteleiro. *PODIUM Sport, Leisure and Tourism Review*, *4*(1), 92–107.

Delmas, M. (2001). Stakeholders and competitive advantage: The case of ISO 14001. *Production and Operations Management*, *10*(3), 343–358.

Drosos, D., Chalikias, M., Skordoulis, M., Kalantonis, P., & Papagrigoriou, A. (2017). The strategic role of information technology in tourism: The case of global distribution systems. *Tourism, Culture and Heritage in a Smart Economy*, 207–219. Springer Proceedings in Business and Economics.

Eusebio, C., Kastenholz, E., & Breda, Z. (2014). Tourism and sustainable development of rural destinations: A stakeholders' view. *Revista Portuguesa Délelőtt Estudos Regionais*, *36*, 13–22.

Fishbein, M., & Ajzen, I. (1975). *Belief, Attitude, Intention, and Behavior: An Introduction to Theory and Research*. Reading, MA: Addison-Wesley.

Fletcher, R. (2019). Ecotourism after nature: Anthropocene tourism as a new capitalist "fix". *Journal of Sustainable Tourism*, *27*(4), 522–535.

Floridi, L. (2015). *The Online Manifesto: Being Human in a Hyperconnected Era*. Cham, Heidelberg, New York, Dordrecht and London: Springer.

Font, X., Torres-Delgado, A., Crabolu, G., Palomo Martinez, J., Kantenbacher, J., & Miller, G. (2023). The impact of sustainable tourism indicators on destination competitiveness: The European tourism indicator system. *Journal of Sustainable Tourism*, 31(7), 1608–1630.

Font, X., Walmsley, A., Cogotti, S., McCombes, L., & Häusler, N. (2012). Corporate social responsibility: The disclosure–performance gap. *Tourism Management*, *33*(6), 1544–1553.

Gasparini, M.L., & Mariotti, A. (2023). Sustainable tourism indicators as policymaking tools: Lessons from ETIS implementation at destination level. *Journal of Sustainable Tourism*, *31*(7), 1719–1737.

Glavic, P., & Lukman, R. (2007). Review of sustainability terms and their definitions. *Journal of Cleaner Production*, *15*(18), 1875–1885.

Goffi, G., & Cucculelli, M. (2018). Explaining tourism competitiveness in small and medium destinations: The Italian case. *Current Issues of Tourism*, *22*, 2109–2139.

Gössling, S. (2020). Technology, ICT and tourism: From big data to the big picture. *Journal of Sustainable Tourism*, *29*(5), 849–858.

Grant, J. (2007). *The Green Marketing Manifesto*. Chichester: Wiley.

Hanna, P., Font, X., Searles, C., Weeden, C., & Harrison, C. (2018). Tourist destination marketing: From sustainability myopia to memorable experiences. *Journal of Destination Marketing & Management*, *9*, 36–43.

House, F (2017). Freedom on the Net 2017. Manipulating Social Media to Undermine Democracy. Retrieved from: https://freedomhouse.org/report/freedom-net/freedom-net-2017.

Hsieh, Y.C. (2012). Hotel companies' environmental policies and practices: A content analysis of their web pages. *International Journal of Contemporary Hospitality Management*, *24*(1), 97–121.

Johnston, P., Everard, M., Santillo, D., & Robert, K. (2007). Reclaiming the definition of sustainability. *Environmental Science and Pollution Research, 14*(1), 60–66.

Jones, P., Hillier, D., & Comfort, D. (2014). Sustainability in the global hotel industry. *International Journal of Contemporary Hospitality Management, 26*(1), 5–17.

Jones, R.J., Reilly, T.M., Cox, M.Z., & Cole, B.M. (2017). Gender makes a difference: Investigating consumer purchasing behavior and attitudes toward corporate social responsibility policies. *Corporate Social Responsibility and Environmental Management, 24*(2), 133–144.

Jørgensen, M.T. (2017). Reframing tourism distribution-activity theory and actor-network theory. *Tourism Management, 62*, 312–321.

Jovicic, D.Z. (2019). From the traditional understanding of tourism destination to the smart tourism destination. *Current Issues in Tourism, 22*(3), 276–282.

Kastenholz, E., Eusebio, C., & Carneiro, M.J. (2018). Segmenting the rural tourist market by sustainable travel behaviour: Insights from village visitors in Portugal. *Journal of Destination Marketing & Management, 10*, 132–142.

Khan, H.R., Lim, C., Ahmed, M., Tan, K., & Bin Mokhtar, M. (2021). Systematic review of contextual suggestion and recommendation systems for sustainable e-tourism. *Sustainability, 13*, 8141.

Khan, M.R., Khan, H.U.R., Lim, C.K., Tan, K.L., & Ahmed, M.F. (2021). Sustainable tourism policy, destination management and sustainable tourism development: A moderated-mediation model. *Sustainability, 13*(21), 12156.

Kornilaki, M., & Font, X. (2019). Normative influences: how socio-cultural and industrial norms influence the adoption of sustainability practices. A grounded theory of Cretan, small tourism firms. *Journal of Environmental Management, 230*, 183–189.

Lee, J.S., Hsu, L.T., Han, H., & Kim, Y. (2010). Understanding how consumers view green hotels: How a hotel's green image can influence behavioral intentions. *Journal of Sustainable Tourism, 18*(7), 901–914.

López-Gamero, M.D., Molina-Azorin, J.F., & Claver-Cortes, E. (2011). The relationship between managers' environmental perceptions, environmental management and firm performance in Spanish hotels: A whole framework. *International Journal of Tourism Research, 13*(2), 141–163.

Macedo, S. R., & da Silva, F. R. (2016). Tourist destination and accommodation: The role of travel agencies and tour operators. *Navus-Revista De Gestao E Tecnologia,6*(5), 115–126.

Malone, S., McCabe, S., & Smith, A.P. (2014). The role of hedonism in ethical tourism. *Annals of Tourism Research, 44*, 241–254.

Marin, L., & Ruiz, S. (2007). 'I need you too!' Corporate identity attractiveness for consumers and the role of social responsibility. *Journal of Business Ethics, 71*(3), 245–260.

Martin Martin, J.M., Jimenez Aguilera, J.D.D., & Molina Moreno, V. (2014). Impacts of seasonality on environmental sustainability in the tourism sector based on destination type: An application to Spain's andalusia region. *Tourism Economics, 20*(1), 123–142.

Martinez, P., Herrero, A., & Gomez-Lopez, R. (2019). Corporate images and customer behavioral intentions in an environmentally certified context: Promoting environmental sustainability in the hospitality industry. *Corporate Social Responsibility and Environmental Management, 26*(6), 1382–1391.

Martinez, P., & Rodriguez del Bosque, I. (2013). CSR and customer loyalty: The roles of trust, customer identification with the company and satisfaction. *International Journal of Hospitality Management, 35*, 89–99.

Matute-Vallejo, J., Bravo, R., & Pina, J.M. (2011). The influence of corporate social responsibility and price fairness on customer behaviour: Evidence from the financial sector. *Corporate Social Responsibility and Environmental Management*, 18(6), 317–331.

McLoughlin, E., & Hanrahan, J. (2019). Local authority sustainable planning for tourism: Lessons from Ireland. *Tourism Review*, *74*(3), 327–348.

Moran, M., & Hunt, B. (2005). *Search Engine Marketing, Inc.: Driving Search Traffic to Your Company's Web Site*. Lebanon, IN: IBM Press.

Morgan, N. (2012). Time for 'mindful' destination management and marketing. *Journal of Destination Marketing & Management*, *1*, 8–9.

Navio-Marco, J., Ruiz-Gómez, L.M., & Sevilla-Sevilla, C. (2018). Progress in information technology and tourism management: 30 years on and 20 years after the internet-Revisiting Buhalis & Law's landmark study about eTourism. *Tourism Management*, *69*, 460–470.

Navio-Marco, J., Ruiz-Gomez, L.M., & Sevilla-Sevilla, C. (2019). Progress in wireless technologies in hospitality and tourism. *Journal of Hospitality and Tourism Technology*, *10*(4), 587–599.

O'Connor, P., & Frew, A. (2002). The future of hotel electronic distribution: Expert and industry perspectives. *Cornell Hotel and Restaurant Administration Quarterly*, *43*, 33–45.

Otero, A. (2007). La importancia de la visión de territorio para la construcción de desarrollo competitivo de los destinos turísticos. *Cuadernos de Turismo*, *19*, 91–104.

Peloza, J., & Shang, J. (2011). How can corporate social responsibility activities create value for stakeholders? *A systematic review. Journal of the Academy of Marketing Science*, *39*(1), 117–135.

Qi, G., Zeng, S., Li, X., & Tam, C. (2012). Role of internalization process in defining the relationship between ISO 14001 certification and corporate environmental performance. *Corporate Social Responsibility & Environmental Management*, *19*, 129–140.

Rex, E., & Baumann, H. (2007). Beyond ecolabels: What green marketing can learn from conventional marketing. *Journal of Cleaner Production*, *15*(6), 567–576.

Rodriguez-Oromendia, A., Reina-Paz, M.D., & Sevilla-Sevilla, C. (2013). Environmental awareness of tourists. *Environmental Engineering and Management Journal*, *12*(10), 1941–1946.

Segittur (2020). Destino Turístico Inteligente. Informe de Diagnóstico y Plan de Acción de Gijón/Xixón. Retrieved from: www.gijonturismoprofesional.es/files/shares/Corporativo/DTI/IDPADTI%20GIJON%20DEF.pdf.

Segittur (2022). Guía buenas prácticas de sostenibilidad para destinos turísticos inteligentes. Retrieved from: https://www.segittur.es/sala-de-prensa/informes/guia_buenas_practicas_sostenibilidad_destinos_inteligentes/.

Sen, S., & Bhattacharya, C.B. (2001). Does doing good always lead to doing better? Consumer reactions to corporate social responsibility. *Journal of Marketing Research*, *38*(2), 225–243.

Sevilla-Sevilla, C., Mendieta-Aragon, A., & Ruiz-Gomez, L.M. (2023). Drones in hospitality and tourism: A literature review and research agenda. *Tourism Review*. doi: 10.1108/TR-11-2022-0557.

Sevilla-Sevilla, C., Mondejar-Jimenez, J., & Reina-Paz, M.D. (2019). Before a hotel room booking, do perceptions vary by gender? The case of Spain. *Economic Research-Ekonomska Istrazivanja*, *32*(1), 3853–3868.

Sevilla-Sevilla, C., Navio-Marco, J., & Ruiz-Gomez, L.M. (2020). Environment, tourism and satellite technology: Exploring fruitful interlinkages. *Annals of Tourism Research*, *83*, 102841.

Souza, R.G., Rosenhead, J., Salhofer, S.P., Valle, R.A.B., & Lins, M.P.E. (2015). Definition of sustainability impact categories based on stakeholder perspectives. *Journal of Cleaner Production*, *105*, 41–51.

Spenceley, A. (2019). Sustainable tourism certification in the African hotel sector. *Tourism Review*, *74*(2), 179–193.

Su, L., Huang, S., & Huang, J. (2018). Effects of destination social responsibility and tourism impacts on residents' support for tourism and perceived quality of life. *Journal of Hospitality & Tourism Research, 42*, 1039–1057.

Su, L., & Swanson, S.R. (2017). The effect of destination social responsibility on tourist environmentally responsible behavior: Compared analysis of first-time and repeat tourists. *Tourism Management, 60*, 308–321.

Teng, Y.M., Wu, K.S. & Liu, H.H. (2013). Integrating altruism and the theory of planned behaviour to predict patronage intention of a green hotel. *Journal of Hospitality Tourism Research, 39*(3), 299–315.

Tudorache, D.M., Simon, T., Frenț, C., & Musteață-Pavel, M. (2017). Difficulties and challenges in applying the European tourism indicators system (ETIS) for sustainable tourist destinations: The case of Brasov County in the Romanian Carpathians. *Sustainability, 9*(10), 1879.

Turker, D. (2009). Measuring corporate social responsibility: A scale development study. *Journal of Business Ethics, 85*(4), 411–427.

UN(2019). World Economic Situation and Prospects 2019. Retrieved from: https://www.un.org/development/desa/dpad/publication/world-economic-situation-and-prospects-2019/.

UNWTO (2022). Retrieved from: https://www.unwto.org/sustainable-development.

Valentin, A., & Spangenberg, J.H. (2000). A guide to community sustainability indicators. *Environmental Impact Assessment Review, 20*(3), 381–392.

Villarino, J., & Font, X. (2015). Sustainability marketing myopia: The lack of persuasiveness in sustainability communication. *Journal of Vacation Marketing, 21*(4), 326–335.

Viñals, M.J., & Teruel, L. (2021). The perspective of environmental sustainability in master and doctoral tourism studies in Spain. *Cuadernos de Turismo, 47*, 599–601.

Von Bergner, N.M., & Lohmann, M. (2014). Future challenges for global tourism: A Delphi survey. *Journal of Travel Research, 53*(4), 420–432.

World Economic Forum (2020). Clean Skies for Tomorrow: Sustainable Aviation Fuels as a Pathway to Net-Zero Aviation. Retrieved from: https://www3.weforum.org/docs/WEF_Clean_Skies_Tomorrow_SAF_Analytics_2020.pdf

World Travel & Tourism Council (2019). World, Transformed. Megatrends and Their Implications for Travel & Tourism. Retrieved from: https://wttc.org/Portals/0/Documents/Reports/2019/World%20Transformed-Megatrends%20and%20their%20Implications%20for%20Travel%20and%20Tourism-Jan%202019.pdf?ver=2021-02-25-182733-437.

World Tourism Organization (UNWTO) (2005). *Indicators of Sustainable Development for Tourism Destinations A Guidebook – Indicadores de desarrollo sostenible para los destinos turísticos.* Guía práctica (Versión española). UNWTO, Madrid, doi: 10.18111/9789284408382.

Xiang, Z., Pan, B., Law, R., & Fesenmaier, D.R. (2010). Assessing the visibility of destination marketing organizations in google: A case study of convention and visitor bureau websites in the United States. *Journal of Travel & Tourism Marketing, 27*(7), 694–707.

Yadav, R., Kumar, A., & Swaroop, G. (2016). The influence of green marketing functions in building corporate image. *International Journal of Contemporary Hospitality Management, 28*(10), 2178–2196.

Yuksel, F., Bramwell, B., & Yuksel, A. (1999). Stakeholder interviews and tourism planning at Pamukkale, Turkey. *Tourism Management, 20*, 351–360.

Zhang, J., & Dimitroff, A. (2005). The impact of webpage content characteristics on webpage visibility in search engine results (Part I). *Information Processing and Management, 41*(3), 665–690.

# 5 Water Use in Tourism

*Amelia Pérez Zabaleta and Ester Méndez Pérez*

## Introduction

Chapter 7 explains how the management of water resources has a profound impact on the success and sustainability of tourism destinations, with particular reference to destinations suffering from water scarcity. It compares a variety of regulatory approaches to water management, with regard to both government regulation and self-regulatory practice in the private tourism industry.

## Background

### *Main Focus of the Chapter*

This chapter serves a dual purpose: on the one hand, to highlight the value of water as a resource for the tourism industry, and on the other hand, to demonstrate that it is possible to reconcile water economy with tourism development through efficient management. Water, apart being an attraction for tourism, is also a vital resource for its development. The water footprint for tourism activities is particularly important in Spain due to the evident water scarcity resulting from the climate crisis. Frequent drought episodes, coupled with the increasing pressure of tourism in specific areas and seasons throughout the year, demand the adoption of efficient water management systems that are compatible with the growth of the tourism sector. The city of Benidorm serves as a prime example of this reconciliation between these two realities: water scarcity and tourism activity.

Water is one of the most in-demand tourist attractions, and the tourism industry is a major consumer of water. The water–tourism nexus is undeniable. Accordingly, the integrated management of the tourism industry and water resources is of vital importance to guarantee the sustainability of both the resource and tourist activity.

The goal of sustainable tourism is to ensure the development of this economic activity alongside the conservation of the environment and natural resources, as has been emphasized by international organizations such as the United Nations (UN), the World Tourism Organization (UNWTO) and the Organisation for Economic Co-operation and Development (OECD).

DOI: 10.4324/9781003369967-8

The chapter first analyses the nexus water–tourism, in order to subsequently highlight what sustainability means for both the tourism industry and water. The characteristic features of water use and consumption in this industry are then presented. Lastly, the main impacts of tourism on the resource are detailed, as well as the key damage limitation measures that enable water conservation for the development of this economic activity.

***Water–Tourism Nexus***

Human activity, and particularly tourism, leads to a deterioration in the quantity and quality of natural resources such as water. This is because it exerts pressure on the use of resources and waste production, especially during holiday periods and in high seasons.

Many authors and researchers have emphasized the idea that water is one of the fundamental resources for the development of tourism (Gabarda-Mallorquí et al., 2017; Gössling and Peeters, 2015; Morrison and Pickering, 2013). In line with the definition given by Grellier et al. (2017), blue spaces are outdoor environments—whether natural or manmade—that prominently feature water and are accessible to people physically (being in or near the water) or virtually (being able to see, hear or otherwise sense the water).

In addition to their tourist value, these spaces are keenly appreciated for the benefits they provide for both physical and mental health. Notable among these benefits are the fact that they encourage physical activity, help restore levels of physical and emotional health (by reducing stress and anxiety, and improving mood and psychological well-being) and lower the risk of diseases, such as cardiovascular disease. Furthermore, blue spaces can contribute to a healthier environment and improve air quality. In relational terms, they also promote social interaction.

On the one hand, water is itself a natural attraction for tourists—beaches, rivers, waterfalls, lakes and so on. But, in addition to the beautiful landscapes of "blue spaces", it offers an opportunity for other types of recreational and sports activities, such as swimming or fishing. Water also has therapeutic properties that promote well-being and relaxation. Hot springs and hydrotherapy treatments are water-related experiences that are popular with tourists. Water is also a key element in adventure tourism, with activities such as rafting, kayaking and surfing attracting a certain segment of the tourist market. In short, tourists value public or private infrastructures that require water, such as gardens and fountains, or sports facilities such as golf courses, which are also provided in hotel accommodations, campsites or other types of tourist accommodation.

There is such a value placed on the quality of blue spaces (Ballesteros-Olza et al., 2020) that these destinations are categorized and differentiated; for example, beaches can be labelled as "blue flag". The Blue Flag dates back to 1985, when the French branch of the Foundation for Environmental Education created an award for marinas and boats to recognize their consideration of the environment. Two years later, in 1987, the European Commission (1987) established the "Blue Flag" label, applicable to all coastal European countries, under the title "Clean beaches and ports of Europe", to encourage compliance with the European Community's Bathing Water Quality Directive. Today, participating municipalities number more than 2,000 in 45 countries on five continents.

Tourists' strong appreciation for blue spaces and their associated quality awards means that water scarcity, water restrictions and any deterioration in quality negatively affect their image and appeal. On top of this, tourism is the economic activity that is most exposed to the effects of global warming. While it is true that agriculture covers more area and is also highly exposed to climate change, tourism has a much greater socio-economic impact in an increasingly globalized and urban world (Miró and Olcina, 2020). This problem is particularly acute when it comes to sun and sand destinations, which are in hot climates with high levels of water stress.

Water stress occurs when the demand for water exceeds the quantity available for a period, or when its use is restricted by low quality.

The Sixth Assessment Report of the Intergovernmental Panel on Climate Change (IPCC) (2022) projects with high confidence an increase in the severity and frequency of droughts, the severity and frequency of heatwaves and extreme rainfall, along with a reduction in annual amounts of precipitation. Climate change is already influencing investment, planning and operations in the tourism industry, while transforming the competitiveness, sustainability and geography of destinations (Roson and Sartori, 2014; Scott, 2021).

Furthermore, tourism places greater demands on water than people do at home. There is no doubt that tourists consume more water in showers and bathrooms, and more water is consumed in the preparation of their meals and in the production of food and energy. There is evidence that tourism increases total water consumption, although it is still difficult to make comparisons (Gössling, 2015; Gössling et al., 2012).

In addition to the availability of water, water quality is also a key factor when considering the impacts of tourism. Polluted water jeopardizes certain tourist activities and damages the public image of a destination. An example of this was examined by Kent et al. (2002), who addressed water scarcity and water quality problems on the Balearic island of Mallorca (Spain). These problems generated negative publicity in the German press and, as a result, a significant reduction in the number of German tourists.

The extensive academic literature on the water–tourism link is centred around three main lines of research (García et al., 2020): (A) Studies based on the measurement of direct or indirect water use in the tourism industry. (B) Studies that analyse the sustainability implications of water use, including water scarcity, competition for scarce resources between tourism and other economic sectors or local populations, as well as the transfer of water use between countries and continents as a result of global tourism flows. (C) Literature on water management, which includes all the actions that can help reduce water demand.

As such, the efficient management of water resources becomes a strategic necessity to guarantee the continuity of this important economic segment (Silva and Mattos, 2020).

There is thus an urgent need to promote water-saving measures, in order to ensure the future availability and sustainability of this resource. The need is especially acute given the rise in demand due to growing numbers of tourists, as well as a decline in availability as a result of the effects of climate change (more intense and frequent droughts, reduced precipitation, etc.). The near future portends a situation of growing tension and conflicts between social and economic sectors in response to increasing competition for this resource (Gössling, Hall and Scott, 2015) and its overexploitation (UNEP, 2019).

### *Water–Tourism Sustainability*

One of the challenges facing the tourism industry is sustainability and the protection of the environment and natural resources. UNWTO defines sustainable tourism as tourism that takes full account of its current and future economic, social and environmental impacts, addressing the needs of visitors, the industry, the environment and host communities. That is, the goal of sustainable tourism is to ensure the development of this economic activity alongside the conservation of the environment. It is about to meeting the needs of today's tourists and host regions, while protecting and fostering opportunities for the future.

The concept thus leads to the "management of all resources in such a way as those economic, social, and aesthetic needs can be fulfilled while maintaining cultural integrity, essential ecological processes, biological diversity, and life support systems" (UNWTO, 2020). Article 3 of the Global Code of Ethics for Tourism states that "All the stakeholders in tourism development should safeguard the natural environment with a view to achieving sound, continuous and sustainable economic growth geared to satisfying equitably the needs and aspirations of present and future generations".

Among the 17 Sustainable Development Goals (SDGs) that form the core of the 2030 Agenda, one is related specifically to water; namely, SDG 6 "Ensure access to clean water and sanitation". It may appear that this goal has been achieved in developed countries, where tourism plays a leading role. In fact, at the UN 2023 Water Conference, the UNWTO highlighted the role of tourism in safeguarding water resources in order to contribute to the achievement of SDG 6.

The main focus of responsible tourism is to promote environmental, socio-economic and cultural sustainability, and to minimize the negative impacts of tourism on communities and ecosystems. By so doing, it contributes to the achievement of several SDGs, including SDG 6.

Both water and tourism are closely related to the rest of the SDGs. For example, water is related to SDG 2, which deals with food security. Also notable is its relationship with SDG 3, because it contributes to health and disease prevention. Both goals are closely related to tourism, as noted previously. There are also links to SDG 8, SDG 9, SDG 11 and SDG 12, as without water the sustainable economic growth of towns and cities cannot be guaranteed, an element which is vital for tourism. Moreover, it is fundamental for action to fight climate change and to ensure environmental sustainability in terrestrial and aquatic ecosystems (SDG 13, SDG 14 and SDG 15).

In turn, tourism can contribute directly or indirectly to the achievement of various goals: SDGs 8 and 11, since tourism should generate decent work and economic opportunities for local communities, promoting sustainable economic development while preserving cultural heritage; SDG 12, by fostering and encouraging the efficient use of resources and minimizing the negative impacts of the industry, on both the demand side and the supply side; and SDG 13, in terms of climate action, responsible tourism can and should reduce emissions by promoting low-carbon tourism products and practices, fostering the use of sustainable transport, improving energy efficiency in tourist establishments and raising awareness among stakeholders about the problems generated by climate change. Furthermore, when responsible tourism contributes to the conservation of marine and coastal ecosystems—for example, by promoting responsible diving practices that protect coral reefs and marine life—it enables the achievement of SDG 14. Responsible tourism should also aim to promote biodiversity and protect natural habitats by introducing organized visits to protected areas and natural reserves, generating tourist satisfaction with the information received about the need to preserve the environment, and thereby contributing to the achievement of SDG 15. The conservation of water—in terms of both quantity and quality—in tourism environments basically depends on the joint management of water and tourism, focusing on both the supply and demand sides.

### *Water Use and Consumption in Tourist Destinations*

The viability and sustainability of any tourist destination ultimately depends on an adequate (in terms of both quantity and quality) water supply; indeed, water is a key factor in the tourism life cycle model (Rico et al., 2009).

In contrast to other economic sectors such as industry and agriculture, which account for almost 90% of global water consumption, tourism's water use represents less than 1%, a figure that may seem insignificant at a global level if we do not account for the concentration of tourism activities in space and time (Silva and Mattos, 2020). If we consider this figure with reference to the contribution tourism makes to global GDP, it is relatively lower, since tourism's contribution has dropped from 10% of global GDP in 2019 to 5.5% in 2022.

Although water consumption may vary depending on certain factors such as size, location and number of facilities, tourist establishments generally consume water for the supply guests use in the rooms; for the laundry service and cleaning the facilities; for aquatic leisure areas, such as swimming pools; for gardens and landscaped areas; and for the not inconsiderable consumption of water in restaurants and kitchens, which can account for between 30% and 40% of total consumption in hotel establishments, as shown in Table 5.1.

With regard to the average consumption in litres per room, this figure ranges between 1,001 and 4,001 per room per day at a global level, which takes into account the average at European level (consumption of between 450 and 8,001 per room per day) and the minimum and maximum values recorded (range of results) (Table 5.2).

*Table 5.1* Distribution of Water Consumption in Hotels

| *Activities* | *Percentage of Total Consumption (%)* |
|---|---|
| Supply to rooms | 25/35 |
| Laundry | 10/15 |
| Swimming pools and spas | 5/15 |
| Watering gardens and green areas | 10/15 |
| Kitchens and restaurants | 30/40 |
| Cleaning and sanitation | 10/15 |
| Cooling systems | 5/10 |
| Maintenance and operations | 5/10 |

*Source:* Own elaboration based on the UNWTO World Tourism Barometer (2019).

*Table 5.2* Average Consumption in Litres per Room per Day

| *Global Level* | *Range of Values* | *European Level* |
|---|---|---|
| 100/400 | 85/2335 | 450/800 |

At global level, the average cost per room of hotel establishments' water consumption is between 1% and 3% of the total cost per room.

The scarcity of water and the degradation and pollution of water sources represent a pending challenge, one which is particularly critical in hot tourist destinations that suffer from water stress, limiting the capacity of the destination to meet tourists' demand for water. This lack of water can affect the quality of the existing water, exacerbating the health risks faced by guests.

In general, water scarcity can compromise the long-term sustainability and viability of tourist destinations. It is therefore essential to implement efficient management measures and promote education for all stakeholders aimed at raising awareness about responsible use of the resource.

Moreover, tourism peaks in summer, which causes a spatio-temporal mismatch between tourist activities and water availability (Liu et al., 2021; Ricart et al., 2020).

For these reasons, water use in tourist destinations has been the subject of discussion in the academic literature in recent years. Based on an exhaustive review of water consumption in the tourism industry, significant differences can be observed depending on the type of tourist establishment, the tourist activities during the stay (sailing, golf, swimming, etc.), the geographical location and the management system in place (Deyà-Tortella and Tirado, 2011; García et al., 2020; Gössling et al., 2012).

Direct use of water ranges between 84 and 2,425 litres per tourist per day in accommodations (including the use of water in the rooms, for watering gardens and swimming pools), with extra activities adding between 10 and 875 litres per overnight stay per guest (Deyà-Tortella and Tirado, 2011; Gössling et al., 2012; Hadjikakou et al., 2013; Page et al., 2014).

In addition to the direct use of water, indirect use is also relevant; that is, the water that is used in the production of goods and services. Both direct and indirect uses of water are covered by the concept of the **water footprint**, which measures the amount of water used in the production of goods and services. The aim of the water footprint is to demonstrate the dependence of a company or supply chain on water, as well as whether regulations protect water resources; in short, it uses data to identify the impact of production or economic activity on water resources.

There are three components of the water footprint. The green water footprint is an indicator that refers to the water which comes from rainfall and does not go to runoff or groundwater, or become wastewater, but is temporarily held in the soil (depending on evaporation). The blue water footprint is an indicator that covers the consumptive use of fresh water (surface water or groundwater) for the production of goods or the provision of a product. Lastly, the grey water footprint refers to the water that is polluted as a result of certain processes. It is defined as the volume of fresh water that is needed to assimilate the pollutant load based on concentrations under natural conditions and the environmental water quality standards or legislation in effect (Figure 5.1).

It is interesting to see the distribution between direct and indirect use in the water footprint. The proportion of direct water use in Travel and Tourism is significantly lower. While in 2019 it represented 0.2% of the global total, by 2021 it had halved to 0.1% of the global total. Much of this industry's water use is indirect, through its supply chain, while agriculture and food production account for two-thirds of the entire Travel and Tourism water footprint.

Comparing continents, between 2010 and 2019 the Travel and Tourism industry in Europe and Africa reduced their direct water use. In Europe, the direct use of water dropped by 8%, and in Africa by 6%.

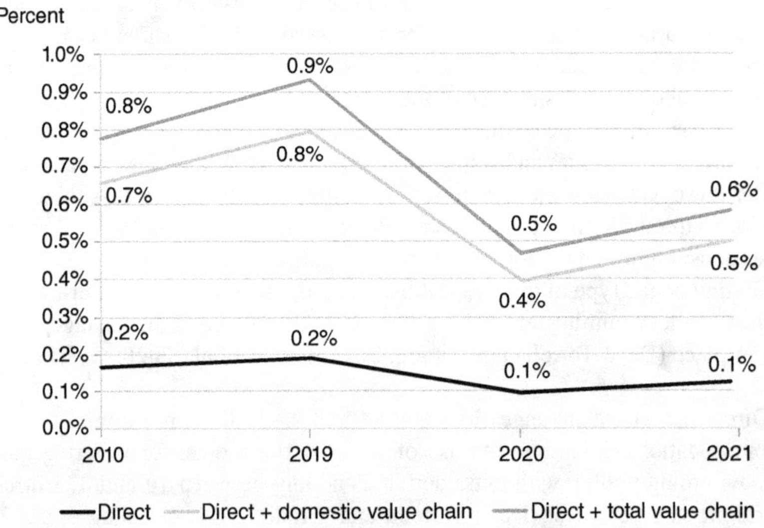


*Figure 5.1* Distribution of travel and tourism water use (water footprint).

*Source:* WTTC.

Another important data point is the water intensity of Travel and Tourism per unit of GDP. This has decreased since 2010, in both direct and indirect use. In 2010, this industry used 0.57 $m^3$ of water for every dollar contributed to the world economy. In 2019, this figure dropped by 19% to 0.46 $m^3$ of water for every dollar contributed to the world economy.

Furthermore, tourism is highly concentrated in space and time, which means that the share of tourism in water consumption varies significantly, as became clear during the lockdown periods of the COVID pandemic. According to García et al. (2020), water consumption in the highly tourism-oriented municipalities in the Balearic Islands of Spain decreased by an average of 58% during the lockdown period from June to September 2020 compared to the same period in 2019. The decrease in water consumption was only 14% for less tourist-focused municipalities.

Likewise, there are important differences between local residents' consumption and tourists' consumption, especially in developing countries (Cole, 2012). Several studies quantify tourists' per capita water consumption as being between two and three times greater than the local demand for water in developed countries (García and Servera, 2003; UNEP, 2019; WTO, 2004), and up to 15 times greater in developing countries (Gössling, 2000).

Furthermore, it is generally recognized that water consumption by tourists is not only high, but also strongly seasonal, which can exacerbate situations of scarcity and conflicts with other users (Torregrosa et al., 2010).

Such knowledge, data and indicators are fundamental for improving management. The city of Valencia (Spain) is the first city in the world to officially certify the water footprint of tourism, which in 2019 (the last full year before the pandemic) was calculated at 74.23 $hm^3$. At that time, Valencia received 2.18 million tourists annually, who stayed an average of 2.42 nights, in addition to 2.24 million day trippers and 435,000 cruise visitors.

To calculate the water footprint, ten components were analysed: (1) travel to the tourist destination by tourists, cruise passengers and day trippers; (2) regulated tourist accommodations; (3) food, restaurants; (4) leisure, entertainment; (5) purchases made by tourists, visitors and cruise passengers; (6) tourism products and services and tourism management (energy consumption); (7) internal transport (public transport and rental vehicles); (8) public services, urban solid waste and waste treatment; (9) public water services; and (10) tourism products and services and tourism management.

The direct water footprint accounted for 16% of the total, and includes the direct water consumption of the city, which is distributed as follows: 0.23% for water services, indicating the high level of water efficiency of the city, the highest in Europe with 87% real efficiency thanks to remote meter reading; and 13.78% for consumption in tourist accommodation.

As for the indirect water footprint, it accounted for the remaining 84% (62.38 $hm^3$). It is associated with the goods and services linked to tourism. The food/leisure/entertainment and shopping chapter accounts for the lion's share, with 68.75%.

Per tourist, a daily average of 0.315 $m^3$ was recorded. However, there was a difference depending on the type of tourist: day trippers left a water footprint of 5.45 $m^3$; tourists, 23.35 $m^3$; and cruise passengers had the greatest impact, with 24.88 $m^3$.

***Impacts of Tourism on Water Resources and Mitigation Measures***

Tourism thus exerts an undeniable pressure on local and regional water resources, and its impacts can be summarized as follows:

- Overexploitation of water resources. As mentioned above, the direct use of water by tourists is additional to (and greater than) the use by residents. As a result, it can cause water stress. Overexploitation affects both surface water and groundwater sources. When surface waters are insufficient, the next step is to turn to the exploitation of aquifers to cope with the demand for water from the different tourist services and activities.
- Pollution. Tourism can create water pollution through the discharge of untreated wastewater, the use of chemical products in facilities, sailing, etc. In some tourist destinations, wastewater treatment infrastructures turn out to be insufficient to meet the increase in population due to tourist arrivals, or they can be inadequate due to obsolescence, resulting in the discharge of untreated wastewater, which affects public health and the environment.
- Degradation of aquatic systems. The development of infrastructure can alter ecosystems such as rivers, lakes and wetlands, adversely affecting biodiversity.
- Furthermore, the pressure on water resources can contribute to climate change through the greenhouse gas emissions generated by tourism activities such as transport. This can cause alterations in precipitation patterns, contributing to the intensity of extreme events such as droughts and floods, which are especially impactful in destinations suffering from water scarcity.

***Solutions and Recommendations***

To mitigate these impacts, it is necessary to implement measures aimed at conserving both the quality and quantity of water. There is thus a need for measures aimed at achieving greater efficiency in the use of water and ensuring the appropriate treatment of wastewater.

In this context, the lines of action taken to achieve a more sustainable and water-efficient tourism industry have gone in two directions. The first, focusing on the demand side, seeks to mitigate the impact of climate change and encourage tourists' sustainable use of resources. The second, focused on the supply, calls for the tourism industry itself to take responsibility and implement the necessary measures to achieve a more efficient use of water. In this way, the tourism industry seeks to move towards a new model of the circular economy—more sustainable, efficient and environmentally friendly.

In order to mitigate the impact of this economic activity on water resources, the following measures are being implemented and should be encouraged in all tourist destinations, with the involvement of all stakeholders:

- Technology and digitalization. Incorporating technology and more efficient water use practices will contribute to a reduction in consumption in tourist establishments. For example, the installation of water-saving devices to reduce the flow rate in taps, dual-flush cisterns, efficient irrigation systems, and, above all, the establishment of a management plan and a detailed study of consumption, can improve the profitability of the business and lead to substantial water savings. Thanks to smart meters, digitalization can facilitate improved management by tourist establishments. These meters offer real-time information on water consumption, which can be translated into specific saving strategies; for example, by setting precise schedules for watering gardens or using the laundry. Water quality control and monitoring are other tasks that can be tackled with the support of digital technologies; for instance, reporting information about water leaks allows immediate action to substantially reduce losses. Likewise, innovation in desalination, rainwater recovery and water reclamation is essential—especially desalination in beach tourism. In short, research, development and innovation are key factors for this industry.
- Reusing and recycling water can help boost use efficiency. The installation of wastewater treatment system, technologies that allow the treated water to be reused for non-potable purposes, such as irrigation or cleaning, is a good management measure, helping to ensure both the efficient use of resources without overexploitation, and the conservation of ecosystems and mitigation of environmental degradation.
- The aforementioned measures are in line with the requirements of the **circular economy**. The European Commission, in its report on the transition towards green tourism in the European Union, puts the focus on the circular consumption of resources (reducing, recycling and reusing are key), with a special emphasis on water—improving treatment and the application of technologies for reuse. The circular economy entails the transition to new business models in the tourism industry. These models should be considered a starting point for the protection of the environment, natural resources and ecosystems, in order to ensure the necessary and proper quantity and quality of water supply for the development of this economic activity.
- Given the above, we highlight the investment required in technology and new water use management models, as well as in building and upgrading infrastructure to ensure less waste and fewer leaks.
- Regulation and intervention by public administrations is essential. (Gössling, 2015) point out the importance of improving the regulation of water use in sun and sand tourist destinations, in order to allocate related responsibilities to all stakeholders. The diversity of interests—sometimes conflicting—makes it difficult to design useful policies. Regulation is key to improving water management

in tourist areas, as it has been shown that awareness and water-saving devices are not enough to compensate for the growing water needs associated with tourism.

- Taxes can help to raise awareness and make reparations, as long as the tax system of the region or the country can ensure the income generated is reinvested in improving resource use. An example of environmental taxation is the "ecotax" that was introduced in the Balearic Islands for tourists who stay overnight in tourist establishments. Similarly, it may be worth introducing a block tariff system that discourages excessive consumption. Its impact will depend on the characteristics of the tariff system.
- Regulation, control and intervention measures involving public-private collaboration could incorporate incentives, certifications, awards or other actions aimed at introducing sustainability criteria into water management. Public-private collaboration is needed to achieve this as it allows alliances to be established with the aim of developing and implementing integrated water resources management policies and programmes, requiring all stakeholders actors to adopt responsible practices in the use of the resource.
- Awareness-raising, education and best practices. Information and awareness-raising initiatives aimed at tourists and industry stakeholders are crucial. Therefore, the public administration and other stakeholders must take action to improve knowledge about the implications of tourism for water conservation. In terms of good practices, there are a number of measures that almost all tourist establishments can implement, which can contribute substantially to a real water-saving strategy: for example, reducing the washing of towels and sheets, responsible cooking and cleaning, promoting a zero kilometre kitchen, or using fewer utensils and dishes.

While such strategies implemented in the tourism industry have a direct impact on water savings, the basis for ensuring the successful coupling of tourism and water sustainability lies in research, technology, best practices and citizen awareness.

The industry is constantly adapting and moving towards an increasingly sustainable and responsible tourism model, as the survival of the industry largely depends on it.

## Conclusion

The link between water and tourism cannot be denied. The impacts of water on tourism and vice versa require coordinated management to guarantee environmental and water resource sustainability, as well as the sustainability of tourism.

The benefits of blue spaces for tourism are incontrovertible, not only because of their contribution as ecosystems, but also because they support tourists' well-being and physical and mental health. The impact of the tourism industry in terms of water consumption is a major challenge that calls for integrated management of water resources and tourist activities.

It is therefore necessary to reduce the water footprint of the tourism industry and introduce measures that minimize the negative impact of the industry on water resources. Such measures can include reducing water consumption, improving water-use efficiency, boosting investment, fostering the circular economy, and ensuring the involvement of all industry stakeholders, on both the demand side (tourists) and the supply side (operators and tourist establishments), as well as public administrations.

## Future Research

The importance of the tourism sector in many countries around the world and its link to water has been highlighted by the large number of studies addressing the issue. However, in recent years, water scarcity and the need for better management have led to a redoubling of analytical efforts to maintain the quality and quantity of the resource. All of this, in accordance with the requirements of the 2030 Agenda.

The tourism sector is and must be committed to the new requirements that allow its survival, as well as that of ecosystems and, therefore, the introduction of all measures and improvements to reduce its water footprint is key.

### Case Study

***Water Management in Tourist Destinations Suffering from Water Scarcity: The Case of Benidorm (Alicante)***

The Mediterranean is a key area for both tourism and water scarcity. The region accounts for about a third of international tourist arrivals and further substantial growth is expected (Fosse et al., 2021). Water scarcity is at its most severe in summer, when rainfall levels are low and agricultural and urban water demand is high. Tourism also peaks in summer, resulting in a lack of spatiotemporal synchronization between tourist activities and water availability (Liu et al., 2021; Ricart et al., 2020).

A leading tourist destination in the Mediterranean is the city of Benidorm (Alicante), a classic Spanish sun and sand holiday destination that, despite covering only 1% of the total surface area of the coastline of the Valencian Community, has consolidated its position in recent years as the fourth most visited tourist destination in Spain. It has 180,000 habitual inhabitants, of which 70,000 are registered residents, and a floating population that reaches 500,000 in the months with the greatest influx of visitors. The municipality registered more than 11.4 million overnight stays in 2019. However, although the population has increased by 44% in the last 25 years, the water supply has shrunk by 18%.

There are major challenges to the success of Benidorm as a tourist destination, as it suffers from extreme recurrent droughts. Therefore, the goal of water management is to guarantee the supply, by reducing water losses, lowering water consumption to a minimum and introducing elements of efficient management in collaboration with the hotel industry. According to the experts, the data demonstrate the effective water management that has been carried out in this municipality, which makes Benidorm one of the most sustainable cities in Spain in terms of water resources.

The figure shows the evolution of the water supplied since 2004 in comparison with the technical performance of the water network, an indicator that measures the relationship between the volume of water registered in the network and the volume supplied to the system. While this indicator should be complemented with others—because it does not take into account the length of the network, its state of conservation or unregistered water—it constitutes an element of service efficiency worth accounting for. In this case, in which the evolution of this parameter is monitored in the specific location of Benidorm, where supply conditions remain the same over time, the indicator is useful (Figure 5.2).

Known as the "Manhattan of the Mediterranean", in 2019 Benidorm was named Spain's first Smart Tourist Destination by The Institute for Spanish Tourism Quality. This distinction was awarded in recognition of Dinapsis Benidorm, the innovation centre for sustainable territorial management launched by Hidraqua, the company that supplies water to the city. Benidorm is a clear example of tourism sustainability projects harnessing digitalization, which are already becoming a reality. In 2023, Dinapsis Benidorm is working on a project involving an application to calculate the floating population, where several tourism variables are correlated with water service variables (remote meter readings of water consumption, ammonia entering the WWTP, etc.). As such, the Benidorm sustainable tourism experience, which is already beginning to be sold as a successful model for other coastal cities, is closely linked to water.

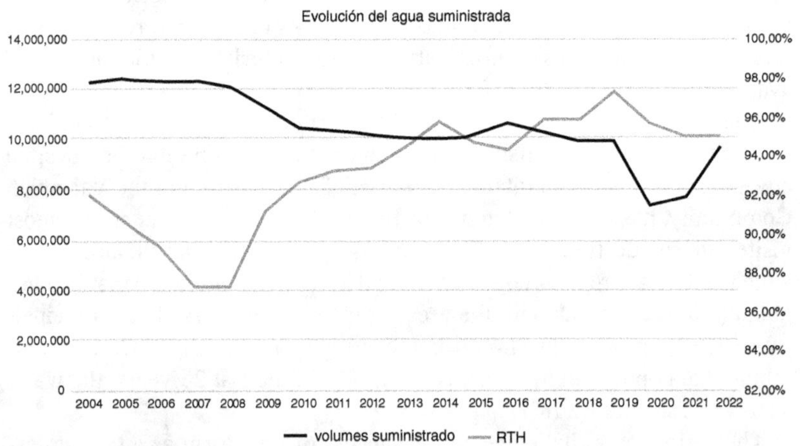


*Figure 5.2* Evolution of the water supplied since 2004, compared with the increase in the technical performance of the water network.

*Source:* Hidraqua (2023).

**Case Questions (Three Questions)**

1 What are the impacts of tourism on water resources?
2 What measures are necessary for the conservation of water resources and ecosystems in tourist destinations?
3 Identify and describe the challenges water scarcity poses to the tourism industry.

**Key Terms and Definitions**

**Water Consumption**: Refers to the water that is consumed and does not return to its original source. This includes the water that is evaporated or incorporated into production and is not available for reclamation.

**Water Footprint**: Indicator that measures the consumption of freshwater throughout the production and consumption processes. It considers both direct and indirect consumption. To calculate it, the total amount of water consumed, evaporated and contaminated throughout the entire process is summed.

**Water Sustainable Management**: It's one of the pillars of the Sustainable Development Goals (SDGs). It refers to systems that ensure a balance between the consumption of freshwater and its natural replenishment. These are systems that enable the responsible use and control without compromising it for future generations.

**Circular Economy**: Production and consumption model aimed to minimize the use of resources and the production of waste, keeping production inputs circulating as long as possible. The use of renewable energies and the elimination of toxic products that hinder land regeneration are also characteristic features of this model.

**Blue Spaces**: Outdoor spaces where water plays a significant role in the landscape. Blue spaces are defined in analogy to urban green spaces where vegetation is the primary feature. They encompass both natural elements such as rivers, beaches, etc., as well as artificial features like fountains, ponds…

**Hydric Stress**: Situation in which the demand of water exceeds the available supply under specific circumstances. For instance, hydric stress is referred when certain areas at certain times of the day experience water shortages, which can hinder such as garden irrigation.

## References

Ballesteros-Olza, M., Gracia-de-Rentería, P., & Pérez-Zabaleta, A. (2020). Effects on general health associated with beach proximity in Barcelona (Spain). *Health Promotion International*, 35(6), 1406–1414. https://doi.org/10.1093/heapro/daaa013. PMID: 32105314.

Cole, S. (2012). A political ecology of water equity and tourism: A case Study from Bali. *Annals or Tourism Research*, 39, 1221–1241. https://doi.org/10.1016/j.annals.2012.01.003.

Deyà-Tortella, B., & Tirado, D. (2011). Hotel water consumption at a seasonal mass tourist destination. The case of the island of Mallorca. *Journal of Environmental Management*, 92(10), 2568–2579. https://doi.org/10.1016/j.jenvman.2011.05.024.

European Commission (1987). 7 July 1987. https://ec.europa.eu/commission/presscorner/detail/en/IP_88_434.

Fosse, J., Kosmas, I., & Gonzalez, A. (2021). *The future of Mediterranean tourism in a (post) covid world* (Eco-med briefing 01/21). Eco-union. https://doi.org/10.5281/zenodo.4616983.

Gabarda-Mallorquí, A., García, X., & Ribas, A. (2017). Mass tourism and water efficiency in the hotel industry: A case study. *International Journal of Hospitality Management*, 61, 82–93. http://doi.org/10.1016/j.ijhm.2016.11.006.

García, C., & Servera, J. (2003). Impacts of tourism development on water demand and beach degradation on the Island of Mallorca (Spain). *Geografiska Annaler: Series A, Physical Geography*, 85, 287–300. https://doi.org/10.1111/j.0435-3676.2003.00206.x.

García, C., Mestre-Runge, C., Moran-Tejeda, E., Lorenzo-Lacruz, J., & Tirado, D. (2020). Impact of cruise activity on freshwater use in the Port of Palma (Mallorca, Spain). *Water*, 12, 1088. http://doi.org/10.3390/w12041088.

Gössling, S. (2000). Sustainable tourism development in developing countries: Some aspects of energy use. *Journal of Sustainable Tourism*, 8(5), 410–425. http://doi.org/10.1080/09669580008667376.

Gössling, S. (2015). New performance indicators for water management in tourism. *Tourism Management*, 46, 233–244. https://doi.org/10.1016/j.tourman.2014.06.018.

Gössling, S., & Peeters, P. (2015). Assessing tourism's global environmental impact 1900–2050. *Journal of Sustainable Tourism*, 23(5), 639–659. https://doi.org/10.1080/09669582.2015.1008500.

Gössling, S., Hall, C. M., & Scott, D. (2015). *Tourism and Water*. Channel View Publications, Bristol, United Kingdom.

Gössling, S., Peeters, P., Hall, C. M., Ceron, J.-P., Dubois, G., Lehmann, L. V., & Scott, D. (2012). Tourism and water use: Supply, demand, and security. An international review. *Tourism Management*, 33(1), 1–15. https://doi.org/10.1016/j.tourman.2011.03.015.

Grellier, J., White, M. P., Albin, M., et al. (2017). BlueHealth: A study programme protocol for mapping and quantifying the potential benefits to public health and well-being from Europe's blue spaces. *BMJ Open*, 7, e016188. https://doi.org/10.1136/bmjopen-2017-016188.

Hadjikakou, M., Chenoweth, J., & Miller, G. (2013). Estimating the direct and indirect water use of tourism in the eastern Mediterranean. *Journal of Environmental Management*, 114, 548–556. https://doi.org/10.1016/j.jenvman.2012.11.002.

IPCC (2022). Climate change 2022: Impacts, adaptation, and vulnerability. *Contribution of working group II to the sixth assessment report of the intergovernmental panel on climate change* (H.-O. Pörtner, D. C. Roberts, M. Tignor, E. S. Poloczanska, K. Mintenbeck, A. Alegria, M. Craig, S. Langsdorf, S. Löchke, V. Möller, A. Okem, & B. Rama, eds.). Cambridge University Press, 3056. https://doi.org/10.1017/9781009325844.

Kent, M., Newnham, R, & Essex, S. (2002). Tourism and sustainable water supply in Mallorca: A geographical analysis. *Applied Geography*, 22(4), 351–374. https://doi.org/10.1016/S0143-6228(02)00050-4.

Liu, H., Jiang, Y., Zhu, H., Chen, Y., Lyu, W., Luo, W., & Yao, W. (2021). Analysis of water resource management in tourism in China using a coupling degree model. *Water Policy*, 23(3), 765–782. https://doi.org/10.2166/wp.2021.155.

Miró, J., & Olcina, J. (2020). Cambio climático y confort térmico. Efectos en el turismo de la Comunidad Valenciana. *Investigaciones Turísticas*, 1, 1–30. https://www.researchgate.net/publication/346602623_Cambio_climatico_y_confort_termico_Efectos_en_el_turismo_de_la_Comunidad_Valenciana.

Morrison, C., & Pickering, C. (2013). Perceptions of climate change impacts, adaptation and limits to adaption in the Australian Alps: The ski-tourism industry and key stakeholders. *Journal of Sustainable Tourism*, 21(2), 173–191. https://doi.org/10.1080/09669582.2012.681789.

Page, S., Essex, S., & Causevic, S. (2014). Tourist attitudes towards water use in the developing world: A comparative analysis. *Tourism Management Perspectives*, 10, 57–67. https://doi.org/10.1016/j.tmp.2014.01.004.

Ricart, S., Arahuetes, A., Villar, R., Rico-Amorós, A. M., & Berenguer, J. (2020). More water exchange, less water scarcity? Driving factors from conventional and reclaimed water swap between agricultural and urban–tourism activities in Alicante, Spain. *Urban Water Journal*, 16(10), 677–686. https://doi.org/10.1080/1573062X.2020.1726408.

Rico, A., Olcina, J., & Sauri, D. (2009). Tourist land use patterns and water demand: Evidence from the Western Mediterranean. *Land Use Policy*, 26(2), 493–501. https://doi.org/10.1016/j.landusepol.2008.07.002.

Roson, R., & Sartori, M. (2014). *International Journal of Climate Change Strategies and Management*, 6(2), 212–228. https://doi.org/10.1108/IJCCSM-01-2013-0001.

Scott, D. (2021). Sustainable tourism and the grand challenge of climate change. *Sustainability*, 13, 1966. https://doi.org/10.3390/su13041966.

Silva, K. B., & Mattos, J. B. (2020). A spatial approach for the management of groundwater quality in tourist destinations. *Tourism Management*, 79. https://doi.org/10.1016/j.tourman.2020.104079.

Torregrosa, T., Sevilla, M., Montaño, B., et al. (2010). The integrated management of water resources in Marina Baja (Alicante, Spain). A simultaneous equation model. *Water Resources Management*, 24, 3799–3815. https://doi.org/10.1007/s11269-010-9634-8.

UNEP (2019). Annual report. https://www.unep.org/annualreport/2019/index.php.

UNWTO (2019). Word tourism barometer. https://www.e-unwto.org/doi/epdf/10.18111/wtobarometereng.2019.17.1.1?role=tab

UNWTO (2020). Global Code of ethics for tourism. https://www.unwto.org/global-code-of-ethics-for-tourism.

WTO (2004). Available at: www.world-tourism.org/market_research/facts/menu.html (accessed 10 August 2004).

# 6 Smart Tourism

## Advancing Sustainable 'Smart' Tourism by a Critical Analysis of the Disparity between the Factors Driving Consumer Decision-Making and the Offerings and Challenges within the Industry

*Saira Sultana*

### Introduction

This chapter assesses the advancement of smart tourism, aimed at offering convenient and timely travel services while potentially enhancing the operational efficiency and visitor capacity of tourist destinations. Additionally, this research scrutinizes the factors influencing consumer decision-making and existing challenges within this domain, with the aim of fostering advancement and forecasting potential future developments and trends.

### Background

Smart tourism has rapidly evolved with advances in technology and changing traveller preferences. Smart tourism involves the use of technology-driven and data-driven solutions to enhance the efficiency, sustainability, and overall experience of tourism for both travellers and destinations (Ye, Ye, and Law, 2020). It encompasses various aspects of the tourism industry, including planning, booking, transportation, accommodation, and on-site experiences. The primary objectives of smart tourism include enhancing the overall visitor experience, ensuring the sustainability of destinations, and facilitating efficient resource allocation. These efforts collectively contribute to the long-term viability and competitiveness of tourist destinations (Mehraliyev, Choi, and Köseoglu, 2019).

Sustainable tourism involves a conscientious and considerate approach to travel and tourism, to reduce adverse effects on the environment, culture, and society, while simultaneously optimizing advantages for local communities and ensuring the ongoing sustainability of destinations (Bressan and Pedrini, 2020) Pandemic in 2019, affected the business model and affected the tourism industry (Scott, Gössling, and de Freitas, 2020). The sustainable tourism sector has received significant attention. However, a deeper understanding of the factors that influence consumer decision-making when choosing their next city break and the underlying causes is necessary (Mehraliyev, Choi, and Köseoglu, 2019). Mainly, what drives

DOI: 10.4324/9781003369967-9

consumers' decision-making when choosing a travel destination? That's why it's crucial to examine psychological aspects for comprehending consumer travel motivations (Yoo, Yoon, and Park 2018), destination preferences influenced by consumer demographics like age, gender, income, and family size (Gretzel et al., 2020), and how social media platforms influence consumer perceptions and aspirations (Xiang et al., 2017).

Furthermore, as outlined by Fermani et al. (2020), sustainability has a substantial impact on individuals' daily behaviours, shaping not only their emotional states but also their social interactions. Anticipating tourists' preference for sustainable hospitality businesses, rather than just environmental practices, can be achieved by examining their sustainable values (Fermani et al., 2020). Therefore, it is important to promote progress and predict future trends by incorporating sustainability into the SMART tourism concept. While the tourism industry has potential, it lacks systematic implementation, with emerging issues.

On the other hand, health and mobility are crucial in the tourism sector, especially with the increased focus on wellness and accessible travel due to the COVID-19 pandemic (Darcy, McKercher, and Schweinsberg, 2020). Urban tourism requires the delivery of an outstanding overall tourism experience and the maintenance of high-quality standards in the town centre (Grah, Dimovski, and Peterlin, 2020). Then, wellness tourism revolves around travel experiences crafted to improve physical, mental, and emotional well-being (Smith and Puczkó, 2020). However, the hospitality sector is presently facing significant challenges and prospects as a result of the swift progress of information and communication technology (Lei et al., 2020). Therefore, the industry faces challenges in meeting consumer preferences, and identifying essential components contributing to sustainable 'smart' tourism involves a comprehensive examination of factors influencing consumer choices. This chapter will address two primary aspects concurrently: first, the exploration of consumer preferences, and second, the examination of the industry's capabilities and hurdles in meeting these preferences through the implementation of a SMART tourism experiment.

## Factors Influencing Consumer Decision-Making

When consumers are contemplating their upcoming holiday destination, various fundamental elements come into consideration. These elements shape their decision-making process regarding where to go for their travels. Grasping these elements can assist businesses within the travel and tourism sector in formulating precise marketing strategies and offerings. Here are the primary underlying factors that influence consumer decision-making when picking a vacation destination:

### *Psychological Factors*

Consumer decisions are often driven by their motivations for travel, such as relaxation, adventure, exploration, cultural experiences, or personal growth (Yoo, Yoon, and Park, 2018). The Plog tourist motivation model (1974) suggested the psychographic characteristics of travellers and how these characteristics influence consumer travel preferences and behaviours. There are three main kind of tourist based on the psychographic traits: such as, allocentrics, midcentrics, and psychocentrics.

Allocentrics are considered by a sense of adventure (Plog, 1974; Yoo, Yoon, and Park, 2018). This kind of tourist have a strong desire for new and exotic experiences. They do prefer to take risk however tend to enjoy exploring unfamiliar destinations. They seek for unique and authentic experiences with off-the-beaten-path destinations, also open to trying different cuisines and activities. This categorical tourist usually are engaged in adventure tourism, backpacking, and cultural immersion. They can easily adopt new tends and often travel independently (Plog, 1974; Yoo, Yoon, and Park, 2018).

Midcentrics are in the middle of allocentrics and psychocentrics on the adventure-seeking spectrum (Plog, 1974; Yoo, Yoon, and Park, 2018). They value comfort and familiarity but are moderately open to new experiences. They search for destinations that offer both adventure and comfort. They prefer familiar facilities and expediencies, but open to trying some new experiences. Overly crowded destinations are avoided by this category of tourist, however they do prefer somewhat popular. They frequently travel with a combination of planned activities and opportunities for spontaneity (Plog, 1974; Yoo, Yoon, and Park, 2018).

Psychocentrics are considered with the preference for familiarity, routine, and comfort (Plog, 1974; Yoo, Yoon, and Park, 2018). They avoid to take any risk and may feel uncomfortable in unfamiliar or challenging environments. They prefer security so usually choose the well-known and tourist-friendly destinations for relaxation. They look for familiar cuisine and activities. They consider the package tours, all-inclusive resorts, and guided experiences which have predictability and convenience (Plog, 1974; Yoo, Yoon, and Park, 2018).

Plog's model has been used to suggest that people's psychological characteristics have a significant impact on their travel motivations and choices. The influence it has had on tourism marketing and destination planning is evident. It aids businesses and destination managers in tailoring their offerings and marketing strategies to specific travellers (Cruz-Milan, 2018). Those looking to attract allocentrics might focus on adventure activities and authentic cultural experiences, while those looking to attract psychocentrics might focus on safety, comfort, and familiar amenities. As travellers age or their life circumstances change, they may transition from allocentrics to psychocentrics. For instance, a person may be more adventurous in their youth (allocentric) but tend to become more risk-averse and comfort-seeking as they age (psychocentric) (Yoo, Yoon, and Park, 2018). Therefore, it is an essential consideration that any business can take into account when developing smart tourism with a focus on sustainability.

### *Personal Factors*

The previous research has proven that destination choices can be influenced by consumer age, gender, income, and family size (Gretzel et al., 2020). Families with children may choose family-friendly destinations, while young adults may choose more adventurous destinations. Furthermore, the type of destination a consumer chooses can be determined by their personal interests, such as outdoor activities, history, culture, or food. However, travel health considerations have had a significant impact due to the recent COVID-19 pandemic (Sigala, 2021). The health and environment of travellers are becoming more of a concern. Travel options that are both eco-friendly and sustainable are becoming more popular (Scott, Gössling, and de Freitas, 2020).

The traveller's journey through the stages of the travel process is shaped by their anticipations regarding the destination. Priorities and social values are being adjusted by people in the post-COVID-19 normalcy (Gretzel et al., 2020). Therefore, identifying tourists' expectations before travelling is crucial. Moreover, identifying the preferences of tourists, establishing their attitudes, and determining their expectations for post-COVID-19 destination selection is key. Studies have indicated that in the post-pandemic era, tourists are more inclined towards urban tourism as their top choice, with cultural tourism and travelling with family members following closely behind. This shift in preferences also reflects a greater emphasis on adopting a conscientious and helpful approach to travel (Gretzel et al., 2020; Scott, Gössling, and de Freitas, 2020).

***Social Factors***

Holiday destination choices can be heavily influenced by the recommendations and experiences shared by friends, family, and peers. Consumer perceptions and desires can be influenced by the portrayal of destinations on social media platforms. Destination choices, travel planning, and experiences are greatly influenced by social media platforms and peer recommendations (Xiang et al., 2017). The image of a destination is shaped by social interactions and discussions among friends and family, which in turn influences travel choices (Miguéns et al., 2016). Tourists tend to double-check their preliminary ideas on social media, regardless of where they get them. The direct and immediate impact of social media is due to the fact that tourists receive approval from social media prior to making a decision. Usually, it occurs during the trip and, on occasion, right in front of the product/service they are about to choose. Social media either endorses or disapproves this choice by suggesting that tourists should not proceed or consider other options. Before the final stage of tourists' decision-making process, namely, before choosing or purchasing, this role takes place (Xiang et al., 2017). The online ecosystem for hospitality and tourism is vast, complex, and diverse. From transaction-based online travel agencies (aka OTAs) like Expedia and Bookings.com to community-based online review platforms like LonelyPlanet, Tripadvisor, and Yelp, reviews are incorporated as electronic word-of-mouth (Gligorijevic, 2016). While they all have a common goal of assisting consumers in making decisions, they are all part of social media. Trusted and shared social knowledge is the way to achieve it (Xiang et al., 2017).

The sociocultural and economic systems of these platforms are complex. Deutsch and Gerard (1955) formulated the dual-process theory within the field of psychology in order to elucidate how social factors impact the psychological processes of individuals, as highlighted by Filieri (2015). Informational and normative influences are two types of influences on individual judgements that are considered in the theory. The receiver's judgement of the relevant content of a message is the basis of informational influence. The quality of information within a message is shaped by informational influence, encompassing factors like relevance, the credibility of the source, and the volume of information involved. Normative influence is a compulsion for individuals to comply with the expectations of others, whether they are implicit or explicit in the choices of a reference group (Deutsch and Gerard, 1955; Filieri, 2015). Dual-process theory asserts that consumers can rely on others' perceptions and judgements for reliable evidence of reality. Hence, it can be anticipated that when multiple consumers share a similar perception of a product (like a hotel), their perceptions will subsequently influence one another. In online word-of-mouth (e-WOM) contexts, users can utilize specific cues to gauge how various customers have assessed a particular product or service. Many websites rely on the collective evaluations from all consumers who have reviewed a product within a specific category to determine the star ratings of products (for example, hotels in a specific location). Consumers often refer to these ratings to gain insights into a product's quality. (Deutsch and Gerard, 1955; Varkaris and Neuhofer, 2017)

Social media holds the capacity to serve as a resource for travellers, enabling them to discover attractions to explore and activities to engage in at their destination. Influencers on social media platforms have become a popular marketing strategy for promoting travel destinations and products, influencing tourists' perceptions (Chung and Law, 2019). The importance and widespread use of social media in various aspects of tourists' decision-making require a thorough and holistic comprehension of its role, as tourists' information search and decision-making processes are impacted by social media platforms (Zeng and Gerritsen, 2014). This influence occurs within the information search and evaluation phases of tourists' decision-making. Song and Yoo (2016) provided evidence that social media significantly influences tourists' decision-making in the pre-purchase phase. The use of social media has a positive impact on tourists' purchase decisions, with a positive impact on their willingness, intention, and decision-making. According to researchers, tourism companies have paid a considerable amount of attention to the consideration and purchase stages. Social media can exert a substantial influence on consumers during the assessment and post-purchase phases (Filieri, 2015). Earlier studies have underscored the significance of social media in shaping tourists' attitudes and behavioural intentions when planning trips. However, the substantial role and impact of social media in the decision-making process are particularly noteworthy in real, lived travel experiences rather than hypothetical travel scenarios (Memon et al., 2015). This is mainly because social media analytics stands to gain from the expansion and potential of the hospitality and tourism sector, which appears to be an ideal domain for its application.

### *Consumer Sustainable Tourism Preferences*

Travel websites, blogs, social media, travel agencies, and recommendations are used by consumers to gather information about potential holiday destinations. Consumers evaluate different destinations based on factors such as cost, accessibility, safety, and attractions. What is most important to consumers is which aspects of a destination are most important, whether it's culture, nature, relaxation, or adventure (Chang, McAleer, and Ramos, 2020). The final decision can be influenced by how easy it is to book accommodations, flights, and activities. After completing a trip, consumers reflect on their experiences, which could have an impact on their future travel choices. Personal interests, values, and external influences are among the factors that influence consumer preferences and priorities in tourism (Fermani et al., 2020).

Gregori et al. (2013) have divided sustainable tourism in three types. During the formation phase, the destination's target allows for ample growth in attendance and arrivals, without compromising the environmental and social balance of a territory. Then, a sustainable destination is one that harmonizes tourism activity with ecological considerations. This approach ensures that the influx of tourists does not compromise sustainability while simultaneously offering economic and social advantages to the area, along with environmental well-being. And finally, The development of tourism aims to enhance visitor numbers and the tourist presence in destinations that might otherwise be financially unviable (Gregori et al., 2013; Fermani et al., 2020).

Sustainability affects people's everyday actions, influencing not just their emotions but also their social interactions. The values of sustainability can be used to predict specific actions, such as reducing energy consumption and pollution. Predicting tourists' preference for a sustainable hospitality establishment over environmentally conscious practices is contingent on their alignment with sustainable values. Initially, sustainable tourists were considered a more lenient type of eco tourists because eco tourists meant different types of tourists who cared about protecting the environment. Four types of eco tourists were identified by (Millar and Baloglu, 2011; Fermani et al., 2020), this spectrum of travellers includes those who are driven to explore pristine and less-visited locations, as well as those who incorporate eco-friendly destinations into more conventional tours. For instance, there are individuals who value the idea of being guests in another culture, society, environment, and economy, recognizing the distinctiveness of their travel experiences (Shamsub and Lebel, 2016). Additionally, sustainable tourists are individuals who adhere to a set of guidelines promoting sustainable behaviour. They are aware that their actions can positively impact the environment and make adjustments accordingly. Their goal is to support the host economy by purchasing locally made items such as crafts and food (Shamsub and Lebel, 2016).

### *Post Pandemic Transformation*

The pandemic is seen as a chance for transformation in the growing dialogues and research within the tourism sector regarding COVID-19 (Mair and Wittmer, 2019). The industry must strive to recuperate, envision a new normal, and reshape the future economic landscape (Mair and Wittmer, 2019). Researchers should not limit themselves to COVID-19 as the sole context for replicating existing knowledge in measuring and forecasting tourism effects (Gössling, Scott, and Hall, 2019).

Ongoing research delves into the extensive societal repercussions of COVID-19 on the tourism and hospitality sectors, with a focus on potential alterations in people's lifestyles and daily routines during these challenging times (Fermani et al., 2020). The COVID-19 period has been regarded as a time marked by resilience, wherein fear can stimulate novel decision-making and encourage a re-examination of our values and driving forces (Niewiadomski, 2020). In the aftermath of COVID-19, tourists are displaying a disinclination to participate in mass tourism, opting instead for the principles of slow tourism, which prioritize building connections with local communities, prolonged stays, and more immersive and enriching travel encounters (Fermani et al., 2020).

Prospective tourists are likely to exhibit a newfound interest in aspects such as destination hygiene, medical facilities, and population density (comprising both local residents and tourists) when making travel-related choices. This presents an opportunity to re-evaluate tourism planning and development to ensure sustainability. Specifically, to understand the potential shifts in global hospitality practices resulting from the pandemic, it's crucial to maintain a focus on sustainability. This approach will assist industry professionals in tailoring their products and services for the post-COVID-19 recovery phase (Gretzel et al., 2020).

Furthermore, given the intricate and interconnected socio-cultural, economic, psychological, and political ramifications of COVID-19 of this scale, conventional predictive models may not be effective as we anticipate explicit trajectories rather than relying solely on historical trends. Moreover, there is ample evidence to suggest that both the tourism industry and research have matured significantly, offering substantial knowledge on how to investigate and adeptly formulate and execute strategies for crisis recovery and response as demonstrated by (McKercher and Chon, 2004) and building resilience to confront future crises (as evidenced by Hall, Prayag, and Amore, 2017).

Nonetheless, there remains a gap in comprehending how crises can serve as catalysts for industry transformation. By recognizing the structural contradictions and paradoxes unveiled through disruptions in socio-economic life, it becomes possible to harness crises as opportunities for positive and transformative change. It is widely acknowledged that effective crisis management is essential before, during, and after a crisis. To achieve a transformative outcome, it is crucial to address the impact and implications of COVID-19 on three key stakeholders: tourism demand, tourism operators, and destinations, in addition to policymakers. The industry should leverage transformative research on COVID-19 to reimagine and implement a human-centred operational environment that prioritizes values of sustainability and well-being (Gretzel et al., 2020).

## Tourism Industry Offerings

### *Accessible Tourism*

Recent studies have emphasized the significance of health and mobility within the tourism sector, particularly due to the heightened interest in wellness and accessible travel experiences in the aftermath of the COVID-19 pandemic. Destination choices may be influenced by health concerns, including physical abilities and any medical limitations. Furthermore, decisions can be influenced by whether consumers are travelling alone, with family, friends, or in groups (Darcy, McKercher, and Schweinsberg, 2020)

The aging population has resulted in an increase in senior travellers who may have unique health and mobility needs. It is crucial to understand their needs (Chen and Chen, 2019) and accessible tourism is essential. The aim of accessible tourism is to ensure that travel experiences are inclusive and accommodating for individuals with disabilities or limited mobility (Darcy, McKercher, and Schweinsberg, 2020). Accessible tourism is a type of tourism that involves cooperative efforts among various stakeholders. It facilitates the independence and fair treatment of individuals with diverse accessibility needs, encompassing mobility, vision, hearing, and cognitive aspects, by providing universally designed tourism products, services, and surroundings. This definition promotes a lifelong approach, ensuring that people, including those with permanent and temporary disabilities, seniors, and individuals with obesity, families with young children, and those seeking safer and socially sustainable environments, can benefit from accessible tourism provisions (Darcy, McKercher, and Schweinsberg, 2020).

### *Urban Tourism*

Then again, urban tourism is the term for tourists visiting cities to explore their cultural, historical, architectural, entertainment, and economic attractions (Grah, Dimovski, and Peterlin, 2020). The broader tourism industry is influenced by the unique features and amenities that cities offer, making urban tourism a significant component.

Urban tourism encompasses a diverse array of attractions and activities within a compact, intriguing, and appealing environment, rather than focusing solely on any individual component. What typically matters is the entirety and the excellence of the overall tourism experience and the quality of the town centre as a whole (Grah, Dimovski, and Peterlin, 2020).

In urban tourism destinations, it is a common challenge that buildings, establishments, public transportation, streets, and squares are often not accessible for individuals with disabilities. Travelling can become intricate and challenging even if some elements are accessible, as there may be disruptions in the accessibility chain. Numerous inconveniences such as thresholds, the absence of elevators, or adapted restrooms can easily transform an otherwise pleasant journey into a substantial challenge. Hence, it is evident that ensuring seamless travel experiences is crucial (Yuan et al., 2018).

For instance, providing accessibility information that empowers people to make informed decisions about their travel plans is indispensable. Innovative technologies like real-time virtual reality and 3D photo imaging offer a wealth of information, enabling individuals to select destinations and facilities that match their accessibility needs and even serving as a virtual gateway to places for those with physical limitations. Establishing the objective of delivering a comprehensive accessible experience, considering both the perspectives of stakeholders and visitors, is of utmost importance in enhancing customer care and satisfaction (Yuan et al., 2018). This approach may not always align with cost-effective practices and can place significant demands on resources and manpower. Nevertheless, disabled travellers often prioritize factors beyond cost when deciding where and how to travel. Once trust is established and previous experiences have been satisfactory, disabled travellers tend to value the additional effort and are more likely to return to specialized operators and providers that offer a comprehensive and accessible experience. This not only fosters loyalty but also can lead to increased profitability.

### *Wellness Tourism*

During the 21st century, tourism has experienced notable structural transformations that have redirected its attention towards meeting the growing and varied needs of travellers. Beyond passive beach vacations, modern tourists are keen on participating in various sports and physical activities, have a growing appreciation for nature, and place high value on access to recreational amenities and services (Andreu, Font-Barnet, and Roca, 2021).

Wellness tourism, in particular, centres on travel experiences designed to enhance physical, mental, and emotional well-being. This encompasses activities such as spa visits, fitness retreats, and mindfulness getaways (Smith and Puczkó, 2020).

Wellness entails the proactive engagement in actions, decisions, and ways of living that contribute to an all-encompassing state of well-being. It is an active commitment that necessitates intentions, choices, and actions as individuals endeavour to attain peak health and well-being, as opposed to a passive state. Wellness is intricately connected to holistic health, which comprises multiple facets such as physical, mental, environmental, spiritual, emotional, and social well-being, all of which should harmoniously coexist (Kelly, 2018; Azara et al., 2018).

In the realm of sustainable and integrated wellness tourism, these three components can lead to vacations that blend enjoyment, acts of kindness, and meaningful experiences, providing a well-rounded holiday (Smith and Diekmann, 2017). Moreover, wellness tourism has the potential to stimulate economic growth.

It is in harmony with health routines and has the potential to revitalize traditional wellness methods, which may lead to a decrease in the occurrence of mental health problems. This, in turn, contributes to the preservation of natural and cultural assets, supports environmental conservation, and promotes sustainable tourism, ultimately aiming to enhance the quality of life and social capital (Andreu, Font-Barnet, and Roca, 2021).

## Industry Challenges

### *Travel Mobility and Technology*

Organizations can boost their ability to adapt, strengthen their resilience, and improve their prospects for survival through innovation (Bressan and Pedrini, 2020). Individuals with mobility challenges can take advantage of mobile applications and platforms that offer details about accessible amenities (Sarkar, 2017).

Innovative consumers tend to expect substantial advantages from new innovations and are more inclined to adopt novel products and services swiftly and extensively compared to others. They become a valuable market segment for companies by offering feedback and generating revenue through their adoption of these new offerings (Tussyadiah, 2016). The convenience and ease of modern travel have been significantly enhanced by mobile devices, enabling travellers to complete various transactions related to travel products and services (Ozturk et al., 2016). Hotel guests are increasingly tech-savvy and utilize a range of devices, including smartphones, mobile phones, tablets, and laptops, while on their travels. They use these devices for activities such as pre-checking into their hotel rooms, browsing the internet, and making hotel-related purchases and service requests during their stay (DeFranco et al., 2017).

The internet's rapid development in recent decades has greatly simplified and accelerated information dissemination and sharing, significantly impacting the hospitality sector (Fang et al., 2017; Pestek and Sarvan, 2020). Mobile technologies enable the gathering and immediate sharing of valuable real-time information, thereby enriching the tourism experience (Corrêa and Gosling, 2021). A variety of transactions, including tourism planning and development related products and services, can be completed using mobile devices. Both convenience and ease for travellers have been introduced by it (Ozturk et al., 2016). Travel websites can be accessed instantly by tourists using smartphones to obtain various types of information. The significance of on-the-go functionalities for travellers is underscored by Kim and Law (2015), who stress the importance of accessing weather updates, accommodation details, attractions, and transportation information from anywhere (Kim and Law, 2015). The adoption of 'on-the-go' technology has seen a surge among consumers in the hotel industry, driven by innovative advancements. This includes the utilization of smartphones and related apps for making hotel reservations worldwide (Peng et al., 2017).

Initially, the first-generation hotel mobile websites primarily offered information about the hotel's location, amenities, and facilities. However, with the advent of new hotel mobile websites and mobile applications, travellers can now access comprehensive details such as room availability, star ratings, cancellation policies, and pricing. Furthermore, services like all-inclusive packages, bed and breakfast options, and free Wi-Fi can influence travellers to make on-the-go room bookings (Ozturk et al., 2016).

Mobile technologies have introduced an efficient communication channel within the tourism and hospitality industry (Li et al., 2020). Innovative methods that evoke positive emotions have a notable impact on visitors' experiences, leading to heightened satisfaction and loyalty (Villacé-Molinero et al., 2021). Within the realm of mobile technology, mobile applications play a pivotal role. In order to develop superior practices for mobile commerce (m-commerce) (Morosan and DeFranco, 2016), it is crucial to comprehend how consumers form intentions to utilize hotel apps for accessing products, services, and information while staying at hotels (Morosan and DeFranco, 2016).

However, merely embracing such innovations is insufficient for gaining a competitive advantage, especially as technology becomes increasingly accessible (Kim and Law, 2015). The hotel industry can derive significant benefits from mobile apps, which can be employed to enhance the effectiveness of promotional materials (Peng et al., 2017).

### *Luxury and Sustainability*

Some travellers prioritize luxury and exclusivity, seeking upscale accommodations, gourmet dining, and personalized experiences (Dubois and Ceron, 2021). A majority of previous research in the field of sustainability has contended that luxury and sustainable development are incongruent in the luxury hospitality sector (as seen in Chu, Tang, and Luo, 2016). Sustainability efforts may pose challenges to the economic performance of luxury hospitality companies (Peng and Chen, 2019). Barber and Deale (2014) have argued that addressing environmental concerns in a luxury hotel setting can potentially have adverse effects on the comfort and opulence aspects of the hospitality experience. The primary emphasis in this field continues to be on delivering opulent experiences to guests (Harkison, Hemmington, and Hyde, 2018). The emotional response and purchase intentions of consumers can be influenced by the hedonic and symbolic value of a hotel's offerings (Peng and Chen, 2019).

To uphold the prestigious status of their businesses, managers may employ practices that have the potential to negatively impact the natural environment. For example, they may choose to replace towels in guests' rooms, even if those towels have not been used (Peng and Chen, 2019). Luxury hospitality is associated with status-oriented consumption, albeit with some distinctions compared to the context of luxury goods (Yang and Mattila, 2017). In the realm of hospitality, the concept of luxury often aligns with the idea of more, such as having more staff, more amenities, more services, and more lavish decor, which can run counter to the principle of less, including reduced consumption, minimized waste, and the conservation of resources, typically associated with environmentally conscious behaviour (Line and Hanks, 2016). Managers of luxury hotels might be apprehensive that adopting environmentally friendly practices could potentially undermine the perceived luxury value and brand image of their establishments (Song et al., 2012).

Line and Hanks (2016) investigated the degree of belief in two categories: environmental beliefs and luxury beliefs related to environmentally friendly hotels. The extent to which an individual believes that choosing green hotels leads to a reduction in luxury experiences is what we refer to as their definition of luxury beliefs about green hotels. The choice to engage with sustainable hospitality services is influenced by both consumer attitudes and behaviours. Based on the idea that there is a trade-off between the concept of luxury and the potential advantages of environmentally conscious consumption, consumers' attitudes towards staying in an eco-friendly hotel show an inverse relationship with their beliefs about luxury. This connection also applies to their inclination to choose eco-friendly hotels and their readiness to spend extra for such accommodations. It's important to highlight that this observed pattern is mainly noticeable in urban tourism spots, rather than in nature-oriented destinations. Similarly, Peng and Chen (2019) illustrated that consumers tend to postpone their decision-making regarding whether to stay in a luxury hotel during their vacation if that luxury hotel adopts new green practices. This delay can have an adverse impact on their desire to book a stay at a luxury hotel. This hesitation stems from the concern that customers express concerns about the ability of upscale hotels to consistently deliver high-quality amenities, uphold their distinctive reputation, evoke a feeling of luxury, and prove to be a sound financial choice (Peng and Chen, 2019).

Luxury hotels may exhibit reluctance in embracing new eco-friendly measures because of concerns about how consumers will react (Peng, Chen, and Hung, 2020). Dolnicar, Cvelbar, and Grün (2017) have pointed out the challenges associated with promoting pro-environmental behaviours in the tourism and hospitality industry due to the inherently pleasurable nature of these experiences. Athwal et al. (2019) have identified conflicts between green building practices and the expectations and satisfaction levels of hotel guests, emphasizing the apparent incompatibility between luxury and sustainability within the hospitality context.

However, recent research (Peng and Chen, 2019) suggests that conveying sustainability initiatives has the potential to generate a positive impact on the reactions of high-end hospitality consumers. This favourable viewpoint emphasizes that upscale hotels have strong incentives to operate with an environmentally conscious approach. Significantly, sustainable travel options are becoming more sought after by millennials, women, and educated consumers (Hassan and Lee, 2015). Furthermore, advancements in technology are facilitating the adoption of green practices that not only enhance hotel cost efficiency but also generate revenue (Perramon et al., 2022). Consumers may indeed view luxury goods that incorporate sustainability-related elements in a positive light (Amatulli et al., 2017; Athwal et al., 2019).

In general, some scholars have argued that luxury and sustainability are not as contradictory as commonly perceived. They share a common focus on rarity. However, Amatulli et al. (2017) put forth the view that luxury pertains to high-quality products that are measurably scarce due to their utilization of scarce materials and exceptional craftsmanship, whereas sustainable development aims to conserve natural resources by curtailing excessive material consumption that surpasses the world's recycling capacity. Sustainable luxury encompasses various aspects of design, production, and consumption that prioritize environmental and ethical considerations or both. Its objective is to rectify various perceived shortcomings within the luxury industry, such as issues related to animal welfare, environmental impact, and human exploitation, as highlighted by Athwal et al. (2019).

On the contrary, budget-conscious travellers prioritize affordability and may seek accommodations and destinations that offer cost-effective options (Xiang et al., 2017), making sustainability more appealing in such cases (Figure 6.1).

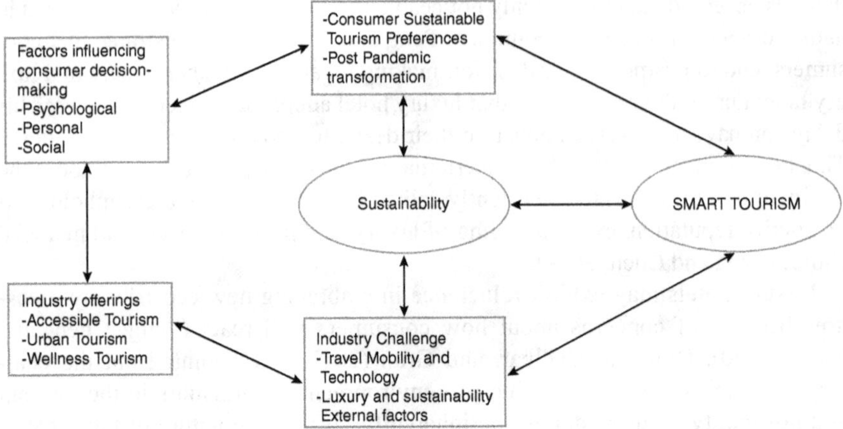


*Figure 6.1* Developing sustainable 'smart' tourism by analysing the gap between consumer underlying factors of the decision-making process and industry offerings and challenges.

## Solutions and Recommendations

### *External Factors*

#### *Economic Stability*

Travel budgets can be influenced by economic stability and exchange rates in influencing destination choices. The tourism industry is heavily influenced by economic conditions, as travellers' decisions are often influenced by factors such as income levels, exchange rates, and overall economic stability. Consumers' ability to travel and their spending on tourism-related activities are influenced by economic conditions, including personal disposable income (Bieger et al., 2018). The cost of international travel can be impacted by exchange rate fluctuations, which can affect destination choices and travel behaviour (Gössling et al., 2019). Short-term and long-term effects on tourism demand can be significant due to economic downturns and crises (Mair and Wittmer, 2019). Both tourism supply (e.g., availability of tourism services) and demand (e.g., travellers' ability to take vacations) are influenced by economic conditions, which affect employment levels (Li et al., 2020). Stimulus packages and tourism promotion are government policies that can play a critical role in mitigating negative impacts on the tourism industry during economic crises (Navarro-Pantin et al., 2019). Economic growth in tourism destinations can be stimulated by investing in transportation and tourism infrastructure (Dwyer et al., 2018). Travellers' price sensitivity and choice of destinations and travel products can be influenced by inflation rates and overall price levels (Vicente, 2020).

*Political Stability and Safety Concerns*

The decision-making process may be influenced by political stability and safety concerns in potential travel destinations. Geopolitical factors have the potential to have a significant impact on the tourism industry by influencing travel patterns, destination choices, and overall tourism demand. International tourism can be directly impacted by geopolitical tensions and government policies related to travel restrictions and visa requirements (Hall, 2020). The decision of tourists to visit or avoid certain areas can be significantly influenced by political instability, conflicts, and security concerns in a destination (Yang et al., 2019). International political interactions can influence tourism flows, as changes in trade or diplomatic relations between countries can encourage or discourage travel (Li et al., 2019). Geopolitical conflicts or disputes that involve a destination can have a negative impact on its image and discourage tourists (Kim and Law , 2019). By limiting economic exchange and travel, economic sanctions and embargoes can have a ripple effect on various industries, including tourism (Yang et al., 2019). Tourism flows can be greatly affected by geopolitical factors, such as international cooperation and border controls during health crises like COVID-19 (Sigala, 2021). To attract tourism investment and infrastructure development, political stability is a crucial factor (Hosany and Witham, 2019). The complex interaction between geopolitics and the tourism industry is emphasized by the external factors mentioned above. Highlighting the necessity for industry stakeholders and policymakers to monitor and adjust to geopolitical changes that can affect tourism dynamics.

*Online Eco-system*

Hospitality and tourism enterprises must gain a comprehensive understanding of the online landscape to make informed decisions when selecting various technological channels, engaging with different user segments, and managing power dynamics within the online ecosystem for interaction with both current and potential customers (Stare and Križaj, 2018). This is particularly crucial when responding to both positive and negative reviews (Sparks et al., 2013; Xie et al., 2014). Social media platforms can serve as valuable tools for monitoring indications of service failures and areas requiring improvement. Constructing online reviews with ample information can enhance their effectiveness as persuasive means of communication (Yoo, Sigala, and Gretzel, 2016). This is especially true when the content is rich in subject matter and conveys strong yet meaningful emotions that align with the rating and content of the review. Implementing strict rules to filter out undesirable inputs may not be the most efficient approach to ensuring the quality of reviews. Therefore, businesses should explore effective methods to encourage customers to share their consumer experiences in a meaningful and constructive manner (Stare and Križaj, 2018).

*Ecological and Social Problems*

The viability of any tourism destination faces threats from ecological and societal challenges (Rees, 2017). A decline in competitiveness damages the image of the targeted destination. These challenges impact the quality of life within the local community through two key factors: the relationship between tourists and residents and the development of the tourism industry itself (Fermani et al., 2020). For instance, the role of transportation has been substantiated in contributing to social and economic issues related to sustainability. The Bruntland Commission established an interconnection among the three dimensions of sustainability (environmental, economic, and social) to address issues like the decline of community in neighbour hoods and the reduced productivity of rural lands (May and Crass, 2007). Hence, there is a specific need for a measurement tool to assess the attitudes of residents in a particular location towards sustainable tourism, in order to fulfil the sustainability requirement in tourism.

Furthermore, contemporary tourism is perceived as a social and cultural experience by both millennials and post-millennials. It fosters social interaction and the construction of identity, giving new significance to their choices as tourists. Vacation planning is customized to their preferences and expectations, with a growing interest in innovative tourism practices and niche offerings (Veiga et al., 2017). To address future risks and expedite the shift towards sustainability, destinations play a pivotal role in building resilience, adaptability, and agility. Although health and safety considerations have become the foremost priorities in the aftermath of the pandemic, the need for sustainable and robust businesses is even more accentuated in the recovery period. Millennials show a keen interest in selecting sustainable travel options, whereas Generation Z may be less inclined to accept higher costs (Monaco, 2018). Sustainable travel experiences are becoming increasingly sought after by consumers of all age groups. Regardless, younger generations, such as millennials, will play a pivotal role in driving substantial changes in behaviour and attitudes regarding the reasons and methods of travel. To endure, the hospitality industry must innovate by adopting eco-friendly practices (Fermani et al., 2020).

One starting point for the hospitality industry is to assess tourists' perceptions regarding their preference for accommodations that prioritize sustainability features. Many travellers aspire to minimize their environmental footprint by choosing destinations and accommodations that adhere to sustainable practices (Lovelock and Sasser, 2017).

## Future Research Directions

In this chapter, consumer preferences and the determinants influencing their choices when selecting a travel destination have been investigated. The industry's capacity and challenges in aligning with these preferences have also been assessed through the execution of a SMART tourism experiment. To advance the initial efforts in smart tourism development, with a specific emphasis on sustainability, it is beneficial to explore various areas that require further research.

To formulate strategies and implement actions aimed at nurturing resilience in support of sustainable tourism practices, with a focus on fostering sustainable growth for destination viability and attracting mindful travellers.

To identify the importance of upholding the principles of the rule of law and effective governance to ensure political stability, as well as to acknowledge the need for investments in tourism safety and security measures, which encompass well-trained and well-equipped law enforcement agencies, surveillance systems, and emergency response teams.

To explore the digital landscape for a more profound insight into the technologies and emerging patterns within the sustainable smart tourism sector.

## Conclusion

Indeed, smart tourism has captured significant attention from both industry practitioners and scholars in recent years. The evolution of smart tourism, driven by technological advancements and changing traveller preferences, has ushered in transformative changes within the tourism sector. In order to establish and comprehend this emerging concept, this chapter has set out to delineate several fundamental elements contributing to the development of sustainable 'smart' tourism. This goal was accomplished through a detailed examination of the factors shaping consumer decision-making and the challenges and opportunities inherent in the industry.

Specifically, this chapter explores consumer preferences, the industry's capabilities, and the obstacles faced when implementing smart tourism solutions. By gaining a comprehensive understanding of these two fundamental aspects, it is possible to navigate the dynamic terrain of the tourism industry more effectively. This, in turn, provides a roadmap for the future, where the convergence of technology, sustainability, and traveller satisfaction can create genuinely intelligent and resilient tourism experiences.

### Case Study

Bliss holiday is a travel company who is trying to establish the business in the tourism sector based on sustainability and smart urban tourism. The company has applied intensive approach as a part of their corporate social responsibility and focused on reducing emissions and carbon footprint to reach a sustainability goal. However, the company has comprehended the urban destination that they are offering on the holiday package, needs alteration by focusing in consumer diverse need to give them a sustainable travel experience. The company also realized that some methods of travel can be extremely harmful to the environment. They took it as their responsibility to ensure that sustainable options are always considered. The company found that they cannot help to reduce something if they do not know they have to reduce. They asked for government support to be more resourceful and to identify that they need to create more awareness to the public. For example, they introduce the option of group travel, which would allow more safety to the consumer and more carbon-friendly options while on the trip, such as considering a travel as a group by using same transport rather than single vehicles. The company has also taken into consideration few important issues such as terrorist threats, health and safety issues among others. In the ongoing pandemic situation, company has analysed the potential risks while travelling internationally during this pandemic include risk of COVID-19 infection while on-trip, lockdown imposed in destination, quarantine requirements on return to home country. Therefore, the business struggled form the fact by measuring the COVID-19 safety regulations. Although global cities have swallowed impact of the pandemic, with nationwide lockdowns forcing the closure of a number of industries and social distancing measures gave a pause to companies' productivity.

However, COVID-19 vaccination rollout programmes, gradual ease of lockdown restrictions allowed once again to rebuild the business position. As countries around the world seek to 'build back better' in the aftermath of the COVID-19 pandemic, the company realized that the need for sustainable urban development has become more important. Bliss have received the government support to strengthen infrastructure and build community resilience with the Next Generation EU Recovery Plan, which will finance the European Green Deal; where, the European Commission has established a set of policies to achieve carbon neutrality across the EU by 2050. The cutting of $CO_2$ emissions on public transport is the part of the programme and this initiative has predominantly effective for the company to support sustainable urban development goals, As such, as a responsible organization they managed efficiently as its duty of care. However, the regulations have undoubtedly impacting the trip experience for the consumers. To engage consumer in travel, Bliss travel has showcased the importance of their duty of care to the customers and implemented the idea into the travel policies.

As government taking initiatives to implement several new strategies and policies for the sustainable urban transformation, the idea of smart cities has become ever more popular that any travel business need to consider. According to the United Nations (UN) estimation, the global urban population increases by 1.5 million people each week and by 2050, around 70% of the world's population will live in cities. And the possible estimation are 60%–80% of the world's GDP will be generated by the cities. This is a clear indication that urban areas are significant to any impact of the economic growth.

What is a smart city? No single definition can explain the concept, but the basis of the concept relies upon the urban settings, which utilize technology to respond to the challenges faced by cities and people to make become safer, more sustainable, and more efficient. Smart technologies can reduce buildings environmental impact since it focus on few key smart city applications. For example; smart buildings which might have different functions, from offices and apartments to schools and hospitals, but broadly has the same needs: energy, water, waste, lighting, ventilation, connectivity, parking, traffic management, security, and emergency services. Then again, smart transportation, electrification and automation could possibly drive the development of smart city transportation. Smart technology can reduce a building's carbon footprint, create more reliable, sustainable, and resilient power grid infrastructure by focusing on people's needs. Therefore, the government investment and focus has shifted for renewable energy, biotechnology, manufacturing, media, and entertainment. Government initiatives in the modernization of cities could be beneficial for tech companies focusing in advanced, state-of-the-art products to improve the efficiency of the urban landscape. The inspiring offerings are to encourage a return to urban living in the aftermath of the pandemic, stimulating economic growth. However, Bliss has faced the uncertainty whether smart cities signify a truly sustainable urban development goal. It has a growing concerns that smart cities are overly reliant on new technology, a 'one-size-fits-all' solution to urban development without taking into consideration whether it will benefit all age population. Bliss comprehended that smart city strategies start with people, not technology, and use technology and data purposefully to deliver a better and unique travel experience. Bliss has taken steps to evaluate how sustainable urban development can be achieved. Considering finding their position in the strong market for companies offering unique, innovative solutions in this field for consumers that will benefit both sustainability and smart urban tourism. They are planning for a programme which connects governments, academics, and businesses from around the world. Bliss intention is to create a collaborative network where experts can share insights, expertise, and resources. Generally, a supports towards the inclusive urban environments that are not guarded by an individual city's needs but are rather scalable at an international level.

**Case Questions**

1 How has Bliss ensured that sustainable urban development is always considered and achievable?
2 In order to provide unique and innovative solutions for sustainability, what are the behavioural aspects of a consumer that Bliss should take into account?
3 What are the factors that bliss ought to consider when engaging consumers in smart urban tourism? What steps can be taken to make consumers aware of smart urban tourism?

**Key Terms and Definitions**

**Smart Tourism**: The application of technology, data, and creative solutions to enrich the tourism sector and to enhance the travel experience for both tourists and service providers (Ye, Ye, and Law, 2020).

**Sustainability**: The capacity to maintain or continue something for an extended duration while safeguarding the welfare and ecological balance of the environment, society, and the well-being of future generations (Fermani et al., 2020).

**Consumer Psychological Factors**: The various mental and emotional elements that influence an individual's buying behaviour and decision-making process when making purchases or consuming products and services (Yoo, Yoon, and Park, 2018).

**Consumer Personal Factors**: The unique individual characteristics and traits that influence a person's consumer behaviour and preferences (Gretzel et al., 2020).

**Consumer Social Factors**: The external influences and societal aspects that impact an affect an individual's buying choices and consumer conduct (Xiang et al., 2017).

**Urban Tourism**: The phenomenon where tourists visit cities and urban areas as their primary destination or as a significant part of their travel experience (Grah, Dimovski, and Peterlin, 2020).

**Accessible Tourism**: The practice to enhance the accessibility and enjoyment of travel and tourism for individuals with disabilities or unique requirements (Darcy, McKercher, and Schweinsberg, 2020).

**Wellness Tourism**: A form of tourism that places emphasis on enhancing the overall health and well-being of individuals, encompassing physical, mental, and emotional aspects (Andreu, Font-Barnet, and Roca, 2021).

**Travel Mobility and Technology**: The intersection of travel and transportation with innovative technological advancements (Sarkar, 2017).

**Luxury and Sustainability**: The integration of high-quality, luxurious experiences and products with a commitment to environmental and social accountability (Dubois and Ceron, 2021).

## References

Amatulli, C., De Angelis, M., Costabile, M., & Guido, G. (2017). *Sustainable luxury brands: Evidence from research and implications for managers*. Springer.

Andreu, M. G. N. L., Font-Barnet, A., & Roca, M. E. (2021). Wellness tourism—New challenges and opportunities for tourism in Salou. *Sustainability*, *13*(15), 8246.

Athwal, N., Wells, V. K., Carrigan, M., & Henninger, C. E. (2019). Sustainable luxury marketing: A synthesis and research agenda. *International Journal of Management Reviews*, *21*(4), 405–426.

Azara, I., Michopoulou, E., Niccolini, F., Taff, B.D., & Clarke, A. (2018). *Tourism, Health, Wellbeing and Protected Areas*, Wallingford: CABI.

Barber, N. A., & Deale, C. (2014). Tapping mindfulness to shape hotel guests' sustainable behavior. *Cornell Hospitality Quarterly*, *55*(1), 100–114.

Bieger, T., Wittmer, A., & Wittmer, A. (2018). Economic conditions and tourism: A structural vector autoregressive analysis. *Tourism Management*, *66*, 34–50.

Bressan, A., & Pedrini, M. (2020). Exploring sustainable-oriented innovation within micro and small tourism firms. *Tourism Planning & Development*, *17*(5), 497–514.

Chang, C. L., McAleer, M., & Ramos, V. (2020). A charter for sustainable tourism after COVID-19. *Sustainability*, *12*(9), 3671.

Chu, Y., Tang, L., & Luo, Y. (2016). Two decades of research on luxury hotels: A review and research agenda. *Journal of Quality Assurance in Hospitality & Tourism*, *17*(2), 151–162.

Chung, N., & Law, R. (2019). The impact of social media and online travel information on travel decision: A study of online travel information adoption. *Tourism Management*, *70*, 28–37.

Corrêa, S. C. H., & Gosling, M. D. S. (2021). Travelers' perception of smart tourism experiences in smart tourism destinations. *Tourism Planning & Development*, *18*(4), 415–434.

Cruz-Milan, O. (2018). Plog's model of personality-based psychographic traits in tourism: A review of empirical research. *In M. A. Camilleri (Ed.), Tourism Planning and destination marketing* (pp. 49–74). Bingley: Emerald Publishing Limited.

Darcy, S., McKercher, B., & Schweinsberg, S. (2020). From tourism and disability to accessible tourism: A perspective article. *Tourism Review*, *75*(1), 140–144.

DeFranco, A. L., Morosan, C., & Hua, N. (2017). Moderating the impact of e-commerce expenses on financial performance in US upper upscale hotels: the role of property size. *Tourism Economics*, *23*(2), 429–447.

Deutsch, M., & Gerard, H. B. (1955). A study of normative and informational social influences upon individual judgment. *The Journal of Abnormal and Social Psychology*, *51*(3), 629.

Dolnicar, S., Knezevic Cvelbar, L., & Grün, B. (2017). Do pro-environmental appeals trigger pro-environmental behavior in hotel guests? *Journal of Travel Research*, *56*(8), 988–997.

Dubois, A., & Ceron, J. P. (2021). Luxury tourism and its sustainable future. In *The Routledge handbook of transport economics* (pp. 299–314). *Cowie J., Stephen & Ison S.* Routledge.

Dwyer, L., Forsyth, P., & Spurr, R. (2018). Tourism economics and policy: Lessons from Australia. *Tourism Economics*, *24*(6), 655–670.

Fang, J., Zhao, Z., Wen, C., & Wang, R. (2017). Design and performance attributes driving mobile travel application engagement. *International Journal of Information Management*, *37*(4), 269–283.

Fermani, A., Sergi, M. R., Carrieri, A., Crespi, I., Picconi, L., & Saggino, A. (2020). Sustainable tourism and facilities preferences: The sustainable tourist stay scale (STSS) validation. *Sustainability*, *12*(22), 9767.

Filieri, R. (2015). What makes online reviews helpful? A diagnosticity-adoption framework to explain informational and normative influences in e-WOM. *Journal of Business Research*, *68*(6), 1261–1270.

Gligorijevic, B. (2016). Review platforms in destinations and hospitality. In R. Egger et al. (Eds.), *Open Tourism: Open Innovation, Crowdsourcing and Co-Creation Challenging the Tourism Industry* (pp. 215–228). Berlin, Heidelberg: Springer.

Gössling, S., Scott, D., & Hall, C. M. (2019). *Tourism and water: Interactions, impacts, and challenges*. Channel View Publications.

Grah, B., Dimovski, V., & Peterlin, J. (2020). Managing sustainable urban tourism development: The case of Ljubljana. *Sustainability*, *12*(3), 792.

Gregori, G. L., Pencarelli, T., Splendiani, S., & Temperini, V. (2013). Sustainable tourism and value creation for the territory: Towards a holistic model of event impact measurement. *Calitatea*, *14*(135), 97.

Gretzel, U., Fuchs, M., Baggio, R., Hoepken, W., Law, R., Neidhardt, J., Pesonen, J., Zanker, M., & Xiang, Z. (2020). e-Tourism beyond COVID-19: A call for transformative research. *Journal of Information Technology & Tourism*, *22*(2), 187–203.

Hall, C. M. (2020). *Tourism and regional development: New pathways*. Routledge.

Hall, C. M., Prayag, G., & Amore, A. (2017). *Tourism and resilience: Individual, organisational and destination perspectives* (Vol. 5). Channel View Publications.

Harkison, T., Hemmington, N., & Hyde, K. F. (2018). Creating the luxury accommodation experience: Case studies from New Zealand. *International Journal of Contemporary Hospitality Management*, *30*(3), 1724–1740.

Hosany, S., & Witham, M. (2019). Exploring the social capital of political stability: A case study of Mauritius. *Journal of Destination Marketing & Management*, *12*, 135–145.

Hassan, A. M., & Lee, H. (2015). Toward the sustainable development of urban areas: An overview of global trends in trials and policies. *Land Use Policy*, *48*, 199–212.

Kelly, C. (2018). 'I Need the Sea and the Sea Needs Me': Symbiotic coastal policy narratives for human wellbeing and sustainability in the UK. *Marine Policy*, *97*, 223–231.

Kim, H.H. and Law, R. (2015). Smartphones in tourism and hospitality marketing: A literature review. *Journal of Travel & Tourism Marketing*, 32(6), 692–711.

Lei, S. I., Ye, S., Wang, D., & Law, R. (2020). Engaging customers in value co-creation through mobile instant messaging in the tourism and hospitality industry. *Journal of Hospitality & Tourism Research*, *44*(2), 229–251.

Lovelock, B., & Sasser, S. (2017). *Essentials of services marketing*. Pearson.

Li, J., Li, X., & Wang, D. (2020). Tourism economics research: A review and bibliometric analysis. *Current Issues in Tourism*, *23*(4), 402–421.

Li, X., Wang, D., Li, Y., & Li, A. (2019). Tourism and economic growth: A panel Granger causality analysis in the frequency domain. *Annals of Tourism Research*, *75*, 65–83.

Line, N. D., & Hanks, L. (2016). The effects of environmental and luxury beliefs on intention to patronize green hotels: The moderating effect of destination image. *Journal of Sustainable Tourism*, *24*(6), 904–925.

Mair, J., & Wittmer, A. (2019). Assessing the impacts of economic crises on tourism: Evidence from Switzerland. *Journal of Travel Research*, *58*(6), 1045–1059.

May, T. and Crass, M. (2007). Sustainability in transport: Implications for policy makers. *Transportation Research Record*, 2017(1), 1–9.

McKercher, B., & Chon, K. (2004). The over-reaction to SARS and the collapse of Asian tourism. *Annals of Tourism Research*, *31*(3), 716.

Mehraliyev, F., Choi, Y., & Köseoglu, M. A. (2019). Progress on smart tourism research. *Journal of Hospitality and Tourism Technology*, *10*(4), 522–538.

Memon, I., Chen, L., Majid, A., Lv, M., Hussain, I., & Chen, G. (2015). Travel recommendation using geo-tagged photos in social media for tourist. *Wireless Personal Communications*, *80*, 1347–1362.

Miguéns, J., Baggio, R., & Costa, C. (2016). Social media and tourism destinations: TripAdvisor case study. *Aslib Journal of Information Management*, *68*(3), 277–293.

Millar, M., & Baloglu, S. (2011). Hotel guests' preferences for green guest room attributes. *Cornell Hospitality Quarterly*, *52*, 302–311.

Monaco, S. (2018). Tourism and the new generations: Emerging trends and social implications in Italy. *Journal of Tourism Futures*, *4*, 7–15.

Morosan, C., & DeFranco, A. (2016). Modeling guests' intentions to use mobile apps in hotels: The roles of personalization, privacy, and involvement. *International Journal of Contemporary Hospitality Management, 28*(9), 1968–1991.

Navarro-Pantin, D., Rojas-Méndez, J. I., & Plaza, B. (2019). The impact of government measures on domestic tourism demand: An econometric evaluation. *Current Issues in Tourism, 22*(14), 1681–1695.

Niewiadomski, P. (2020). COVID-19: From temporary de-globalisation to a re-discovery of tourism? *Tourism Geography, 22*, 651–656.

Ozturk, A. B., Bilgihan, A., Nusair, K., & Okumus, F. (2016). What keeps the mobile hotel booking users loyal? Investigating the roles of self-efficacy, compatibility, perceived ease of use, and perceived convenience. *International Journal of Information Management, 36*(6), 1350–1359.

Peng, N., and Chen, A. (2019). Luxury hotels going green—The antecedents and consequences of consumer hesitation. *Journal of Sustainable Tourism*, 27(9), 1374–1392.

Peng, N., Chen, A., & Hung, K. P. (2020). Dining at luxury restaurants when traveling abroad: incorporating destination attitude into a luxury consumption value model. *Journal of Travel & Tourism Marketing, 37*(5), 562–576.

Peng, H., Zhang, J., Lu, L., Tang, G., Yan, B., Xiao, X., & Han, Y. (2017). Eco-efficiency and its determinants at a tourism destination: A case study of Huangshan National Park, China. *Tourism Management, 60*, 201–211.

Perramon, J., Oliveras-Villanueva, M., & Llach, J. (2022). Impact of service quality and environmental practices on hotel companies: An empirical approach. *International Journal of Hospitality Management, 107*, 103307.

Pestek, A., & Sarvan, M. (2020). Virtual reality and modern tourism. *Journal of Tourism Futures, 7*(2), 245–250.

Plog, S. C. (1974). Why destination areas rise and fall in popularity. *Cornell Hotel and Restaurant Administration Quarterly, 14*(3), 13–16.

Rees, W. E. (2017). Ecological footprints and appropriated carrying capacity: What urban economics leaves out. *Environment and Urbanization, 2*, 66–77.

Sarkar, A. (2017). Mobile applications as an inclusive tool in the tourism and hospitality industry: A state-of-the-art analysis. *Journal of Hospitality and Tourism Technology, 8*(2), 225–240.

Scott, D., Gössling, S., & de Freitas, C. R. (2020). Preferred tourism destinations for a post-COVID-19 world. *Tourism Geographies, 22*(3), 491–501.

Shamsub, H., & Lebel, L. (2016). Identifying tourists with sustainable behaviour: A study of international tourists to Thailand. *Journal of Environmental Management and Tourism, 3*, 26–40.

Sigala, M. (2021). Tourism and COVID-19: Impacts and implications for advancing and resetting industry and research. *Journal of Business Research, 117*, 312–321.

Sparks, B. A., Perkins, H. E., & Buckley, R. (2013). Online travel reviews as persuasive communication: The effects of content type, source, and certification logos on consumer behavior. *Tourism Management, 39*, 1–9.

Smith, M. K., & Diekmann, A. (2017). Tourism and wellbeing. *Annals of Tourism Research, 66*, 1–13.

Smith, M., and Puczkó, L. (2020). Post-Socialist Tourism Trajectories in Budapest: From Under-Tourism to Over-Tourism. In Slocum S. & Klitsounova V. (Eds.)*, Tourism Development in Post-Soviet Nations* (pp. 109–123). London: Palgrave Macmillan.

Song, H. J., Lee, C. K., Kang, S. K., & Boo, S. J. (2012). The effect of environmentally friendly perceptions on festival visitors' decision-making process using an extended model of goal-directed behavior. *Tourism Management, 33*(6), 1417–1428.

Song, S., & Yoo, M. (2016). The role of social media during the pre-purchasing stage. *Journal of Hospitality and Tourism Technology*, *7*(1), 84–99.

Stare, M., & Križaj, D. (2018). Evolution of an innovation network in tourism: Towards sectoral innovation eco-system. *Amfiteatru Economic*, *20*(48), 438–453.

Tussyadiah, I. P. (2016). Factors of satisfaction and intention to use peer-to-peer accommodation. *International Journal of Hospitality Management*, *55*, 70–80.

Varkaris, E., & Neuhofer, B. (2017). The influence of social media on the consumers' hotel decision journey. *Journal of Hospitality and Tourism Technology*, *8*(1), 101–118.

Veiga, C., Santos, M. C., Águas, P., & Santos, J. A. C. (2017). Are millennials transforming global tourism? Challenges for destinations and companies. *Worldwide Hospitality and Tourism Themes*, *9*, 603–616.

Vicente, J. (2020). Seasonal variation in the price elasticity of demand for tourist accommodations. *Tourism Economics*, *26*(6), 943–958.

Villacé-Molinero, T., Fernández-Muñoz, J. J., Orea-Giner, A., & Fuentes-Moraleda, L. (2021). Understanding the new post-COVID-19 risk scenario: Outlooks and challenges for a new era of tourism. *Tourism Management*, *86*, 104324.

Xiang, Z., Du, Q., Ma, Y., & Fan, W. (2017). A comparative analysis of major online review platforms: Implications for social media analytics in hospitality and tourism. *Tourism Management*, *51*, 51–65.

Xie, H. J., Bao, J., & Kerstetter, D. L. (2014). Examining the effects of tourism impacts on satisfaction with tourism between native and non-native residents. *International Journal of Tourism Research*, *16*(3), 241–249.

Yang, W., & Mattila, A. S. (2017). The impact of status seeking on consumers' word of mouth and product preference—A comparison between luxury hospitality services and luxury goods. *Journal of Hospitality & Tourism Research*, *41*(1), 3–22.

Yang, Y., Wang, D., & Li, X. (2019). Tourism and terrorism: A macroeconomic analysis of the impact of terrorism on tourism. *Tourism Management*, *70*, 368–378.

Ye, B. H., Ye, H., & Law, R. (2020). Systematic review of smart tourism research. *Sustainability*, *12*(8), 3401.

Yoo, K.H., Sigala, M.,and Gretzel, U. (2016). Exploring TripAdvisor. In: Egger, R., Gula, I., Walcher, D. (Eds) *Open Tourism. Tourism on the Verge* (pp. 239–255). Berlin, Germany: Springer.

Yoo, C. K., Yoon, D., & Park, E. (2018). Tourist motivation: An integral approach to destination choices. *Tourism Review*, *73*(2), 169–185.

Yuan, J., Deng, J., Pierskalla, C., & King, B. (2018). Urban tourism attributes and overall satisfaction: An asymmetric impact-performance analysis. *Urban Forestry & Urban Greening*, *30*, 169–181.

Zeng, B., & Gerritsen, R. (2014). What do we know about social media in tourism? A review. *Tourism Management Perspectives*, *10*, 27–36.

## Section 3

# Innovations, Technologies and Branding in Tourism

# 7 Immersive Technology in Tourism

## Applications of Immersive Technologies and Impacted Tourism Experience

*Carmen Arroyo*

### Introduction

In recent years, immersive technologies such as augmented reality (AR) and virtual reality (VR) have improved the users' travel and hospitality experience. This article aims to analyze how immersive technologies will eventually change the future of tourism and the use of these immersive technologies especially in the international tourism. It provides first a definition of immersive technologies, their different types, the costs, health and safety implications and an analysis of the benefits and how the future will be shaped in the travel industry by the progressively adoption on these new technologies.

### Background

Immersive augmented reality (AR) and virtual reality (VR) technologies have been widely used in the travel industry. Previous studies have independently investigated the effective applications of AR/VR in tourism from different perspectives and contexts (Pratisto et al., 2022). Thus, the presence of AR/VR has been defined as a main feature in tourism. Attendance affects tourist experience both directly and indirectly through the mediators of value perception and psychological response. In addition, simulation type and social interaction positively moderated the effect of visit on tourism experience, previous visit had a negative moderating effect, while the effect of experience type was not significant. These findings contribute to the development of AR/VR in tourism (Yung and Khoo-Lattimore, 2019).

The COVID-19 pandemic has catalyzed the digitization of the global economy, which has led to an increase in the number of IT users worldwide. The changes also affected the international tourism industry. Immersive technologies allow users to get to know the tourist, quickly find the desired attraction before purchasing a travel service in a foreign place or even visit different countries without leaving home. An AR/VR travel experience relates to a cognitive, emotional or behavioral response which triggers certain touchpoint experiences during the journey through AR/VR technologies trip that affect the overall travel experience (Flavian et al., 2019). Although some literature reviews offer many constructive observations based on only a few studies, a number of systematic reviews have discussed the general

DOI: 10.4324/9781003369967-11

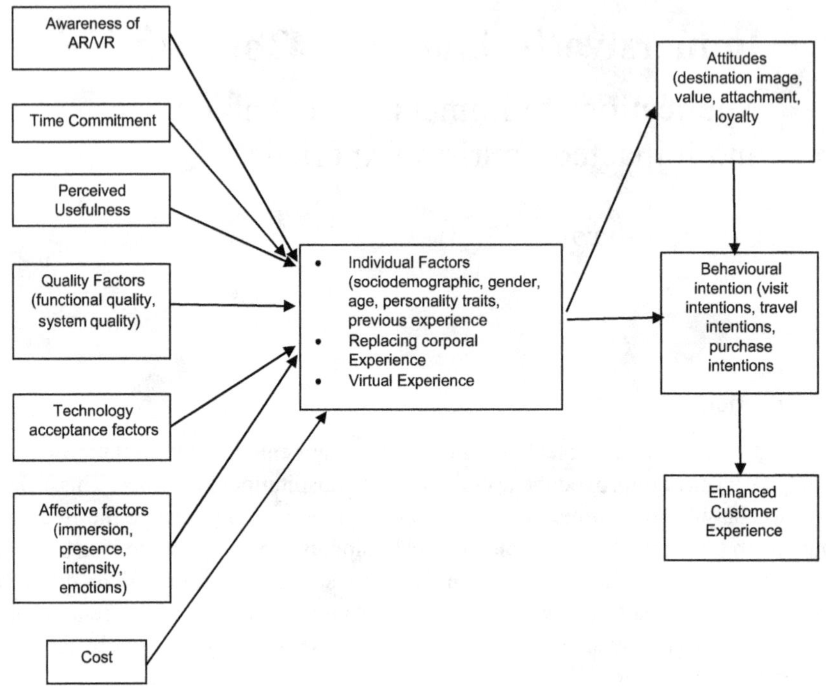


*Figure 7.1* Conceptual framework.

mechanisms of AR/VR affects the tourist experience (Yung and Khoo-Lattimore, 2019). In addition, some similar technical AR/VR attributes such as presence and immersion were explored of their specific research contexts and objectives. From entertainment to education, immersive technologies for healthcare manufacturing are poised to change the game. Among some of the advantages of immersive technologies, their ability to create realistic and immersive experiences to the users is one of the most critical one. Virtual reality, for example, can transport users to different worlds and allow them to interact with objects and people in ways that were previously never done. This encompasses immense possibilities for industries like entertainment and tourism. This has implications for industries like healthcare and retail, and furthermost in the travel sector, where AR can be used to enhance the customer experience (Flavian et al., 2019) (Refer to Figure 7.1).

However, there are also concerns about the impact of immersive technologies on society. These technologies could conduct isolation and progressed addiction, as people become less connected to the real world and more connected to the virtual world. There are also concerns about the potential for these technologies to be used for unethical purposes, such as manipulating people's perceptions or invading their privacy. The evolution of these immersive technologies is inevitable and it will be important to address these patterns and ensure that they are used in an ethical manner (Yung and Khoo-Lattimore, 2019).

## What Are Immersive Technologies?

Immersive technologies are a group of innovative technologies with potential changing the way people interact with the world around them. These techniques create an immersive experience that allows you to feel part of something different when you use it in the world. They are a set of technologies that allow users to experience something computer generated environment as if it were real. Virtual reality (VR), augmented reality (AR) and mixed reality (MR) are the key technologies. VR creates a fully immersive experience simulating the physical presence of the user in a computer-generated to the environment. AR overlays digital information with the real world, while MR combines the two to create a hybrid environment.

There are several types of immersive technologies, each with its unique characteristics and applications. The following are the most common types of immersive technologies: – Virtual Reality (VR) – VR is a fully immersive experience that engages the user computer environment generated. It requires headphones and sometimes accessories such as gloves or controllers; Augmented Reality (AR) – AR superimposes digital information over the real thing in the world AR is often used in mobile apps, advertisements and games; Mixed Reality (MR) – MR combines VR and AR to create a hybrid environment. MR is often used in education, training and entertainment; Haptic Technology – Haptic technology uses vibration and other physical senses and creates a more immersive experience. It is often used in games, medical education and rehabilitation (Pratisto et al., 2022).

### *There Are, However, Some Challenges*

One of the most significant challenges of immersive technologies is their cost. The hardware required for virtual and augmented reality experiences can be expensive, making it difficult for many individuals and organizations to afford. For example, high-end virtual reality headsets can cost several hundred pounds, and that doesn't even include the cost of a powerful computer to run them. Additionally, the cost of their developing could be very high, requiring skilled professionals and specialized software.

Another significant challenge of immersive technologies is accessibility. Not everyone has access to the necessary hardware or software to experience immersive technologies. This can create a divide between those who have access to these technologies and those who do not, potentially exacerbating existing inequalities. Additionally, some individuals with disabilities may experience serious limitations in their use due to physical impediments or sensory sensitivities. Immersive technologies can also raise health and safety concerns. Prolonged use of virtual reality headsets can cause motion sickness, eye aches, and other health issues. Additionally, these experiences could be risky and dangerous if they distract users from their surroundings, leading to accidents or injuries. It is essential to ensure that immersive experiences are designed with user safety in mind and that users are educated on how to use them safely. Finally, there are technological limitations to consider when it comes to immersive technologies. For example, current hardware may not be powerful enough to support highly detailed and complex virtual environments. Additionally, there may be limitations in the software used to create immersive experiences, such as limited interactivity or a lack of support for certain types of content. As technology continues to evolve, these limitations may be addressed, but they are still a significant challenge for immersive technologies (Pratisto et al., 2022).

In conclusion, while immersive technologies have the potential to revolutionize many industries, there are significant challenges and limitations that must be considered. These include cost, accessibility, health and safety concerns, and technological limitations. By addressing these challenges, immersive technologies can become more widely adopted and accessible, leading to a more immersive and engaging future.

## Tourism and Travel

Immersive technologies can transform the travel and tourism industry. VR and AR can allow to create virtual tours of destinations that allow travelers to explore and experience the place before you visit. It can be especially useful for people who have mobility problems or people who cannot travel for other reasons.

### *Virtual Reality for Travel Experiences*

Virtual reality refers to interactive images or videos that allow the viewer to explore 360 degrees of the whole scene. Unlike a normal captured video image, the VR production captures every part of the location. On the way industry, virtual reality can capture the unique and attractive way. This is achieved with special cameras, equipment and software. The finished content can then be viewed either with a VR headset or a regular computer or mobile device. Many people assume that VR content can only be viewed with dedicated VR headsets, but it is not so. Although VR is more immersive when viewed this way, it is can also be viewed on any device, including mobile phones. Surrounding technologies can also do this used to improve the personal travel experience (Pratisto et al., 2022). For example, AR can provide travelers with information about sightseeing or historical sites to get to know the city. VR can also be used to create immersive experiences, e.g. simulate a hot air balloon ride across the city.

VR is undoubtedly a powerful marketing tool as it does capture places in a memorable way. The users have the feeling of "being there", whereas regular videos or pictures cannot achieve this and will only work in demonstrating what the destination offers, they do not provoke an emotional response from the users. VR has the ability to make people imagine themselves in a place that never saw before or dreamt they could visit. Users are at the heart of the whole scene that makes them purchase and wish to be there at any cost.

A VR travel video works like a regular video. They can be viewed on social media or websites, but unlike a regular video, the user can explore the entire scene while the video is playing. VR travel videos are shot using special cameras called omni-radiators. These cameras capture the object from every angle simultaneously. After filming, the footage is taken back to the studio where it is stitched together to create a VR travel video. There are two types of VR travel videos: monoscopic VR travel video and stereoscopic VR travel video.

Monoscopic VR tourism videos can be viewed on standard devices such as mobile phones and computers. The viewer can click or drag across the screen to rotate the field of view, similar to turning the head to explore a scene.

Stereoscopic VR videos for tourism are produced for VR headsets and cannot be viewed on a standard device. Although they take longer to produce and are usually more expensive, they offer a more immersive travel experience. These videos have head tracking, which allows the user to move their head to realistically explore the environment (Maslova and Belov, 2022).

### *Travel Experiences with Virtual Reality*

Virtual reality travel experiences usually refer to VR travel videos made for VR headsets. The goal of these virtual travel experiences is to create a feeling as close as possible to the actual destination. At the forefront of 360 VR, virtual reality travel experiences offer the user something truly unique and memorable. The number of travel agencies and travel companies using this technology is constantly increasing and they promise a bright future for this industry. VR headsets usually provide the user with the most realistic virtual reality travel experience. VR headsets use special software that tracks the movement of the user's head. This allows the user to get to know the destination as in real life. Currently, the number of owners of VR headsets is growing rapidly. This growth in headsets is largely due to the gaming market, where the technology is being pushed hard. In addition, all major internet platforms such as Google, Facebook and Amazon are investing heavily in VR headsets and VR content, promising a bright future for this industry (Zeng et al., 2022). One of the most common VR headsets in tourism is used by travel agencies themselves. They can offer potential customers in-store virtual travel experiences that completely change the meaning of visiting a travel agency.

Instead of showing visitors brochures and computer screens, travel agencies can offer their customers a virtual experience. This approach can also be used effectively at trade shows and events, quickly attracting the interest of the general public. With VR, travel brands stand out and offer users an experience they won't forget. Many travel companies have adopted VR technology and used it to improve sales and increase brand visibility. With virtual hotel tours, users can get to know the hotel and its surroundings in a more immersive way than ever before. Hotel interiors and exteriors can be captured in extreme detail with high-resolution cameras. The images are then combined (stitched) into a complete 360-degree interactive tour where the user can choose which room to explore. VR hotel tours are typically single-view, meaning they can be viewed on any device, including mobile and desktop. The tours can then be uploaded to websites and social media for prospective customers to view at any time. They can also be stereoscopic if the situation and budget allow. This can lead to a more realistic and immersive experience.

Unlike standard hotel photos, these tours allow users to imagine themselves in the space. Such immersion helps create unique brand engagement and leave a lasting impression on the user (Zeng et al., 2022).

### Benefits of Virtual Reality in Travel

The advantages of virtual reality in tourism include e.g. • The user can imagine that he/she is at the destination, allows the user to explore the scene at will, it creates memorable and unique experiences for the user, the opportunity for travel companies to differentiate themselves from others, providing travel experiences for those who cannot travel, and finally reduce the impact of tourism on vulnerable destinations.

### The Future of VR Travel

At Immersion VR, we don't see travel VR anywhere. On the contrary, the use of VR for travel has increased. We cannot predict how this space will evolve or what new VR travel technologies will be developed. However, we can observe emerging trends in the industry. Some of the more common trends in VR travel are related to VR aviation and landmarks, especially hotel virtual hotel tours, senior experiences and virtual reservation interfaces. These trends are growing rapidly and many travel companies are using VR as defined above to improve travel packages and promote new destinations. Popular attractions often suffer from environmental problems associated with too many tourists. By making VR experiences out of these landmarks, the number of users can be controlled, thus reducing environmental impact. The virtual booking interface is another very recent development in travel VR. Users can book their vacation using a VR headset. The entire booking process takes place in virtual reality. Everything from choosing a hotel to paying for a vacation happens while the user is experiencing VR. Although it has limited applications, we see travel agencies and companies using this approach to increase conversions. One of the areas where virtual reality can replace travel is for those who cannot travel, especially the elderly. When people think of VR, they don't usually think of the elderly. But being able to provide them with travel experiences that wouldn't otherwise be possible can be very rewarding.

#### *Will Virtual Reality Replace Travel?*

VR is great for creating immersive moments, but it still can't replace 24/7 full immersion in real space. In a recent survey by European travel agency Italy4Real, 81% of adults said that VR cannot replace travel. 92% said that visiting a VR destination is not the same as in real life. In addition, 77% considered it important for them to taste local food – (https://www.mediapost.com/publications/article/302947/virtual-reality-not-seen-as-substitute-for-travel.html). Other disadvantages of VR are smells and the general atmosphere created by people and animals. While VR technology is advancing rapidly, it's pretty safe to say that virtual reality won't be replacing travel anytime soon.

**International Travel**

One of the main trends in the era of COVID-19 has been the acceleration of digitization of the global economy. This was manifested, among other things, in the growing popularity of immersive technologies. As a key feature of XR technologies, immersion provides a suitable environment for tourism companies to promote tourism services. Today, some countries are reimposing coronavirus restrictions (e.g. China) and others are strengthening entry requirements for certain categories of tourists (EU citizens from Russia). In addition, in the current geopolitical uncertainty, inflation and the cost of living are rising around the world. Therefore, in the near future, the use of virtual tours may become the only way for the general public to get the desired travel experience, if the situation does not change for the better. Therefore, it is necessary to find out the most promising directions of use of immersive (VR, AR, MR) technologies in international tourism (Pratisto et al., 2022). The purpose of the section is to analyze the use of immersive (XR) technologies in international tourism. In the context of this study, it is important to: (1) identify the role of XR technologies in international tourism; (2) analyzes the technological aspect of the implementation of AR, VR, MR, the advantages and disadvantages of these technologies; (3) provides general recommendations for the application of environmental technologies in the field of international tourism. VR is a fully virtual environment created through 3D modeling; AR is a real environment with virtual objects; MR is a real environment with interactive virtual objects. Augmented reality is a term that combines these three concepts. Immersions, on the other hand, combine XR technologies that mimic the physical world through digital environments, creating a sense of immersion (Kornilov and Popov, 2020). The restrictions related to the coronavirus encouraged digitization worldwide and at the same time increased the dependence of the population on digital technology. This is reflected in an increase in the number of Internet users by 5.03 users and an increase in the number of unique mobile phone users to 5.34 billion people by mid-2022. Under these conditions, the XR industry ecosystem is enriched. Although technological limitations hinder the current industrial development of XR, the XR market is approaching a tipping point of rapid development as technology giants expand their adoption in the XR industry, targeting next-generation computing platforms. The AR market was valued at $12.45 billion in 2021 and is expected to reach more than $36 billion by 2026. The virtual reality consumer and enterprise market was valued at $11.97 billion in 2022 and is expected to grow to 15.81 billion USD by the end of 2023. Virtual tourism is becoming a solution to the problem of social distance. In addition, travel companies can use extensive technology even after the COVID-19 pandemic, which promotes a positive image of the tourist destination. There are two main reasons for this. First, augmented reality creates a sense of presence that engages the user and can increase trust in the destination and provide a comprehensive understanding of the destination (Tsai, 2022). Second, as a way of pre-inquiry about tourist destinations, XR creates a first impression of a tourist destination, which is why some hotels offer advance online tours of the hotel and rooms. Based on the first impression, the user makes the final purchase decision.

As a result, immersive technologies have been successfully combined with one of the most important motivations for travel consumption – the desire for a unique experience. Immersive technologies offer interactivity, trips to the past, and even models of the future and alternate realities. Examples from the gaming industry (3D content, not immersive VR) include trips to Ancient Greece ("Odyssey"), Ancient Egypt ("Origins") and Scandinavia ("Valhalla"). These tours are offered in the latest games in the "Assassin's Creed" series. Although the gaming sector of the global economy is not directly related to tourism, the experience gained on these trips undoubtedly created an interest in the history of the players and their tourist destinations (the modern versions of them). As for virtual tours, there are those. However, there are some downsides to XR tours. First, there may be a difference between the image of the XR application and the natural landscape of the tourist destination. It is not necessary to reflect the real world as a whole to create a satisfying travel experience, but the difference should not create cognitive dissonance. Second, XR does not provide the interpersonal interaction that is an integral part of international travel (Rauscher et al., 2020). Therefore, tourism companies must find a balance between the distance inherent in XR-related tourism products and adequate research on the culture, worldview and historical experience of the local population (Kornilov and Popov, 2020). VR developers should emphasize content with interactive and immersive features, using immersive factors such as storytelling to engage visitors and deepen their cultural understanding, as pride mediates the link between VR experiences and cultural diffusion. It has also been proven that authentic experiences of tourism-related VR activities significantly influence potential tourists' assumptions and beliefs (cognitive response) and feelings and emotions (emotional response). However, there are more factors influencing the intention to visit a tourist destination presented in a VR tour, such as the attachment of users to such technologies (Oncioiu and Priescu, 2022). We will now focus on the technical part of augmented reality technologies and the prospects of implementing immersive technologies related to this aspect in tourism.

### *Building AR Reality*

There are two approaches to building augmented reality: the first involves using the coordinates of the user's location (calculated by mobile devices using GPS receivers, gyroscopes, etc.) and the second uses a marker. A mark is an object of reality and its location is used for the next projection of the virtual object (Maslova and Belov, 2022). There are also two types of devices for AR guidance: wearable devices and wearable devices such as smart glasses. Camera-equipped smartphones and tablets provide powerful computing power to run AR-based applications. In addition, these mobile devices are more accessible to consumers than mobile devices due to their lower prices. In another type of device, smart glasses are equipped with a processor, various sensors and transparent lenses so that digital information covers the physical environment (Pratisto et al., 2022). AR has the potential to transform both indoor (for example through ARCore applications) and outdoor navigation. Tourists can find tourist attractions and suitable accommodation through AR-based virtual tours. There is already an example of Premier Inn's Hub Hotels using AR to turn rooms into city maps that direct the user to the most important attractions in the area. However, despite all the advantages of AR, there is a big problem to be solved. AR applications collect more data than necessary, and vulnerability to cyberattacks puts this private information at risk of being leaked. Pratisto et al. (2022) highlight the following challenges in the development of augmented reality applications in addition to user privacy: lack of interoperability between device platforms, network connection requirement for some AR applications to retrieve data from the server, physical size of AR devices., AR tracking capability while using the camera as a sensor, system feedback, AR application localization, user engagement with a real object or environment, and of course user privacy. Several design principles have been proposed for the successful implementation of a tourism mobile application with AR (Shukri et al., 2017): create a user-friendly user interface (UI) with convenient information presentation; ensures the optimal placement of information, which facilitates the identification of interesting objects; provide training to the user to improve further use of the application and to ensure the realism and interactivity of the 3D environment.

***VR Tours***

The virtual environment is accessed through wearable VR headsets. Today, a constant computer connection is not required to provide sufficient processing power, and there are untethered headsets such as the Oculus Go (Weber-Sabil and Han, 2021). According to Pratisto et al. (2022), the following challenges are associated with the use of virtual reality devices: first, familiarization with the devices (which can be time-consuming), second, the relationship and connection of the physical information of the devices. In addition, one of the biggest obstacles to the mass adoption of VR is the enabling conditions (costs, equipment requirements and knowledge needed to use this technology) (Oncioiu and Priescu, 2022).

Tsai et al. (2022) proposed four mechanisms of VR technology in tourism: limited (which limits the user's mobility; pre-recorded 360-degree virtual tours), semi-restricted (which limits the user's actions), advanced semi-restricted (where visual aid) and is not restricted (the user's movement is not restricted; the user can even ignoring the manual). Immersion in the whole body is very empowering a rich feeling of being in another place.

On the other hand, it makes the user vulnerable to all dangers, threats and social antagonism in the place where the physical body is located.

There are a few suggestions for creating VR tours: – An authentic and believable atmosphere must be created to stimulate a sense of immersion. – One must take care of all the details of the scenario and determine the interactions (if any) in the virtual environment to avoid cognitive dissonance. – User interface elements must be sufficiently compatible with the environment and be hidden in a menu that can be named separately. – Sound has an enhanced impact on the user experience and therefore VR requires high quality sound. – It should be noted that phobias (e.g. acrophobia, claustrophobia, nyctophobia) are strengthened in the virtual environment. Depending on the goals and the entourage of the VR tour, it is possible to use these elements to adjust the mood of the user (Oncioiu and Priescu, 2022).

When implementing mixed reality, MR technology combines the capabilities of a VR headset and an external video camera. MR creates a third dimension where both real and virtual reality exist, are displayed and are interactive. One of the types of MR offered by HoloLens and Magic Leap works with translucent glasses, allowing users to see their immediate surroundings and creating the effect of holograms in the user's peripheral vision. Another type (e.g., Windows MR) fully immerses the user in a computer-generated environment and uses the device's cameras to mirror the immediate environment (Weber-Sabil and Han, 2021). The use of holograms is expected to be the next innovation trend in teleconferencing. However, holograms are limited by expensively furnished rooms at both ends. Therefore, flexible digital travel has an advantage over this mixed reality technology (Tjostheim and Waterworth, 2022). Now that the implementation of all three components of XR has been analyzed, it is possible to identify the main characteristics that XR should have to maximize effectiveness: effective, engaging, flexible and scalable, able to work with change management, simple, physically attractive (Stewart et al., 2020). In the following years, the consumer base of tourism services will be represented by generations Y, Z and alpha. These people cannot imagine their life without digital technology. Therefore, travel companies should start using the surrounding technologies as soon as possible and adopt best practices (Weber-Sabil and Han, 2021). Although digital travel is unlikely to completely replace traditional travel, as XR technologies evolve, both types of travel can evolve side by side, increasing the value of travel brands and creating unique travel experiences. AR, MR, AR improves the data collection processes of tourist destinations, makes it possible to promote cultural and natural heritage without harming historical monuments and our planet, gives disabled people the opportunity to explore the world, etc.

## Solutions and Recommendations

A challenging marketplace combined with soaring consumer expectations means the travel industry faces significant challenges. The platforms, processes and cultures fit for the businesses of tomorrow are not far away. Now is the time to define and transform organizations into a digital-first powerhouse offering unparalleled guest experience, and also, the keys to unlock new revenue streams for the travel organizations. Change is happening first, and these organizations need to stay ahead.

Today, putting customers first is the only way. The clear path forward is to learn how to deliver a world-class customer experience and automate personal moments with data. In the post-pandemic world, travel habits, behavior and expectations have changed. What does this mean for travel and hospitality brands looking to deliver a better experience for their customers? Undoubtedly, immersive experiences give customers a better idea of how a product would look, feel and act in real life, making the purchase decision easier. VR/AR methods allow users to interact with objects and perform tasks that mimic real life. This technique helps the user gain new skills and experience without facing real consequences. Virtual travel can remove many constraints from travelers such as budget constraints, physical constraints or time constraints. Companies only need to invest in VR technologies and immersive travel experiences around the world can be enjoyed. However, virtual travel experiences are still rare because many miss the authenticity of real travel. The idea of a virtual trip where you don't even have to pack your bags can be confusing to some. We cannot deny that the global virtual travel market is growing rapidly due to the inventive use of virtual and augmented reality technology. While VR travel may seem strange to some, virtual reality certainly enhances the travel experience. VR technology has the potential to revolutionize the travel industry by changing the way travelers plan and experience travel. In the future, it can be used for destination marketing, creating personalized travel experiences and enhancing destination experiences through immersive tours and interactive exhibits. This technology can also help travel agencies attract more visitors and encourage them to stay longer. Travel companies can also create customized travel plans and offer travelers more interesting and memorable experiences. As the technology advances, we can expect innovative applications of VR in the travel industry.

Overall, the author's recommendations are to invest in new VR and AR technologies to enable expanding the business and profit revenue. We have only seen a glimpse of what the new technologies will bring to the future, but it is inevitable to see how consumers and travelers opt more and more to experience virtual reality to then purchase their packages and travel destinations. Investing in this new technology enables also to acquire more knowledge, trends through consumers' data that were not available a few decades back.

**Future Research**

A research agenda for virtual tourism should include the use of computer-generated travel experiences that offer travelers the ability to view, immerse, and control their environment. Considering the level of immersion in the virtual environment, it has been proposed that tourists can have affective, cognitive and sensory experiences by visiting virtual attractions, choosing travel transport and accommodation, admiring landscapes and interacting with other virtual tourism providers and tourists. Concepts of co-creation and participation can be evaluated to determine if the design deserves more functionality and interactive features.

The current level of adoption of mobile and web-based applications allows participants to visit virtual target scenarios using smartphones and computers, VR headsets and other augmented reality technologies. The main difference between the virtual experience and traditional hypothetical experimental scenarios is the motivation of the participants to have virtual travel experiences that they cannot have in real life and the level of immersion in the virtual target scenarios. In addition, with the help of mobile technology, it is possible to plan different travel scenarios. The pandemic has made technological communication more common, with consumers buying VR and AR devices by more than 50%; therefore, now is the perfect time to consider innovative data collection and technology adoption in tourism.

The use of virtual travel experiences can advance tourism research in several ways. First, it provides an ideal intangible experience that is difficult to provide in a real environment. It also facilitates the objective measurement of the time dimension of the travel experience at different times before, during and after the virtual trip. Then, it allows learning subjects in a natural virtual environment, considering the level of immersion and realism of the virtual scenarios. Finally, it helps prevent self-reporting by observing actual tourist behavior and collecting sensor and mobile phone psychophysiological responses.

Virtual reality scenarios allow researchers to plan and test results for different target situations by placing peak experiences at different moments and segmenting visitors according to socio-demographic and personality characteristics. The implementation of virtual destinations also has promising implications for destination marketing and management, tourism providers and tourists. First, governments can use virtual target scenarios to pretest policies and marketing campaigns against existing and emerging targets. Second, virtual destinations help the virtual tourism research program use computer-generated travel experiences that offer travelers the ability to view, immerse, and control their environment. Considering the level of immersion in the virtual environment, it has been suggested that tourists get affective, cognitive and sensory experiences by visiting virtual attractions, choosing travel equipment and accommodation, admiring the scenery and interacting with other virtual tourism providers and tourists. Concepts of co-creation and participation can be evaluated to determine if the design deserves more functionality and interactive features.

The current deployment of mobile and web-based applications allows participants to visit virtual target scenarios using smartphones and computers, VR headsets and other augmented reality technologies. The main difference between the virtual experience and traditional hypothetical experimental scenarios is the motivation of the participants to have virtual travel experiences that they cannot experience in real life and the level of immersion in the virtual target scenarios. In addition, it is possible to plan different travel scenarios with the help of mobile technology. The pandemic has made technological communication more common, with consumers buying VR and AR devices by more than 50%. Therefore, now is the best time to consider innovative data collection and technology adoption in tourism. The use of virtual tourism experiences can contribute to tourism research in several ways. First, it provides an ideal intangible experience that is difficult to provide in a real environment. It also facilitates the objective measurement of the time dimension of the travel experience at different times before, during and after the virtual trip. It then allows subjects to be learned in a natural virtual environment, considering the immersion and realism of virtual scenarios. Finally, it helps to avoid self-reporting by observing real tourist behavior and collecting psychophysiological responses from sensors and mobile phones. Virtual reality scenarios allow researchers to plan and test results for different target situations, placing peak experiences at different moments and segmenting visitors according to socio-demographic and personality characteristics.

The implementation of virtual destinations also has promising implications for destination marketing and management, tourism providers and tourists. First, governments can use virtual target scenarios to test policies and marketing campaigns against existing and new targets. Second, virtual destinations help manage visits to overdeveloped destinations by providing opportunities for alternative virtual experiences. Next, virtual tourism offers travel providers new business opportunities in difficult times and creates new market niches for certain customer segments. Virtual destinations can provide opportunities for people who cannot visit real destinations or for disadvantaged groups such as the poor, the disabled or the elderly. Manage visits to overdeveloped destinations by offering options for alternative virtual experiences. Next, virtual tourism offers travel providers new business opportunities in difficult times and creates new market niches for particular customer segments. Virtual destinations can provide opportunities for people who cannot visit real destinations or for vulnerable groups of people, including poor people, people with disabilities or the elderly.

## Conclusion

In conclusion, the article analyzed the use of immersive (XR) technologies in international tourism. It can be said that the role of immersive technologies in people's lives has grown in recent years. Augmented reality is actively changing international travel, and the XR experience often becomes the basis for the purchase decision of travel services. This article examines the benefits of using AR, VR and MR for both travel agencies and tourists. Although the surrounding technologies still have some shortcomings in terms of technological aspects, accessibility and facilitating conditions, the scientific community has already made recommendations to improve the implementation of XR. This article also contains some relevant recommendations.

### Case Study

The Board of Hospitality and Official Tourism purchased a special platform that used VR technologies to promote and network the Western Greece region of Greece. Contemporary researchers are interested in relevant technologies and their role in the tourist experience. The special platform included the creation of a virtual platform promotion training program for foreign tour operators and a competition between foreign tourism professionals playing the tourism elements of the West Greece region. The result is an information-rich domain useful for visualization without limits to explore destination experiences of visitors in the era of COVID-19. The platform offers foreign tourism professionals (travel agencies, tour providers, local governments) an educational opportunity that ultimately leads them to vote for the "best" option. The platform includes a virtual tour focusing on the beaches of the Western Greece region, which was a trend during the COVID-19 pandemic. Through the virtual project, the ten beaches of the Western Greece region are presented through an enriched 360-degree multimedia tour (image and video), the content of which is accessibility, beach characteristics, attractions of the wider region and cultural specifics of the wider area beach area. Combined and multidisciplinary measures implemented to create the platform, especially the creation of a platform hosting multimedia multimodal content (virtual tours, text, videos, etc.) about the beaches of the Western Greece region. In addition to text content, the platform uses 360-degree photos, 360-degree panoramic videos, photos and pop-ups. More specifically, all the 360-degree videos of the beaches have been integrated, resulting in an interactive platform in mixed reality that combines digital and physical information, includes the surrounding environment, completes the real world and offers interaction thanks to play that responded. to participating travel agencies. About 360 videos can be converted to normal videos and especially to project the Western Greece area and all the beaches. Travel agents can use 360-degree panoramic views and use their desktop computers or mobile phones. This is a new style that also adds movement and vibrancy to static 360 shots (like 360 images) to capture the viewer. In the same environment, it is possible to make 360 rotations with all the 360 videos designed on the beaches, which improves the synthetic sensory experiences and reception channels in the human-computer interface. Each virtual environment is enriched with advanced information about the destination. Travel agencies can see the destination in detail by experiencing an additional environment through the object handler, which essentially provides interactively accessible automatic information about the destination in the West Greece region. Destinations were advertised and tourism professionals were trained by a vote of a significant part of the Western Greek region created by the gamified characteristics of foreign professionals.

In addition to the online training competition, a number of tools were available for the implementation of game practices and the further development of content and the training of travel agencies in general and in particular: learning content that can be transferred to different platforms; different presentations of learning content for the studied object; application of personal content according to the user's choices, and complete control system for individual courses to Western Greece Region, addition of interactive quizzes to parts of the courses; playfulness and interactive learning games (gamification) to connect information.

In addition to the virtual immersion platforms of the Western Greece region, the travel agency guides users to the quiz through an online training that worked. Successful are defined as "target experts" who create loyalty conditions and a step before a business deal or partnership. A fully researched virtual immersion platform follows the most important dimensions included in the target information systems models, such as ease of use, data quality, functionality, usability, responsiveness, appearance, design and presentation, and interaction, which are described in various research papers. Evaluation and control form a dynamic process for immersive platforms and occur at all stages of design and implementation. As a process, however, the results were evaluated in terms of users and their behavior within the platform and in their wider communication. In short, the process of creating and implementing a comprehensive platform is a demanding process that, due to its multifactorial nature, requires a strategic approach to procedures. One of the most important advantages of virtual reality in the hotel industry is the ability to increase guest. Virtual reality offers guests the opportunity to explore and experience the hotel before arrival, which gives them a better understanding of the hotel's amenities and layout and can potentially improve guest satisfaction and better reviews. In addition, virtual reality can be used to show the surroundings and attractions near the hotel, giving guests a better idea of what the place has to offer and helping them plan their stay accordingly. Virtual reality can also improve booking times, which can lead to cost savings for hotels. With a virtual tour of the hotel's rooms and amenities, guests can make more informed decisions and better understand their reservation, which can reduce reservation times and improve the use of personal time. In addition, virtual reality can be used as an effective marketing tool for hotels. By providing an immersive and immersive way to experience a hotel, virtual reality can help hotels stand out in a competitive market and attract more guests. Virtual reality can also be used to showcase hotel amenities and facilities in a more interactive and engaging way than traditional photos, which can lead to bookings and increased revenue. According to recent studies, several challenges must be overcome to enable virtual adoption. In hotel industry, the main obstacle is the cost of implementing VR technology, which can be prohibitive for some hotels. The high costs are partly due to the need for specialized equipment and expensive software to create an immersive experience. Additionally, VR technology is still relatively new and integrating it into existing systems and workflows can be difficult. This can make it difficult for hotels to provide a seamless experience to customers accustomed to traditional hotel services. Finally, there is the risk that VR does not meet expectations, which can lead to disappointed guests and negative reviews. Despite these challenges, some hotels are already exploring the potential benefits of VR. For example, Marriott International has experimented with VR technology to offer guests virtual tours of its properties. Similarly, Hilton Worldwide used VR technology to create an immersive experience for guests at the Innovation Gallery72 where visitors can explore new hotel concepts and designs. By harnessing the power of VR, hotels can offer guests unique and immersive experiences that set them apart from the competition.

**Case Questions**

- What were the ultimate benefits of applying the immersive technology/methodology?
- What is the future of these technologies in specific Regions like the Western Greek area? Can VR substitute reality/presence to profit these industries?
- How does compare the VR experience in the Western Greece Region with the Hilton hotels?

**Key Terms**

*Augmented Reality (AR)*: This is an interactive experience that enhances the real world with computer-generated observational data. Using software, apps and hardware such as AR glasses, augmented reality overlays digital content on real environments and objects. In addition, it is an enhanced version of the real physical world created by digital visual elements, sound or other sensory stimuli and delivered by technology This is a growing trend, especially among companies involved in mobile IT and business applications (Yung and Khoo-Lattimore, 2019; Weber-Sabil and Han, 2021).

*Gyroscopes*: A device with a rotating disc or wheel mechanism that exploits the principle of conservation of momentum: the tendency of a system to remain constant when no external torque is applied to it (Wrigley & Hollister, 1965).

*Haptic technology*: This is a technology that can create a tactile experience by directing movements to the user. These techniques can be used to create virtual objects in computer simulation to control virtual objects and improve the remote control of machines and equipment (telerobotics). Haptic devices may include touch sensors that measure the force applied by the user to the user interface. It uses vibration and other physical sensations to create a more immersive experience. It is often used in gaming, medical education and rehabilitation (Pratisto et al., 2022).

*Immersive technologies*: It was brought up with the invention of Sensorama. The movie experience developed by Morton Heilig placed the viewer in a "sensory" theater that contained speakers, fans, scent generators, and a vibrating chair that immersed the viewer in the film. Technology that allows you to see or interact with simulated effects and environments. These range from 360-degree photography and video to virtual and augmented reality (Pratisto et al., 2022).

*Mixed Reality (MR)*: This is a term used to describe a combination of a real environment and a computer environment. Physical and virtual objects can coexist in mixed reality environments and interact in real time. Mixed reality that includes haptics has sometimes been called videohaptic mixed reality. MR combines VR and AR to create a hybrid environment. MR is often used in education, training and entertainment (Siddiqui et al., 2022).

*VR (Virtual Reality)*: This is a simulated 3D environment that allows users to explore and interact with a virtual environment in a way that matches the reality perceived by the user's senses. The environment is created using computer hardware and software, although users must also wear equipment such as a helmet or goggles to interact with the environment. The more users can immerse themselves in a VR environment – and mask their physical environment – the better they can suspend their belief and accept it as real, even if it is fantastical in nature (Yung and Khoo-Lattimore, 2019).

*XR (Extended Reality)*: This is an all-encompassing term that encompasses the current and future developments in augmented reality, mixed reality and virtual reality. Augmented Reality (AR): Matching digital information to the real world (Pratisto et al., 2022).

## References

Flavian, C., Sanchez, A.I., & Orun, C. (2019). Volume 30. Impacts of technological embodiment through virtual reality on potential guests' emotions and engagement. *Journal of Hospitality Marketing & Management*, 1–15.

Kornilov, U. V. & Popov, A. A. (2020). On the terminology and classification of immersive technologies in education. *Problems of Modern Teacher Training*, 68(2): 171–174.

Maslova, U. A. & Belov, U. S. (2022). Augmented reality technologies. *E-scio*, 2(65): 313–322.

Holban Oncioiu Ionica & Iustin Priescu. (2022). "The use of virtual reality in tourism destinations as a tool to develop tourist behavior perspective". *Sustainability,* 1–20.

Pratisto, E. H., Thompson, N. & Potdar, V. (2022). Immersive technologies for tourism: a systematic review. *Information Technology & Tourism*, 24: 181–219.

Rauscher, M., Humpe, A. & Brehm, L. (2020). Virtual reality in tourism: is it "real" enough? *Academica Turistica*, 13(2): 127–138.

Shukri, A., Haslina, A. & Rimaniza, Z. A. (2017). Volume 1891. The design guidelines of mobile augmented reality for tourism in Malaysia. *AIP Conference Proceedings*, 4–34.

Siddiqui, M. S., Syed, T., Nadeem Al Hassan, A., Nawaz, W. & Alkhodre, A. (2022). Virtual tourism and digital heritage: an analysis of VR/AR technologies and applications. *International Journal of Advanced Computer Science and Applications*, 1–13.

Stewart, D., Westcott, K. & Cook, A. V. (2020). Immersive technologies in the enterprise, *Deloitte Insights,* 1, 22–45.

Tjostheim, Ingvar & Waterworth, John A. (2022). The psychosocial reality of digital travel, being in virtual places, 2– 24.

Weber-Sabil, J. & Han, D. I. (2021). Immersive tourism state of the art of immersive tourism realities through XR technology. *Breda University of Applied Sciences*, 2, 1–15.

Wrigley, W. & Hollister, W.M. (1965). The gyroscope: Theory and application. *Science*, 149, 1–2.

Yung, R. & Khoo-Lattimore, C. (2019). New realities: a systematic literature review on virtual reality and augmented reality in tourism research. *Current Issues in Tourism,* 22, 5–47.

Zeng, Y., Liu, L. & Xu, R. (2022). The effects of a virtual reality tourism experience on tourist's cultural dissemination behaviour. *Tourism and Hospitality*, 3, 314–329.

# 8 Destination Branding Authenticity

## Building Relationship Orientation among Visitor Attractions, Tourism and Millennials

*Ilma Aulia Zaim, Annisa Rahmani Qastharin and Agi Agung Galuh Purwa*

### Introduction

This chapter explores the importance of authenticity in destination branding and its impact on building strong relationships with the millennial generation as the largest segment in the tourism industry. The article discusses the challenges faced by Destination Marketing Organizations (DMOs) in attracting and retaining young travellers and provides suggestions and strategies to effectively engage millennials and create authentic yet memorable travel experiences.

### Background

Destination branding is defined as a set of destination marketing activities that incorporate the value creation of the destination identity through a name, symbol, logo, and wordmark, which identifies and distinguishes a destination from other places (Blain et al., 2005). It is also further defined by Zenker and Braun (2010) who defined destination branding as a tourists' association of the place that is embodied through the aims, values, and culture of the place. Destination branding provides many benefits for tourists as well as tourism business providers, particularly in offering destination competitive advantage (Mior Shariffuddin et al., 2023; Murphy et al., 2007).

In the destination branding context, destination brand image is one of the most crucial parts due to its strong influence on tourists' perceptions and behaviour (Beerli & Martin, 2004; Wisker et al., 2023). It is suggested that some attributes of destination brand images can be classified into the physical environment, people, economy, facilities, environment as well as the brand attitudes (Hankinson, 2005; Malik et al., 2022). For example, changing its specific destination image attributes, New York is a city that has successfully transformed its brand image from negative to positive, as it was initially known as a post-industrial city that turned into a vibrant leisure city as well as a business trip destination (Ward, 1998).

According to UNWTO (2016), youth travel has now become the fastest-growing segment of global tourism. The increased number of youth travel growth represents the socio-economic opportunities for local communities, given that young travellers stimulate local communities and businesses as well as encourage the

DOI: 10.4324/9781003369967-12

protection of the environment. This becomes even more particularly relevant in the post-COVID pandemic situation since these young tourist generations have really helped the economy to recover after countries reopened their borders to receive travellers coming from all around the world (Babii & Nadeem, 2021).

Furthermore, destination authenticity is an important factor that directly affects visitors' engagement with the destination, as well as the tourist's intention to revisit and their likelihood to recommend the destination to others (Chen et al., 2020). For Destination Marketing Organizations (DMOs) who are looking to attract millennials and younger generations, destination branding becomes a pivotal tool to communicate the values, credibility as well and promises that the tourism place or destination can offer to its target visitors (Almeyda-Ibáñez & George, 2017). Moreover, for young travellers who are looking for genuine and organic experiences, destination authenticity becomes an essential asset for DMOs to have (Pine & Gilmore, 2008). Destination authenticity encourages tourists' engagement which may potentially further enhance the possibility for tourists to revisit the destination they previously visited (Bryce et al., 2015).

As such, this also shows the significance of destination authenticities in building tourist/visitor relationships, particularly with the aim of creating and maintaining destination branding. Therefore, Visitor Relationship Management (VRM), which is developed from Customer Relationship Management (CRM) can be the answer to help DMOs build and strengthen visitor relationships as well as to assist visitor retention and drive visitor growth. Nonetheless, implementing VRM is a challenge of its own. The study by Pike, Murdy and Lings (2011), for example, indicates that while the management of DMOs recognized the potential of CRM, the implementation of it, is somewhat problematic.

This chapter aims to discuss the role of authenticity in destination branding, which plays a significant role in building relationships between visitor attractions/tourism destinations with the young generation as the biggest travel segment in the tourism industry nowadays. The chapter is divided into three parts. The first part provides a brief overview of destination branding and the importance of destination authenticity. The second part explores some strategies that can be potentially employed to engage millennials and generation Z as the primary target market. Lastly, the third part of this chapter elaborates a case to illustrate the discussed topic.

## Destination Authenticity

According to Chen et al. (2020), the concept of destination authenticity pertains to the extent to which visitors perceive a destination brand as consistent and true to its own identity (consistent), trustworthy and believable to visitors (credible), accountable and ethically upright (responsible), and supportive of tourists in their personal growth and self-expression (facilitates self-discovery). Destination authenticity is a significant factor that directly influences visitors' engagement with the destination brand, their intention to revisit, and their likelihood to recommend the destination to others (Ruixia et al., 2020). It also plays a significant role in influencing tourists' decision-making processes and has the potential to impact their satisfaction and loyalty (Yi et al., 2017). Achieving destination authenticity necessitates striking a balance between preserving the destination's unique identity and heritage while also adapting to evolving tourist preferences and market trends (Cole, 2012).

Authenticity can be attained through several methods, including the preservation of local culture and traditions, the promotion of sustainable tourism practices, and the provision of unique and genuine experiences to tourists. For instance, a destination can promote its local cuisine, arts, and crafts, and encourage tourists to participate in local festivals and events. This process contributes to the development of a sense of place and identity, which is crucial for destination branding (Kumail et al., 2022). On top of that, destination authenticity does not only aid in distinguishing a destination from its competitors but also in attracting tourists in search of authentic experiences. Achieving authenticity necessitates a delicate balance between safeguarding local culture and traditions and catering to the needs and expectations of tourists.

According to studies (e.g. Cetin & Bilgihan, 2016), authenticity in the context of tourism incorporates various dimensions. First, cultural tourism authenticity refers to the degree to which a cultural tourism experience is perceived as original and genuine to the culture and traditions of the locals. According to Cetin and Bilgihan (2016), there are five constructs that are the main determinants of the cultural tourist experience in a destination, such as social interaction, local authentic clues, service, cultural heritage, and cultural challenges. Second, heritage and historical authenticity that incorporate historical sites and landmarks. Park, Choi and Lee (2019) argue that experiencing heritage authenticity creates a strong effect in influencing visitors' intention to revisit the destination.

Several studies discuss the significant role of destination authenticity. For example, a study by Jiménez-Barreto et al. (2020) discusses the influence of online destination brand experience on destination brand authenticity. The study suggests that in an online setting, destination authenticity positively influences visitors' intention towards the destination. Kumar, Kaushal and Kaushik (2023) study the destination authenticity of a heritage tourism destination and suggest that the authenticity of a destination significantly impacts visitors' intention to return and their inclination to provide positive recommendations. Next, Lv and Wu (2021) reveal that sensory experience contributes to the formation of destination brand love, which subsequently leads to higher degrees of visitor satisfaction and loyalty.

Also, it has been suggested that destination authenticity has a significant influence on tourist satisfaction (Shi et al., 2022). Visitors who perceive a destination as authentic are more likely to be satisfied with their experiences, which are also influenced by destination image or branding (Sitepu et al., 2021; Zhang et al, 2018). Furthermore, destination authenticity plays a significant role in influencing tourists' intention to revisit the destination that they have visited before (Maarif et al., 2023; Mohamad et al., 2021). Therefore, it is important for tourism destinations to focus on preserving their authenticity which can help leverage the destination branding and offer unique experiences to tourists to enhance their satisfaction and loyalty.

Moreover, the relationship between destination authenticity and VRM has been explored in a number of empirical studies. The findings of these studies suggest that destination authenticity has a number of positive benefits for VRM, including the previously mentioned increased visitors' satisfaction due to authentic experiences are more likely to meet visitors' expectations, influencing the quality of the experience and providing them with a sense of meaning and purpose (Domínguez-Quintero et al., 2019), and following their increased satisfaction is the increased visitors' loyalty, which is indicated with their intention to revisit and recommend the destination. This is because they have developed a positive emotional attachment to the destination. Alternately, changes to the destination which affect the authenticity can reduce visitors' satisfaction and ultimately, their loyalty (Hu & Xu, 2023).

Another benefit of destination authenticity is the increased visitors' engagement. Visitors who perceive a destination as being authentic are more likely to engage with the local culture and people. This is because they are more interested in learning about and experiencing the real destination, in contrast to the perceived artifice and unnaturalness of daily life (McIntosh, 2004). Visitor engagement, moreover, is important for destination branding. When visitors engage with a destination, they are more likely to feel a sense of connection with it, which leads to increased loyalty and repeat visits. Therefore, it is crucial for Destination Marketing Organizations to focus on creating engaging tourist experiences in order to build a strong destination brand identity and increase loyalty (Borges et al., 2018).

### Destination Branding

Understanding the role of branding for destinations and its different nuances in product and services is essential for formulating effective destination marketing strategies. Unlike product or service branding, destination branding encompasses more myriad elements, for example, physical places like buildings, facilities and venues, intertwined with the complex process of tourists' decision-making journey (Almeyda-Ibáñez & George, 2017; Gartner, 2014). Ritchie and Ritchie (1998) were among the first who introduce the definition and concept of destination branding, which was defined as:

> …a name, symbol, logo, word mark or other graphics that both identifies and differentiates the destination: furthermore, it conveys the promise of a memorable travel experience that is uniquely associated with the destination: it also serves to consolidate and reinforce the recollection of pleasurable memories of the destination experience.
>
> (p. 18)

Blain, Levy and Ritchie (2005) later enriched the concept of destination branding through their research on Destination Marketing Organizations (DMOs).

> Destination branding is the set of marketing activities that (1) support the creation of a name, symbol, logo, word mark or other graphics that readily identifies and differentiates a destination: that (2) consistently convey the expectation of a memorable travel experience that is uniquely associated with

> the destination: that (3) serve to consolidate and reinforce the emotional connection between the visitor and the destination; and that (4) reduce consumer search costs and perceived risk. Collectively, these activities serve to create a destination image that positively influences consumer destination choice.
> (p. 337)

The process of destination branding is intricate due to the multi-faceted nature of destinations. Unlike tangible products, destinations are amalgamations of tangible and intangible elements, experiences, and emotions that tourists encounter throughout their journey. Consequently, destination branding involves not only marketing strategies but also careful management and promotion of the destination's reputation and image (Gartner, 2014). Managing reputations, images, and branding is an intricate task for destinations as they cannot exercise complete control over the perceptions formed by tourists. As a result, Destination Marketing Organizations (DMOs) play a crucial role in constructing destination branding and influencing tourists through various marketing activities (Lund et al., 2018).

In today's digital age, the emergence of social media and digital platforms has revolutionized destination branding and marketing practices. DMOs now have easier access to set marketing activities on social media that help shape a positive destination image and branding for their target audiences (Tiago & Veríssimo, 2014). Social media platforms have become indispensable tools for DMOs, offering direct and instant connections with potential visitors, allowing for user-generated content, real-time updates, and personalized engagement. These platforms serve as dynamic spaces for showcasing the authenticity and uniqueness of a destination, capturing the attention and interest of potential tourists worldwide (Gretzel et al., 2015).

Moreover, DMOs' destination branding activities on social media have shown promise in being more effective and efficient compared to traditional advertising in mass broadcast media (Pike et al., 2011). The interactive nature of social media enables DMOs to engage with potential visitors, respond to enquiries, and provide personalized recommendations. This two-way communication fosters a sense of authenticity and trust, contributing to the establishment of strong and lasting relationships with prospective tourists.

Numerous scholarly investigations have examined branding strategies and their implementation within the tourism industry. Almeyda-Ibáñez and George (2017) conducted a comprehensive literature review that traced the evolution of destination branding. Their study shed light on various strategies and issues related to branding applications in the tourism sector. For instance, they exemplified the strategic branding approach employed by New York City in the 1980s, epitomized by the iconic slogan "I love New York." Subsequently, this image-building marketing approach gained traction and was adopted by prominent global cities such as Spain, Las Vegas, and Pittsburgh (Morgan et al., 2011). Almeyda-Ibáñez and George (2017) further contended that while existing research predominantly focuses on the destination branding approach, it often overlooks discussions regarding managerial solutions (Hankinson, 2005).

In addition to Destination Management Organizations (DMOs), local residents assume responsibility for overseeing the destination's management, maintenance, and promotion. Zouganeli et al. (2012) contend that local stakeholders play a significant role in establishing the authenticity of the destination's image. Consequently, it becomes imperative to formulate an authentic and dependable destination branding strategy aimed at shaping the true identity of the destination, with a focus on ensuring the sustainability of the tourism destination branding. This perspective aligns with Pike, Murdy, and Lings' (2011) suggestion that DMOs should adopt a Visitor Relationship Management (VRM) orientation, emphasizing the image authenticity of the destination and the cultivation of relationships with tourists over purely profit-driven motives, sales expansion, or revenue generation.

## Visitor Relationship Management

According to Buhalis and Law (2008), Visitor Relationship Management is "the systematic and coordinated activities and processes that an organization adopts to manage interactions with visitors throughout the visitor lifecycle, with the goal of improving visitor satisfaction, loyalty, and advocacy, and ultimately enhancing the organization's performance" (p. 98). Visitor Relationship Management (VRM) is Customer Relationship Management (CRM) in the tourism marketing context. The fundamental principle of Customer/Visitor Relationship Management is rooted in the belief that cultivating enduring relationships with specific customers/visitors leads to greater long-term profitability compared to a continuous stream of one-time sales transactions. The expenses associated with acquiring new customers/visitors are significantly outweighed by the costs involved in nurturing and sustaining connections with existing customers/visitors (Kincaid, 2003).

In order to strengthen visitor relationships, one of the most essential matters is to emphasize the authenticity of the tourism destination to the visitors. Tourism plays a pivotal role in fulfilling tourists' necessity for authenticity (Meng & Choi, 2016). As modernity and technological advancement continue to shape societies, individuals often feel a sense of disconnection from the real authentic experience within their everyday lives (MacCannell, 1973). Consequently, during their travels, tourists would actively search for genuine and authentic experiences. The pursuit of authentic experience becomes the driving force behind their travel decisions since tourists seek to explore destinations that offer unique and authentic experiences that give meaning and resonate with their inner values.

The influence of destination authenticity on destination branding and its role in visitor relationship management has been a discussion in various literature. For example, Shi et al. (2022) examine the revisit intentions of tourists who had visited the tourist destination of Guilin in China. They indicate that the authenticity of Guilin, perceived by visitors as a "world-class tourist city and international tourist resort," influences visitors' satisfaction levels and their intentions to revisit Guilin. This, in turn, results in the retention of existing visitors, who potentially become repeat visitors (Figure 8.1).

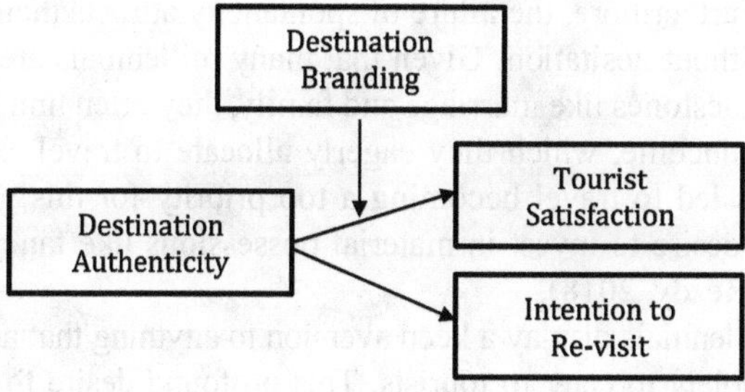


*Figure 8.1* The relationship between destination authenticity and destination branding towards tourist satisfaction and revisit intention.

It is beneficial for DMOs to incorporate destination authenticity into the strategy of Visitor Relationship Management. By focusing on delivering authentic and unique experiences, DMOs can foster strong emotional connections between visitors and the destination. These emotional ties are more likely to endure over time, encouraging visitors to return and engage in word-of-mouth marketing, thus attracting new visitors organically. Moreover, authentic experiences create memorable moments that resonate deeply with visitors, leading to positive reviews and social media mentions, which further enhance the destination's reputation and visibility.

To capitalize on the potential of destination authenticity, DMOs must actively collaborate with local communities and stakeholders. Engaging local residents in the tourism development process ensures that the destination's unique culture and traditions are preserved and showcased authentically. The involvement of local communities in tourism activities can create a sense of pride and ownership, fostering a warm and welcoming atmosphere for visitors.

## Challenges in Building Visitor Relationships

Establishing strong connections with visitors in the tourism sector is not just an optional marketing strategy but has become an essential prerequisite for achieving success in destination branding (Pencarelli et al., 2020). According to UNWTO (2016), the emergence of younger generations, such as millennials and Generation Z, is playing a pivotal role in shaping the growth and transformation of the global tourism economy. These young travellers are characterized by their information-seeking behaviour, unique preferences and travel expectations, and have become a critical segment for Destination Marketing Organizations (DMOs) to understand and take into account in overcoming the tourism trends.

One of the defining features of this young and influential group of travellers is their fascination with the concept of "living like a local," which entails seeking immersive experiences that allow them to engage directly with local businesses, cultures, and residents. Millennials, in particular, have emerged as avid travellers, demonstrating a penchant for extended-stay trips that encourage cultural immersion

(Rezdy, 2018). Furthermore, the allure of spontaneity attracts them to book weekend getaways without hesitation. Given that many millennials are opting to delay traditional life milestones like marriage and family, they often find themselves with more disposable income, which they eagerly allocate to travel experiences. This phenomenon has led to travel becoming a top priority for this generation, often surpassing their desire to invest in material possessions like fancy cars or luxury home products (Rezdy, 2018).

Moreover, millennials display a keen aversion to anything that appears inauthentic or designed solely to cater to tourists. This profound desire for choice and authenticity has been significantly amplified by the advent of personal, transportable technology. With just a few taps on their phones, they can effortlessly explore destinations, connect with like-minded individuals, and book accommodations in remote locations (Fiz, 2018). Millennials are driven by a passion for experiences rather than material possessions, prompting them to be willing to pay more for unique and memorable experiences, further reinforcing their pursuit of authenticity (Sofronov, 2018).

A growing body of research indicates that millennials differ from their older counterparts, such as baby boomers, in their information-seeking behaviour when planning their travels. Rather than relying on traditional sources like travel guidebooks, millennials overwhelmingly turn to the internet, primarily through their electronic devices, to gather information about tourism destinations (Zaim, 2021). Studies have revealed that popular platforms like Google, YouTube, and social networking sites such as Instagram, Facebook, and Snapchat play a crucial role in influencing their decision-making process (Zaim, 2021; Zaim, 2022).

The information search process undertaken by young travellers has a significant influence on their decision-making in travel and can shape their perception of destination authenticity (Kim & Kim, 2020). Specifically, social media platforms play a vital role in enabling younger generations to explore tourism destinations virtually, granting them access to real-time updates that greatly impact their perception of a destination's authenticity and their level of satisfaction with the overall experience (Rahman et al., 2021). Meanwhile, the internet and social media often present a distorted portrayal of the physical reality of a destination as they are often edited to enhance their visual appeal (Kim & Stepchenkova, 2015).

As a result of accessing massive sources from the internet and social media, young generations often form preconceived preferences and expectations when it comes to travel. The exposure to curated and filtered content on social media platforms creates a disparity between their online perception and the reality of the destination. Consequently, young travellers may develop unrealistic expectations based on these filtered representations, leading to disappointment when their actual experience fails to align with the online perception. The mismatched expectations can impede the development of positive visitor relationships and negatively impact visitor satisfaction. These are the challenges that DMOs might face in effectively conveying the authentic experiences of a destination and maintaining its reputation (Uşaklı et al., 2017). Understanding the unique characteristics and preferences of millennials and Generation Z is key to formulating successful marketing strategies that resonate with this demographic.

In light of the importance of visitor relationship marketing (VRM), destination marketing organizations acknowledge the potential benefits it offers in terms of building strong connections with visitors, increasing satisfaction levels, and generating positive word-of-mouth endorsements (Murdy & Pike, 2012). However, there is a gap between the perceived importance of VRM strategies and the actual implementation of these strategies (Murdy & Pike, 2012). DMOs recognize the pivotal role of digital platforms like social media, mobile apps, and review websites, which have eclipsed traditional marketing channels. To effectively execute VRM strategies, DMOs must develop unique content and strategic approaches tailored to these digital channels (Gretzel et al., 2015). Nevertheless, managing various communication channels, promptly responding to visitor feedback, maintaining a consistent brand image, and effectively monitoring the performance of VRM initiatives present formidable challenges for DMOs (Gretzel et al., 2015).

To address these challenges and establish strong visitor relationships, destination marketing organizations must embrace a multifaceted approach that caters to the preferences and expectations of younger travellers. Leveraging user-generated content can build trust and authenticity, as it allows visitors to share their experiences and perspectives on social media and review platforms (Rezdy, 2018). Engaging with travellers through social media interactions and personalized responses can foster a sense of connection and authenticity. Furthermore, DMOs can collaborate with local businesses and influential personalities to craft unique and immersive experiences that align with the desires of young travellers. Curating off-the-beaten-path itineraries, promoting sustainable practices, and highlighting cultural richness can enhance the destination's appeal to millennials seeking authentic experiences (Sofronov, 2018).

Experiential marketing can also play a crucial role in forging deeper connections with visitors. By designing events, workshops, and interactive experiences that reflect the destination's identity, DMOs can create lasting impressions on young travellers, cultivating loyalty and generating positive word-of-mouth endorsements (Pencarelli et al., 2020). Additionally, managing expectations becomes critical to avoiding disappointment among young travellers. Providing transparent and authentic representations of the destination can help align online perceptions with the reality experienced by travellers (Uşaklı et al., 2017).

On top of that, Destination Marketing Organizations can harness the power of data analytics and insights from digital platforms to better understand the preferences and behaviours of young travellers. Data-driven strategies can enable DMOs to deliver personalized content and recommendations, catering to individual preferences and enhancing overall visitor experiences (Murdy & Pike, 2012). This personalized approach can create a sense of exclusivity and align with millennials' desire for unique and tailored experiences (Sofronov, 2018). Incorporating innovative technologies and collaborating with technology companies and startups can also provide DMOs with novel solutions to connect with young travellers. Mobile apps, augmented reality experiences, and virtual tours can offer engaging ways for visitors to interact with the destination even before embarking on their journey, fostering a sense of anticipation and connection (Fiz, 2018).

**Solutions and Recommendations**

Authenticity plays a vital role in attracting and retaining visitors. DMOs should leverage the user-generated content (UGC) features offered by social media platforms, as they have become influential tools for shaping the authentic image of a destination. It is essential for DMOs to encourage visitors to share their experiences, provide honest feedback, and post reviews on UGC platforms. This approach enables the amplification of positive narratives about the destination, fostering trust and strengthening visitor relationships (Marine-Roig & Clavé, 2016). Furthermore, visitor-generated video content on platforms like Instagram reels, TikTok videos, or YouTube vlogs can offer immersive experiences that further enhance the authenticity of the destination.

As digital technology continues to evolve, social media platforms have also emerged as powerful tools for cultivating the authentic image of a destination. Destination Marketing Organizations (DMOs) must make the most of user-generated content type of social media content to establish an authentic connection with their audiences. The dynamic nature of social media has also shaped the way tourists seek travel information and make decisions in terms of the destination to visit. Instead of relying on curated and induced advertising materials from DMOs, nowadays tourists seek guidance and inspiration from their family, friends, or even unknown persons on digital platforms. They seek out genuine stories shared by fellow travellers who have visited the destination, and real-time experiences posted by those who are currently at the destination. In this context, UGC serves as a bank of insights, providing valuable information and organic proof that influences the potential visitors' decision-making process (Fu et al., 2022).

In order to make the most of user-generated content, DMOs should encourage and facilitate visitors to proactively share their experiences on their own various social media platforms like Instagram posts/reels/stories, travel blogs, TikTok videos, Twitter threads, with destination's authentic photos, videos and narrations. DMOs should enhance their social media presence across the internet landscape to actively engage with users, by acknowledging and sharing users' posts about the destinations to create a sense of visitor relationships and potential visitor engagement. Also, to emphasize destination authenticity through the eyes of young visitors, DMOs should create an emotional connection to make these millennials and Gen Z segments feel they are already part of the experience. Video content shared on YouTube, for example, may complement the authenticity of the destination as it can portray genuine and real visitors' reactions and emotions during their visit to the destination, which is more powerful than induced tourism video (Briciu & Briciu, 2020). Another opportunity that user-generated content may generate is its reach. Social media platforms have billions of active users globally, and every single user-generated content can potentially reach a vast number of audiences around the globe. This multiplier effect of user-generated content not only increases the exposure of the destination brand but also allows destination marketers to engage with visitors from diverse cultures and backgrounds.

Nonetheless, DMOs should understand that harnessing the power of UGC does not mean cherry-picking the positive user-generated content, reviews and testimonials. Instead, DMOs should encourage visitors to share their honest experiences, whether it is positive or negative. Honest feedback will help DMOs improve their values and reinforce the destination authenticity. Addressing and acknowledging negative reviews also shows DMOs' commitment to being transparent in terms of visitors' feedback, which demonstrates the commitment to continuous improvement to enhance destinations' reputation.

Besides that, collaborations with local residents as ambassadors or representatives of the destination can significantly impact building visitor trust, making them valuable assets. By selecting local inhabitants who align with the destination's values and target audiences, DMOs can establish authentic connections with existing visitors and increase their motivation to revisit the destination (Bornhorst et al., 2010). Working with these ambassadors, DMOs can create a positive destination image and provide reliable recommendations, ultimately strengthening visitor relationships and encouraging return visits. The importance of these local representatives extends to the aim of retaining visitors. By creating authentic relationships with visitors, destination marketing organizations can increase their intention to revisit the destination after their first visit. Local ambassadors can help DMOs by sharing their insights and stories to inspire visitors to build relationships with the destination and encourage them to advocate the place by recommending the destination to their friends or family members.

Furthermore, one of the key benefits of collaborating with local representatives is their natural ability to organically emphasize the positive image of the destination. Recommendations from local individuals may carry genuine credibility. Tourists may perceive any information coming from the locals as reliable, genuine, and authentic. This visitor perception may foster a positive destination image and influence potential tourists to make decisions to choose the destination and encourage them to explore the place on their own. Besides that, local representatives can share valuable information and reliable recommendations for exploring the destination with the visitors. These locals must have an in-depth understanding of the hidden places, underrated restaurants, lesser-known places and not-so-popular kinds of experiences that are often unknown by mainstream travellers. By emphasizing this local knowledge, DMOs can create personalized itineraries and unique experiences that might suit individual preferences and interests. These personalized DMO suggestions for the visitor do not only enhance the satisfaction of the visitor but also increase the chance of them repeating their visit to the destination and creating positive word-of-mouth recommendations (Chancellor et al., 2021).

Once the ambassadors share their stories and experiences with visitors, the DMO gains valuable input and knowledge from an authentic and neutral perspective. This direct communication allows the DMO to better understand the needs and expectations of visitors and can facilitate continuous improvement and development in the destination offering. In short, the local ambassadors act as intermediaries between the destination and visitors, ensuring that their experience aligns with the destination's values and promises.

To initiate a productive and collaborative relationship with local ambassadors, the DMO must provide training and support activities. Providing easy-to-understand training to ambassadors guarantees they will have the necessary knowledge and skills to represent their destination effectively. This training can cover a variety of subjects, such as local history, cultural nuances, sustainable tourism practices and customer service. DMO can also facilitate networking events and workshops that bring ambassadors together to share experiences, exchange ideas, and foster a sense of community among them. This friendship can further strengthen their commitment to promoting tourism destinations in an authentic and responsible manner.

Furthermore, DMOs can leverage digital platforms to multiply the impact of their local ambassadors. Social media, in particular, provides a way for these ambassadors to share their experiences and further engage with a wider audience. Through engaging blog content and Instagram stories, enlightening YouTube videos, or live sessions, local ambassadors can connect with potential tourists on a global scale. DMOs should actively work with these ambassadors through advertising and social media promotion, providing them with the necessary resources and tools to effectively reach a wider audience.

## Future Research Directions

In light of the many advantages of implementing visitor relationship marketing orientation approaches, several recommendations are put forth for future research directions and explorations. First, given the digital-native nature of the young generation, including millennials and Generation Z, it becomes necessary to examine the role of advanced technologies such as location-based virtual reality (VR) and augmented reality (AR) has the potential to help destination marketing organizations in enhancing visitors experience even if they are not physically present at the destination. Furthermore, an effective integration between offline and online marketing strategies like real-time customer service chatbots in shaping perceptions of destination authenticity helps build relationships between visitors and the destination attractions. Future research can explore how these technologies may enhance the visitors' authentic experience in the destination and engage millennials to revisit the place they have visited before.

Second, future studies can conduct empirical research on millennials and Generation Z's perspectives regarding the authenticity of specific destinations. This will enable a better understanding of the unique viewpoint of this generation, including their motivations for seeking authentic experiences and the impact of destination authenticity on their decision-making process. Additionally, studies can explore the effectiveness of visitor relationship-oriented strategies implemented by Destination Marketing Organizations. This study direction may help DMOs understand which approaches are most effective and considered potential in formulating strategies that enforce relationships with millennials and Generation Z visitors.

Next, given that millennial visitors often seek authentic and more personalized experiences, it would be worthwhile to investigate how destination attractions can involve this young generation in the co-creation of such authentic experiences.

The concept of co-creation, or the joint value creation between destination attractions, destination marketing organizations and visitors to construct unique as well as personalized experiences may include the experience enhancement during visitors' stays at the hotel, creating viral messages or sentiments over social media channels regarding the destination atmospheres and active participation in local events and cultural activities (Buhalis & Sinarta, 2019). This future research topic will shed light on how to create more meaningful content that reflects young generations' experience, which will formulate a better comprehension of the visitor relationship strategy implementation.

Future studies should explore the degree of relationship between destination authenticity, visitor relationship management and sustainable tourism practices as they are intercorrelated (Mateoc-Sîrb et al., 2022). These three concepts are linked, and understanding each role and how they play a role is vital for formulating a holistic approach to implementing a responsible and sustainable destination marketing strategy. For example, examining how the strategy of destination branding can relate to the principle of sustainable tourism, such as promoting the practice of an environmentally friendly approach, and preserving cultural heritage as well as local wisdom and values, can highlight the significance of sustainable tourism in engaging young generations as the potential tourists. Some topics may include underlining the importance of the Environmental, Social and Governance (ESG) concept in the destination marketing organizations programmes. This is to align the destination marketing efforts with Sustainable Development Goals (SDGs) and encourage young generations to support responsible tourism.

Lastly, another promising area of research is investigating how the authenticity of tourism destinations plays a role in maintaining visitor relationship management and contributes to destination loyalty and repeat visitation. Understanding the driving factors of visitor loyalty and revisiting intentions will help destination marketing organizations to improve their strategies and foster long-term relationships with this young generation group of travellers to become repeat visitors in the future. Additionally, exploring the effect of destination authenticity on the level of satisfaction and overall experiences of visitors will provide important insights into how authenticity can play a role as a key differentiating factor in attracting and retaining young visitors.

## Conclusion

This chapter aims to discuss the role of authenticity in destination branding which plays a significant role in building relationships among visitor attractions and the millennial generation as the biggest travel segment in the tourism industry. Understanding how young travellers perceive authenticity is essential for destination managers to create experiences that resonate with this influential segment. By addressing these challenges, destinations can establish meaningful connections and foster long-term relationships with young travellers.

All in all, understanding the distinct characteristics of destination branding is vital for destination marketing organizations that are seeking to attract, engage and retain young generation travellers. The combination of various destination branding elements is necessary to formulate a holistic marketing approach that incorporates destination marketing strategies, destination branding, image management, visitor retention and visitor relationship management. Social media and digital platforms have become powerful tools for destination marketing organizations to enable their strategic effort in shaping positive destination images and engaging with potential visitors. Successful visitor relationship management strategies require the enhancement of destination authenticity and the fostering of visitors' trust and long-lasting relationships with them. Collaborating with local ambassadors is also essential, with the support of the implementation of innovative technologies to further enhance the destination's authenticity and place image. By formulating and implementing good alignment between destination marketing efforts with the young generation's desires and expectations, destination marketing organizations can create meaningful and memorable experiences that resonate with young travellers as their target market, while also ensuring the sustainable growth and development of the destination in the tourism industry.

Building visitor relationships with young travellers requires an in-depth understanding of their unique characteristics and perceptions of destination authenticity. Destination managers must navigate the challenges posed by evolving definitions of authenticity, the influence of social media, the demand for personalized experiences, sustainability expectations, and the desire for local engagement. By aligning their strategies and offerings with the preferences and values of young travellers, destinations can foster genuine connections and build lasting visitor relationships. Embracing authenticity as a core principle and actively involving young travellers in the destination's development process will contribute to their satisfaction and loyalty, ensuring long-term success in an increasingly competitive tourism landscape.

**Case Study**

West Java, one of Indonesia's largest provinces, is home to nearly 49 million people, accounting for about 20% of the country's total population of approximately 273 million. The majority of West Java's residents, around 68%, fall within the 15 to 64 age range, indicating a young and potentially productive workforce that is crucial for economic growth. It comes as no surprise that West Java has emerged as one of Indonesia's most productive and economically competitive provinces.

However, West Java also faces several challenges that require attention. One of these challenges is the digital divide between rural and urban communities, which stems from inadequate rural infrastructure. According to the Information and Communication Technology Development Index of

2021, West Java ranks fifth out of 34 provinces in Indonesia in terms of access and infrastructure for digital technology. This indicates that West Java still has room for improvement in providing digital access and infrastructure to its population. Therefore, accelerating infrastructure development, promoting digital literacy, and fostering technology adoption within the community are crucial factors for achieving sustainable development in West Java, particularly in the tourism sector.

In the realm of tourism, the digital ecosystem of West Java Province focuses on three main priorities. First, reducing the digital divide, especially in tourist destinations located in areas with limited internet connectivity. Second, increasing the digital literacy rate among citizens. And third, developing the potential of tourism villages through digital innovation and collaboration. To harness the power of social media in the tourism sector, West Java has launched several programmes that centre around social media intervention to promote tourism through these new channels.

To achieve these goals, the government is facilitating the installation of base stations to provide internet access to areas without connectivity, often referred to as "blank spot villages." Through this initiative, the number of blank spot villages in West Java has decreased to 359 areas from the previous count of 5,311 villages, leaving less than 10% of villages without internet access. Additionally, the West Java Provincial Government has established 230 free access points, with 41 of them strategically located in tourist attractions. The selection of these locations is based on recommendations from the local government and is aligned with the unique characteristics of each region. For instance, free access points are available at Citepus Beach in Sukabumi Regency due to its popularity as a coastal destination, as well as at Makam Sunan Gunung Jati and Kacirebonan Palace in Cirebon Regency because these areas are renowned for their historical sites. Once the basic infrastructure is in place, the next crucial step is to ensure that the Internet is effectively utilized for sharing and disseminating information about tourist destinations through social media. However, the government faces limitations in terms of available human resources for these responsibilities.

To generate widespread attention for tourist destinations in West Java, the Provincial Government has introduced the "1000 Smiling West Java Ambassadors" programme. Through this initiative, the government handpicks 1000 content creators, particularly residents of tourism villages, to promote their respective villages and serve as ambassadors for their areas. This approach fosters genuine engagement between tourists and the local community. Additionally, the programme involves collaborating with local influencers to amplify exposure through social media platforms. The 1000 Smiling West Java Ambassadors play a crucial role in optimizing the storytelling aspect of the West Java tourism campaign, primarily utilizing popular social media channels such as Instagram, YouTube, and TikTok to share captivating content. This approach aligns with the demographic composition of West Java, which consists mostly of a young population, particularly millennials and Gen Z, who are well-versed in using social media as their primary means of sharing and accessing information. Leveraging social media for exposure proves to be a cost-effective strategy compared to traditional mass media campaigns.

Furthermore, in the ongoing efforts to optimize the tourism sector, West Java is developing the Smiling West Java Application, which serves as a gateway to West Java's events and the West Java Tourism Center (WJTCC). These three digital initiatives share a common vision of enhancing the tourism industry's image in the digital era, with a focus on simplifying access to tourist destinations within the West Java region for all types of tourists, both domestic and foreign. These approaches align with the prevailing digital familiarity among the population in West Java, who are predominantly accustomed to using digital tools provided through portals and applications.

The Smiling West Java Application plays a vital role in building visitor relationship management, enabling the government to track the number of visitors and gain insights for visitor retention plans. This real-time data on tourist perceptions of West Java is crucial for strengthening the brand of West Java tourism. Moreover, the application has the potential to provide an authentic experience of West Java tourism, as all visitors can directly share their experiences within the application for other users to enjoy. This adds to the authenticity and richness of the West Java tourism experience.

**Case Questions (Three Questions)**

1 How have the different digital literacy levels between rural and urban areas in West Java impacted the tourism industry in the areas, and what initiatives have been implemented by the government to overcome these challenges?
2 How are the local ambassadors playing a role in fostering destination authenticity in West Java?
3 How does the Smiling West Java mobile application contribute to visitor relationship management and the enhancement of destination branding, and what advantages does it offer to create positive experiences for young travellers?

**Key Terms and Definitions – Definitions for the Key Constructs**

**Destination Authenticity**: According to Chen et al. (2020), the concept of destination authenticity pertains to the extent to which visitors perceive a destination brand as consistent and true to its own identity (consistent), trustworthy and believable to visitors (credible), accountable and ethically upright (responsible), and supportive of tourists in their personal growth and self-expression (facilitates self-discovery).

**Destination Marketing Organizations**: Organization(s) that promotes a place as an attractive tourism destination for visitors.

**Destination Branding**: A set of marketing effort that incorporates the destination identities, building visitors expectation of a memorable travel experience to reinforce the emotional connection between the visitor and the destination and reduce visitors negative perception to the destination (Blain et al., 2005).

**Visitor Relationship Management**: "the systematic and coordinated activities and processes that an organization adopts to manage interactions with visitors throughout the visitor lifecycle, with the goal of improving visitor satisfaction, loyalty, and advocacy, and ultimately enhancing the organization's performance" (Buhalis & Law, 2008, p. 98).

**Millennials**: A generation that consists of individuals born between 1981 and 1996.

**Generation Z**: A generation that consists of individuals born between 1997 and 2012.

**User-Generated Content**: organic contents created and shared by visitors on their social media or other digital platform channels in the form of photos, videos, reviews, testimonials or feedback.

## References

Almeyda-Ibáñez, M., & George, B. P. (2017). The evolution of destination branding: A review of branding literature in tourism. *Journal of Tourism, Heritage & Services Marketing (JTHSM)*, *3*(1), 9–17.

Babii, A., & Nadeem, S. (2021). *Tourism in a Post-Pandemic World*. IMF Country Focus.

Beerli, A., & Martin, J. D. (2004). Factors influencing destination image. *Annals of Tourism Research*, *31*(3), 657–681.

Blain, C., Levy, S. E., & Ritchie, J. R. B. (2005). Destination branding: insights and practices from destination management organizations. *Journal of Travel Research*, *43*, 328–338.

Borges, A. P., Vieira, E., & Rodrigues, P. (2018, July). Tourist engagement and the identification with the brand of destination: The case of (re)visiting the city of Porto. In *2018 Global Marketing Conference at Tokyo Proceedings*. Global Alliance or Marketing and Management Associations (pp. 525–526).

Bornhorst, T., Ritchie, J. B., & Sheehan, L. (2010). Determinants of tourism success for DMOs & destinations: An empirical examination of stakeholders' perspectives. *Tourism Management*, *31*(5), 572–589.

Briciu, A., & Briciu, V. A. (2020). Participatory culture and tourist experience: Promoting destinations through YouTube. In *Strategic Innovative Marketing and Tourism: 8th ICSIMAT, Northern Aegean, Greece, 2019* (pp. 425–433). Springer International Publishing.

Bryce, D., Curran, R., O'Gorman, K., & Taheri, B. (2015). Visitors' engagement and authenticity: Japanese heritage consumption. *Tourism Management*, *46*, 571–581.

Buhalis, D., & Law, R. (2008). Visitor relationship management: A research agenda. *Journal of Travel Research*, *47*(1), 93–104.

Buhalis, D., & Sinarta, Y. (2019). Real-time co-creation and nowness service: Lessons from tourism and hospitality. *Journal of Travel & Tourism Marketing*, *36*(5), 563–582.

Chancellor, C., Townson, L., & Duffy, L. (2021). Destination ambassador programs: Building informed tourist friendly destinations. *Journal of Destination Marketing & Management*, *21*, 100639.

Chen, R., Zhou, Z., Zhan, G., & Zhou, N. (2020). The impact of destination brand authenticity and destination brand self-congruence on tourist loyalty: The mediating role of destination brand engagement. *Journal of Destination Marketing & Management*, *15*, 100402.

Cetin, G., & Bilgihan, A. (2016). Components of cultural tourists' experiences in destinations. *Current Issues in Tourism*, *19*(2), 137–154.

Cole, S. (2012). Synergy and congestion in the tourist destination life cycle. *Tourism Management*, *33*(5), 1128–1140.

Domínguez-Quintero, A. M., González-Rodríguez, M. R., & Roldán, J. L. (2019). The role of authenticity, experience quality, emotions, and satisfaction in a cultural heritage destination. *Journal of Heritage Tourism*, *14*(5–6), 491–505.

Fiz (2018). Millennial travellers and how they've changed travel for the better. Accessed August 1, 2018, http://www.fiz.com/blog/travel-trends/millennial-travellers/.

Fu, X., Wan, F., & Wu, Y. (2022). Inbound tourists' perception of tourist destination image classified by UGC picture computer program. *Journal of Electrical and Computer Engineering*, *2022*, 1–13.

Gartner, W. C. (2014). Brand equity in a tourism destination. *Place Branding and Public Diplomacy*, *10*(2), 108–116.

Gretzel, U., Sigala, M., Xiang, Z., & Koo, C. (2015). Smart tourism: Foundations and developments. *Electronic Markets*, *25*, 179–188.

Hankinson, G. (2005). Destination brand images: A business tourism perspective. *Journal of Services Marketing*, *19*(1), 24–32.

Hu, Y., & Xu, S. (2023). Repeat tourists' perceived unfavorable changes and their effects on destination loyalty. *Tourism Review*, *78*(1), 42–57.

Jiménez-Barreto, J., Rubio, N., & Campo, S. (2020). Destination brand authenticity: What an experiential simulacrum! A multigroup analysis of its antecedents and outcomes through official online platforms. *Tourism Management*, *77*, 104022.

Kincaid, J. W. (2003). *Customer Relationship Management – Getting it Right*. Prentice Hall PTR.

Kim, M., & Kim, J. (2020). Destination authenticity as a trigger of tourists' online engagement on social media. *Journal of Travel Research*, *59*(7), 1238–1252.

Kim, H., & Stepchenkova, S. (2015). Effect of tourist photographs on attitudes towards destination: Manifest and latent content. *Tourism Management, 49*, 29–41.

Kumail, T., Qeed, M. A. A., Aburumman, A., Abbas, S. M., & Sadiq, F. (2022). How destination brand equity and destination brand authenticity influence destination visit intention: Evidence from the United Arab Emirates. *Journal of Promotion Management, 28*(3), 332–358.

Kumar, V., Kaushal, V., & Kaushik, A. K. (2023). Building relationship orientation among travelers through destination brand authenticity. *Journal of Vacation Marketing, 29*(3), 331–347.

Lund, N. F., Cohen, S. A., & Scarles, C. (2018). The power of social media storytelling in destination branding. *Journal of Destination Marketing & Management, 8*, 271–280.

Lv, X., & Wu, A. (2021). The role of extraordinary sensory experiences in shaping destination brand love: An empirical study. *Journal of Travel & Tourism Marketing, 38*(2), 179–193.

Maarif, L. A., Ratnawati, K., & Dwi Vata Hapsari, R. (2023). The authenticity and social media effect on revisit intention mediated by destination image. *International Journal of Research in Business and Social Science*, 2147–4478, *12*(4), 33–43.

MacCannell, D. (1973). Staged authenticity: Arrangements of social space in tourist settings. *American Journal of Sociology, 79*(3), 589–603.

Malik, G., Gangwani, K. K., & Kaur, A. (2022). Do green attributes of destination matter? The effect on green trust and destination brand equity. *Event Management, 26*(4), 775–792.

Marine-Roig, E., & Anton Clavé, S. (2016). A detailed method for destination image analysis using user-generated content. *Information Technology & Tourism, 15*, 341–364.

Mateoc-Sîrb, N., Albu, S., Rujescu, C., Ciolac, R., Ţigan, E., Brînzan, O., & Milin, I. A. (2022). Sustainable tourism development in the protected areas of Maramureş, Romania: Destinations with high authenticity. *Sustainability, 14*(3), 1763.

McIntosh, A. J. (2004). Tourists' appreciation of Maori culture in New Zealand. *Tourism Management, 25*(1), 1–15.

Mior Shariffuddin, N. S., Azinuddin, M., Hanafiah, M. H., & Wan Mohd Zain, W. M. A. (2023). A comprehensive review on tourism destination competitiveness (TDC) literature. *Competitiveness Review: An International Business Journal, 33*(4), 787–819.

Meng, B., & Choi, K. (2016). The role of authenticity in forming slow tourists' intentions: Developing an extended model of goal-directed behavior. *Tourism Management, 57*, 397–410.

Murdy, S., & Pike, S. (2012). Perceptions of visitor relationship marketing opportunities by destination marketers: An importance-performance analysis. *Tourism Management, 33*(5), 1281–1285.

Mohamad, N., Chandran, N. S., Marasol, N. K., & Syed Ismail, S. I. (2021). Exploring tourists' intention to visit a heritage destination. *Journal of Management & Science, 19*(2), 12–12.

Morgan, N., Pritchard, A., & Pride, R. (2011). Tourism places, brands, and reputation management. In N. Morgan, A. Pritchard, & R. Pride (Eds.), *Destination Brands: Managing Place Reputation* (3rd ed., pp. 3–19). Routledge-Taylor & Francis Group.

Murphy, L., Benckendorff, P., & Moscardo, G. (2007). Destination brand personality: Visitor perceptions of a regional tourism destination. *Tourism Analysis, 12*(5–6), 419–432.

Park, E., Choi, B. K., & Lee, T. J. (2019). The role and dimensions of authenticity in heritage tourism. *Tourism Management, 74*, 99–109.

Pencarelli, T., Gabbianelli, L., & Savelli, E. (2020). The tourist experience in the digital era: The case of Italian millennials. *Sinergie Italian Journal of Management, 38*(3), 165–190.

Pike, S., Murdy, S., & Lings, I. (2011). Visitor relationship orientation of destination marketing organisations. *Journal of Travel Research, 50*(4), 443–453.

Pine, B. J., & Gilmore, J. H. (2008). The eight principles of strategic authenticity. *Strategy & Leadership, 36*(3), 35–40.

Rahman, A., Ahmed, T., Sharmin, N., & Akhter, M. (2021). Online destination image development: The role of authenticity, source credibility, and involvement. *Journal of Tourism Quarterly, 3*(1), 1–20.

Rezdy (2018). Millennials: An in-depth look into the travel segment. Accessed July 2, 2018, https://www.rezdy.com/blog/millennials-depth-look-travel-segment-infographic/.

Ritchie, J. R. B., & Ritchie, R. J. B. (1998, September). The branding of tourism destination: Past achievements and future challenges. In *Presentation Delivered at Annual Congress of the International Association of Scientific Experts in Tourism, Marrakech, Morocco* (pp. 1–31).

Sitepu, E. S., Medan, P. N., & Rismawati, R. (2021). The influence of service quality, destination image, and memorable experience on revisit intention with intervening variables of tourist satisfaction. *International Journal of Applied Sciences in Tourism and Events, 5*(1), 77–87.

Shi, H., Liu, Y., Kumail, T., & Pan, L. (2022). Tourism destination brand equity, brand authenticity and revisit intention: The mediating role of tourist satisfaction and the moderating role of destination familiarity. *Tourism Review, 77*(3), 751–779.

Sofronov, B. (2018). Millennials: A new trend for the tourism industry. *Annals of Spiru Haret University. Economic Series, 18*(3), 109–122.

Tiago, M. T. P. M. B., & Veríssimo, J. M. C. (2014). Digital marketing and social media: Why bother?. *Business Horizons, 57*(6), 703–708.

UNWTO (2016, March 1). Global report on the power of youth travel. https://www.wysetc.org/. Retrieved April 16, 2023, from https://www.wysetc.org/wp-content/uploads/2016/03/Global-Report_Power-of-Youth-Travel_2016.pdf.

Uşaklı, A., Koç, B., & Sönmez, S. (2017). How 'social' are destinations? Examining European DMO social media usage. *Journal of Destination Marketing & Management, 6*(2), 136–149.

Ward, S. V. (1998). *Selling Places: The Marketing of Towns and Cities, 1850–2000.* Routledge.

Wisker, Z. L., Kadirov, D., & Nizar, J. (2023). Marketing a destination brand image to Muslim tourists: Does accessibility to cultural needs matter in developing brand loyalty? *Journal of Hospitality & Tourism Research, 47*(1), 84–105.

Yi, X., Lin, V. S., Jin, W., & Luo, Q. (2017). The authenticity of heritage sites, tourists' quest for existential authenticity, and destination loyalty. *Journal of Travel Research, 56*(8), 1032–1048.

Zaim, I. A. (2021). Young British tourists' tourism-related information sources. In N. Boukas & D. Stylidis (Eds.). *Tourism Marketing in Western Europe* (pp. 26–42). CABI.

Zaim, I. A. (2022). Young Indonesian travellers' information search behaviour. In D. Stylidis, S. Kim & J. Kim (Eds). *Tourism Marketing in East and Southeast Asia* (pp. 114–127). CABI.

Zenker, S., & Braun, E. (2010, June). The place brand centre–a conceptual approach for the brand management of places. In *39th European Marketing Academy Conference, Copenhagen, Denmark,* Copenhagen: European Marketing Academy (pp. 1–8).

Zhang, H. L., Cho, T., Wang, H., & Ge, Q. (2018). The influence of cross-cultural awareness and tourist experience on authenticity, tourist satisfaction and acculturation in world cultural heritage sites of Korea. *Sustainability*, *10*, 927.

Zouganeli, S., Trihas, N., Antonaki, M., & Kladou, S. (2012). Aspects of sustainability in the destination branding process: A bottom-up approach. *Journal of Hospitality Marketing & Management*, *21*(7), 739–757.

# 9 Unleashing Innovation through Internal Branding and Resident Involvement

*Ioana S. Stoica*

## Introduction

This chapter aims to explore of how residents are influencing their place brands through diverse acts of entrepreneurship by investigating (1) the impact of residents' entrepreneurial acts on internal place branding in shaping a place's identity, culture, and reputation and (2) the opportunities associated with residents' entrepreneurship acts.

## Background: Place Branding

Traditionally, place branding was largely perceived as a form of top-down marketing or promotional activity in which the brand managers are taking all the main decisions on what the brand is and how it is communicated. However, recently scholars contend that the concept of 'branding' has grown more inclusive (Kavaratzis et al., 2018), and it is prioritising the incorporation of internal stakeholders in the brand development and communication process (Källström and Siljeklint, 2023). Nowadays' successful place branding strategies are seen as interactive processes between the private and the public sector (Lucarelli, 2019) focused on internal developments or sustainable changes into the place infrastructure, environment, or communities. These developments and changes are often illustrated through internal place branding initiatives that fulfil diverse stakeholders' needs.

Wide parts of literature on place branding initiatives fulfilling stakeholders' needs is focused on highlighting the importance of residents' involvement in place activities and branding (Kavaratzis, 2004; Freire, 2009; Klijn et al., 2012; Zenker and Petersen, 2014; Jeuring and Haartsen, 2017; Govers, 2018; Skinner, 2018; Casais and Monteiro, 2019; Florek and Insch, 2020; El Banna and Stoica, 2021). This is due to the fact that residents have a special connection to the place; they are highly interacting with it and promoting it on a daily basis through day-to-day mundane activities (Govers, 2011; Su et al., 2018). Zenker and Petersen (2014) suggest that when residents participate in the development of a place's brand, the process enables the place to effectively convey and communicate its identity. This usually occurs because residents possess the capability to endorse the brand messages transmitted through official place communication, but also convey their

DOI: 10.4324/9781003369967-13

personal positive narratives and messages of the place that improve a place's image and reputation. In addition to this, we need to regard branding as a complex process or as Casais and Monteiro (2019) mentioned, the process of branding extends above the creation of a logo and extends beyond the realm of visual communication. In Baxter et al.'s (2013) and Boisen et al.'s (2018) views, place branding and reputation are based on place's identity and the reality often created through the experiences of the locals who are the people engaging with the place's offerings regularly. Some of these studies lead us to associate residents' involvement in place activities with internal branding.

The chapter provides an overview of internal place branding through resident entrepreneurship acts. First, it defines internal branding and place attachment in relation to resident involvement in place activities and proposes a conceptual framework for discussing seven areas in which resident entrepreneurship acts brought a significant contribution to a place's image and reputation. Then, the chapter follows up with a discussion of each of the seven areas identified in the framework. The 'Solution and recommendations' section problematises some of the common issues identified in internal branding and resident involvement studies and possible ways to mitigate risks, and the last sections discuss areas of further research and considerations and conclude the chapter.

## Internal Branding through Resident Involvement and Entrepreneurship

From a corporate perspective, internal branding refers to a process of involving internal stakeholders, often employees in the brand development and promotion by having an input on the brand values and brand culture. Internal branding aims to shape a uniform internal and external image of the brand through the involvement of the internal stakeholders (Barros-Arrieta and García-Cali, 2021).

Although, conventional marketing communications for corporate brands cannot be wholly applied to place brands due to numerous place complexities and intricate interconnectedness of interactions and exchanges among multiple stakeholders (Warnaby, 2009; Braun et al., 2014) necessitating to satisfy different stakeholders' requirements (Casais and Monteiro, 2019), internal branding can be seen an ideal process of developing place brands through the help of residents.

For places, internal branding is a collaborative process which aims at creating consistent images of the brand through the involvement of internal stakeholders, often the residents because of their significant role in creating and promoting place associations (Zenker and Erfgen, 2014). Govers (2011) refers to internal branding as a congregation of non-marketing methodologies in which the residents actively contribute to the formation and essence of the place. In the authors' view, this dynamic process possesses significant potential to not only alter the internal perceptions of the place but also impact its external portrayal, thus influencing global place image and reputation. Internal branding for places has often been idealised in the literature through topics like stakeholders' involvement and participatory processes in which place's stakeholders have more power and influence towards a place's worldwide reputation than the marketing initiatives undertaken by local administrators and councils (Seo, 2013; Braun et al., 2014; Braun et al., 2018; Marine-Roig and Ferrer-Rosell, 2018). Often, internal place branding practices analyses the progression of place developmental initiatives utilising grassroots or bottom-up approaches (Hudson et al., 2017), thereby redistributing influence and power from the managers of the brand to the grassroots participants or bottom-up actors, such as the residents.

In this chapter internal branding is discussed in relation through resident involvement and entrepreneurship acts. Resident involvement and entrepreneurship acts refer in this instance to the proactive initiatives and entrepreneurial endeavours undertaken by the individuals within a place to shape, improve and promote their environment and/or communities. These acts involve the active participation of residents in various entrepreneurial activities, such as establishing businesses, organising events, promoting local culture and heritage and developing and implementing sustainable practices within their place. The connection between resident entrepreneurship acts and internal branding lies in the active role that residents play in shaping and promoting the brand identity of their locality from within. By initiating entrepreneurial activities that highlight the distinct cultural, artistic, historical, or natural elements of their community, residents contribute to the establishment of a unique and authentic brand narrative that resonates with both locals and external audiences. Residents engage in these entrepreneurial acts to leverage the unique attributes and resources of their locality, contributing to the overall development and enhancement of the place's identity and reputation. Hence, many studies tried to understand residents' involvement in place activities (Mohammadi et al., 2018; Vaiou, 2018) to be able to propose useful place strategies that are branding the place inside-out or use resident support to shape and communicate authentic place brands (El Banna and Stoica, 2021). In this chapter, the drivers of resident entrepreneurship acts are mainly discussed through three factors: (1) Place attachment; (2) Local Resources; (3) Local authorities support and governance.

**Place attachment** is a concept defined as the bond between people and places but also seen as 'an individual's conscious awareness of the community environment that they are in and involves positive perceptions, emotions, and behaviour towards the community' (Chang et al., 2022: 21). To understand the concept, Altman and Low's (1992) place attachment theory underscores the idea that individuals cultivate both emotional and cognitive attachments to particular locations which significantly shapes their behaviour and attitudes concerning these places, contributing to a multifaceted relationship that goes beyond mere physical surroundings. In the context of residents' involvement, place attachment is linked with place-related behaviour and intentions (Chen et al., 2018) and can act as a driver of local entrepreneurs who are connected to their homes and due to their connections to the place's offerings, they are investing time, efforts and resources into improving their place. As people develop intricate connections with their environment, these connections play a pivotal role in determining their actions and perceptions within those places (Chang et al., 2022). In this chapter, place attachment is seen as one of the main drivers of residents' entrepreneurial acts because it fosters a sense of emotional connection and belonging among residents, encouraging them to actively participate in initiatives that showcase the unique characteristics of the place. As Chang et al. (2022) suggested, residents with deep place attachment are more likely to contribute to activities that highlight the authentic character of the place, thereby strengthening its brand identity and creating an engaging narrative for visitors and other stakeholders.

However, place attachment is not the only factor driving resident entrepreneurial acts or involvement in place initiatives, the local resources available within the place and local authorities support, leadership and government are also an important factors to consider (Insch and Stuart, 2015; Gao et al., 2022). **Local resources** are important to amplify unique selling propositions of the place, empowering residents to contribute to the development of a distinct and authentic brand narrative that highlights the unique features and cultural richness of the place. Leveraging these resources not only enhances the place's brand identity but also fosters a deeper connection between the local community and visitors, creating a memorable and immersive experience for all stakeholders involved. **Local authorities support and governance** are also important to resident entrepreneurial acts as they can enable strategic guidance and facilitate support for residents to access local resources, implement their ideas and fund local initiatives. Collaborative efforts between residents and local authorities create a strong sense of community engagement, leading to a supportive environment for the local population. It is also important to note, that previous studies shows that place governance is linked with active resident participation in the place various activities and can drive positive or negative reactions towards the place (Casais and Monteiro, 2019).

The aforementioned factors can lead residents to initiate, continue and promote various initiatives that require their entrepreneurship skills in seven main areas that are portrayed in Figure 9.1. These entrepreneurial acts can serve as an inspiration for informing internal place branding strategies in various areas and they are discussed in the next section.

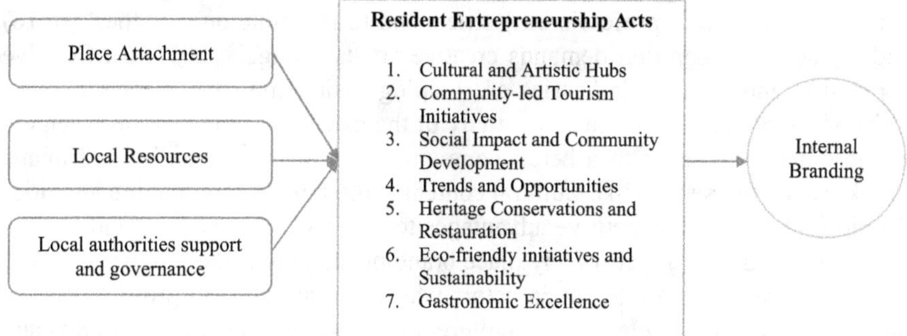


*Figure 9.1* Internal branding through resident entrepreneurship acts.

The next part of the chapter is discussing the seven main themes identified in the framework, exploring how residents, through various entrepreneurship actions are taking ownership of their place brands and shape it in line with their visions, experience and values and thereby unleashing innovation and bringing a positive change to the local culture, economy, and overall place identity and reputation.

## Unleashing Innovation through Resident Entrepreneurship Acts

### *Residents' Involvement in Cultural and Artistic Hubs*

Place reputations can be formed in many ways and one of them is through the residents who are gaining worldwide attention (Ashworth and Kavaratzis, 2011). Famous people who were born or lived in a particular place could add an artistic feel to the place, as it happened to the homes of many artists. For instance, Salzburg, Austria gained renown for its association with Mozart; Bayreuth, Germany, for Wagner, Liverpool, UK, for 'The Beatles', etc. However, this is not something all places are orchestrated to, or residents can control. Instead, residents can transform their home in a cultural or creative destination by establishing and advocating for cultural and artistic hubs. Locals with a taste for art, culture, and heritage can establish cultural centres, run festivals, art galleries, studios, or theatres and this is primarily seen when the locals take ownership of the place brand and identify with its local and cultural identity.

For example, in Bristol, United Kingdom, which is a well-known place for street art (including for Banksy's who started his career in the town) residents have been contributing to enhancing the creative landscape of the town in multiple ways. They have been supporting the artistic vibe of the town by creating public art installations, graffiti arts and art festivals (e.g. Upfest). The impact of the residents was not only building a stronger community and opportunities for the locals, but enhancing local tourism and strengthening the cultural identity of the town (Davies, 2009).

In today's age where globalisation treats to fade away local character of place, it is extremely important to support and enhance local, cultural and artistic hubs as part of the branding strategy of the place, or as Magala (2011) stated: 'without blooming creative industries (journalists, writers, visual artists, designers, composers, musicians, media professionals) cities tend to resemble one another' (p. 16) and a place's 'recognition demands creative artists' (Magala, 2011: 6). Creative expression and creative industries are helping maintaining the place's cultural traits which are adding to the authenticity of the place and residents' involvement in those activities leads to a better communication of place identity and image (Zenker and Petersen, 2014). Cultural content is often used to reimagine identities for cities and bring competitive advantages to destinations, or as Go and Govers (2013) argue, a critical part of any place branding strategy is the representation of culture and residents are the main actors who can represent and promote place's tangible and intangible elements of culture, including customs, traditions, arts and societal norms.

***Residents' Involvement in Community-Led Tourism Initiatives***

Due to their high involvement with the place and having depth knowledge about it, residents can create community-led tourism initiatives that are more credible than top-down promotional activities (Gajdošík et al., 2018). Many scholars contend that the satisfaction of the residents with their surroundings, their emotional connections and their overall positive perception of the place are indispensable factors for effectively promoting the place (Insch, 2010; Campelo et al., 2014; Zenker and Rütter, 2014; Peighambari et al., 2016; Casais and Monteiro, 2019). This leads to the idea that residents do not only need to be included in tourism initiatives, but them themselves can take over the promotions of the place.

Kangjuan et al.'s (2017) suggests that the active participation of residents in place management, policymaking and official promotion leads to more impactful communication for the place brand. According to Casais and Monteiro (2019) the essence of the place is shaped by the interconnections and social ties among the stakeholders and communities within the place. The authors also suggests that the resident-tourist interactions establish the uniqueness of the place, as individuals engage with one another, and these relationships significantly shape tourists perceptions of the place. Moreover, online, residents add credibility to the place brand by offering recommendations and feedback to the place surroundings and offerings through various social media channels and online platforms (Feng et al., 2023), and as Uchinaka et al. (2019) emphasised, often the residents are becoming the principal creators and promoters of place brands for destinations. Destinations are benefiting from resident's participation in social media communication (Chen and Šegota, 2015) and become co-creators of the world-wide place associations (Skinner, 2018). Resident often participate in endorsing favourable destination impressions through word of mouth (Jeuring and Haartsen, 2017) and become committed to destination's branding efforts by advocating for the place brands (Kemp et al., 2012).

It is worth considering that resident's commitment to the place promotion is only seen when they are proud of their place, and identify with the promoted place identity (Choo et al., 2011). For example, the residents of Austin, Texas actively participate in promoting the place through local tourism initiatives and community-led activities that led to the creation of the South by Southwest (SXSW) conference festivals. SXSW became a world-renowned event that celebrates not only music but also film, interactive media, and technology and is a perfect example of how local creatives (artists, musicians) and entrepreneurs work collaboratively to curate diverse showcases, performances, and interactive events that highlight the city's musical, cinematic, and creative identity. Austin's robust music community have a crucial role in creating the music-focused character of SXSW, and showcase the city's authenticity and reputation as the 'Live Music Capital of the World' (Govers, 2018).

Moreover, by interacting with tourists and disseminating their local knowledge and expertise, residents can shape the place's brand as a welcoming and immersive destination (Fan et al., 2019). Smaller communities and family businesses are particularly taking advantage of this personal approach to create lasting impressions for tourists on their visits and promote place's authenticity. Moreover, residents can be involved with cultural exchanges, guided tours, homestay programmes, authentic initiatives or distribution of positive place messages online.

### *Residents' Social Impact and Community Development*

Residents can also have an impact on community development outside the creative or touristic domains and bring positive social changes. Residents can engage in social actions that drive social impact and be part in community development programmes through establishing social enterprises and non-profit organisations that tackle local or global issues. One example of residents' involvement that led to community development is seen in High Line Urban Park, New York City. High Lane is an urban park established on an elevated historic freight rail line above the streets on the west side of Manhattan, as an outcome of community-led initiatives and innovative urban renewal. Local activists played a notable role in transforming the old space by advocating for High Line's preservation and pushing its development into a public urban park by repurposing underutilised space for community's beneficial use. Local residents' involvement, not only offered a second life to the space, but helped foster a sense of pride and ownership of the place among local communities and transformed High Lane into a symbol for community-driven innovation and development in the heart of New York City.

In addition, residents can enhance local initiatives that bring positive worldwide changes. Training programmes or activities to support marginalised groups can foster favourable perceptions and images of the place brand and highlight locals' commitment to inclusion, diversity and social responsibility. Local government action can inform some resident's activities, community development and foster resident social impact. A memorable example of this is the governmental action taken by Bhutan's government to support residents and enforce active participation of residents in various aspects of their country's progress through their 'Gross National Happiness Index' (Govers, 2018). In the context of entrepreneurship, the 'Gross National Happiness Index' initiative encourages businesses that contribute to the well-being of society and the environment, rather than solely focusing on profit. This supports residents to create, join, and promote various entrepreneurial ventures that follow the principles of ethics, social responsibility, sustainability, and/or cultural preservation, leading to the country's overall well-being. Although not initiated by residents, the well-known initiative directs attention towards positive social changes and people's well-being, and Bhutan shape its reputation by having local residents involved in activities focused on well-being such as cultural heritage and natural environment preservation, community-based tourism and traditional handicraft.

### *Residents Leading Trends and Create Opportunities*

Although often residents are treated as the target of place branding and promotion campaigns and initiatives (Stoica et al., 2021), it is crucial to highlight that they can be the driving force of place branding value generation (Källström, 2016).

Källström (2016)'s study, focused on the creation of place values, demonstrates that residents play an essential role in both generating fundamental new place values and actively shaping and evolving existing place values. They are modifying existing place values by testing existing value propositions on a regular basis (e.g. the place value propositions emerged through official promotions done by place managers and authorities or by local business activities) and comparing them with their personal experiences leading to new co-created values of the place brand. This could explain why residents play a vital role in leading trends, and creating opportunities that have a profound impact on local and even global scales. Through their engagement in various spheres, including urban development, entrepreneurship, and community initiatives, residents often drive innovation and contribute to positive change. For example, the residents from the Portland, Oregon, have been engaged in numerous activities in the technology industry that led the place, particularly the Silicon Forest community, to be known globally for its innovation in technology. Residents were unconsciously involved in the internal branding processes through their actions: engineers, innovators, and entrepreneurs work collaboratively to create new businesses and lead technological advancements in the area. Their actions have resulted not only on successful place branding but the whole region's branding being recognised as an emerging technology hub with many successful start-up blooming. By fostering this culture of innovation and collaborative technological progress, the community attracted top global elites and talent in the area and investments, creating jobs and opportunities in the technological ecosystem. All of these solidified the place's reputation and region's identity as a centre for tech entrepreneurs and innovation driven fanatics.

Aside from the developments that residents can bring in the technology industry, there are far many other industries that bloomed through local entrepreneurs innovations that caught global audience's interest. An example of this is Asheville, North Carolina which got its reputation through the local community's involvement in craft brewing industry. The local entrepreneurs from Asheville have pioneered the city's reputation as a craft beer destination by establishing various microbreweries and brewpubs. The community's active involvement in local breweries led to the formation of the city's beer culture which in turn, has led to the development of a robust craft beer scene, attracting tourists and beer aficionados from across the country. Residents' actions in this case fostered a thriving hospitality industry, including pubs, restaurants, and hotels, thereby creating a diverse range of job opportunities for locals. Additionally, the collaborative spirit among residents and brewers has spurred innovation in brewing techniques, flavours, and community engagement, further enhancing Asheville's identity as a vibrant and creative city.

### *Residents Involvement in Heritage Conservation and Restoration*

Braun (2011) argues that the foundation of place brands lies in their historical origins, as opposed to corporate brands that can be quickly established through promotional marketing endeavours. A place's history then is intertwined with its heritage, which is frequently regarded as the primary tool in place branding (Ashworth and

Kavaratzis, 2011) because it has the potential to evoke strong emotions and develop authentic place identities. Entrepreneurial activities concentrated on preserving cultural heritage, narratives, and historical buildings and landmarks can help a place keep its history alive and sometimes even rebrand the place through enriching its history and authentic charm. By converting old structures into boutique hotels, museums, or cultural centres, residents can be involved in activities that will attract heritage tourists and promote the area's unique identity.

Residents' involvement in heritage conservation and restoration often signifies a deep sense of attachment and responsibility towards the preservation of the community's cultural legacy and can also stimulate tourism and promote cultural heritage tourism in the region. By preserving and showcasing historical landmarks, traditional architecture, and cultural artefacts, residents contribute to the development of cultural tourism offerings, attracting visitors who are keen to explore the rich history and cultural significance of the locality. For example, Shirakawa-go is a traditional mountain village in Japan and a UNESCO World Heritage known for its thatched-roof houses that were built, preserved and promoted with the help of residents. Local residents actively participate in cultural preservation efforts, offering guided tours, traditional craft demonstrations, and homestay experiences. All these activities were done mostly independently by local entrepreneurs whose dedication to preserving the village's unique heritage has attracted tourists seeking an authentic glimpse into traditional Japanese rural life (Kuroda, 2019).

### *Residents' Involvement in Eco-Friendly Initiatives and Sustainability*

Places and destinations can also involve residents in their activities when they aim for more sustainable outcomes and eco-friendly initiatives (Gajdošík et al., 2018). Sometimes, the sustainability aspect of the initiatives is deriving from the sense of accountability and responsibility residents have towards their place, its image and its environment. For example, the residents from Germany and Denmark have been leaders in renewable energy adoption by advocating, supporting and creating wind and solar energy projects and supporting sustainable energy production (Sijm et al., 2006). These initiatives are not only empowering local economy but also are minimising greenhouse gas emissions and advocating to the adoption of clean and renewable energy (Sijm et al., 2006).

Another example is the case of the local community in Portland, Oregon, who were driving the city's reputation as a sustainable and eco-friendly hub. Local entrepreneurs in collaboration with local policymakers have actively promoted sustainable practices within the place, fostering a culture of environmental consciousness and green living. The emphasis on sustainable urban development, green infrastructure, and eco-friendly initiatives has not only attracted like-minded residents but has also positioned Portland as a national leader in sustainable living and urban innovation. The city's commitment to eco-consciousness has resulted in the development of green spaces, bike-friendly infrastructure, and a thriving local food scene, all of which have contributed to the city's internal branding as a sustainable urban environment. According to Ind and Holm (2013) place sustainable development requires a place brand's concentration on its stakeholders needs and their relationships and connections, as these various stakeholders may collectively shape the place associations and co-create meaning for the place brand. In the case of Portland, these collaborations between residents, businesses, and policymakers in promoting sustainable living and urban development has not only enhanced the quality of life for Portland's inhabitants but has also drawn attention to the city's unique identity, reinforcing its position as a progressive and environmentally conscious metropolis.

### *Residents' Involvement in Gastronomic Excellence*

Culinary studies have primarily explored gastronomy within the context of tourism, encompassing event, experience, and heritage tourism, however, lately, some studies started to acknowledge its potential in place branding for place-making, focusing on visitors' reactions, their satisfaction with culinary experiences, the perceived authenticity and its contribution to the quality of a place (Visković, 2021). Residents are the custodians of culinary traditions passed down through generations and their commitment to maintaining and sharing family recipes and cooking techniques is helping places getting recognition for their gastronomic excellence. Residents can champion local food movements and supporting locally sources products, advocate for local food and their demands prompted local municipalities to support farmer markets and farm-to-table initiatives developing urban agriculture industry. One example of residents' involvement in gastronomic excellence through internal branding and innovation can be seen in the case of the local food movement in San Francisco, particularly the establishment of the Ferry Plaza Farmers Market. The market, founded by residents, local farmers, and food advocates, has become a pioneering force in the local food movement. Through their collective efforts, they have not only transformed the culinary landscape of the city but also fostered a culture of sustainability, community support, and innovation. The market serves as a medium for local producers to exhibit their goods, fostering a direct connection between consumers and the source of their food. The residents' active involvement has not only led to the market's success but has also enhanced the city's reputation as a hub for culinary innovation and sustainable practices, reinforcing San Francisco's internal branding as a leader in the food industry.

Another example is seen in Vizela, Portugal in which local community fostered an understanding and love for local cuisine through gastronomy events and the cultivation of slow food practices. The local community was supported by the authorities' strategic approach to position the place as a slow city, approach which has progressively laid the groundwork for a gastronomic attraction. The formulation of a signature dish for the city was discovered to contribute to the revival and re-evaluation of local expertise, while the backing of gastronomic culture by the local administration aided in fortifying regional identity, fostering an appealing tourist offering (Emmendoerfer et al., 2023). Consequently, the authors highlight the significance of residents in this developmental process and outlined the prerequisites that could facilitate the preservation and joint creation of typical dishes for tourism.

## Solutions and Recommendations

The previous section explored seven areas where residents could bring essential contributions to shape their place environment and communities. The examples above prove how entrepreneurial acts from various industries and disciplines can create, enhance and communicate a place identity. However, not all places are managing to foster an environment that will allow residents to develop these entrepreneurial acts successfully. There is a common view that internal branding through residents' involvement in place developmental activities can bring numerous advantages to the place brand, but there are also some critics of internal branding and residents' involvement. These critics emerge from the idea that when various participants contribute to the development of a brand, there is a potential for conflicts to arise from divergent perspectives and preferences among the various place stakeholders which might dilute the place messages or create confusing place images. However, it is worth mentioning that 'a city brand should be able to represent the different visions of multiple stakeholders and the authenticity of the place, both engaging internals and externals, who look for that authenticity' (Casais and Monteiro, 2019: 231). Hence, the entrepreneurial activities of the residents should not only follow a single route but represent what the place and local residents have to offer to the world. The entrepreneurial activities of the residents should reflect the multifaceted aspects of the place and its residents, showcasing the diverse offerings and unique characteristics that make the place distinct. As such, it implies that a balanced and inclusive approach is essential in incorporating the perspectives of the local community in the place branding process, thereby ensuring a comprehensive representation of the place's identity and values.

Another critique to internal branding and residents' involvement in such entrepreneurial acts is presented in the way local authorities perceive and treat residents. While residents are depicted as vital components in the establishment and communication of place branding, scholars discuss that the challenge lies in perceiving residents merely as targets of the place initiatives rather than active contributors (Lucarelli, 2019) and often they are not given the power to represent their local communities (Stoica et al., 2021). This one-sided perspective can limit the extent to which residents can participate in shaping the development and representation of their local communities within the broader context of place branding initiatives. Consequently, the potential for residents to actively contribute their insights, experiences, and cultural understanding to the branding process may be overlooked or undervalued. This asymmetrical power dynamic can also result in a lack of empowerment for residents, restricting their ability to authentically represent their communities. Therefore, the branding efforts might fail to fully capture the diverse and nuanced essence of the local culture, heritage, and identity, ultimately leading to a diluted representation of the place.

Often, in place branding practices, the inclusion of residents in the branding process proves more challenging than anticipated. According to Cassinger and Thelander (2018) participatory initiatives and residents' integration in place activities are significant subjects of current debates, but there is a dearth of literature exploring residents' behaviours in these activities. The authors also note that despite the well-meaning intentions, participatory cultures tend to eventually become 'exclusionary due to its selective recruitment of participants and vague aim' (p. 71) especially when these are initiated by local place managers or municipalities. Moreover, the majority of place initiatives are geared towards meeting the needs of external audiences rather than the needs of the local residents (Lichrou et al., 2018) which might make the internal branding process harder, might disengage residents from local activities, or even create oppositions to the local initiatives (Casais and Monteiro, 2019).

Insch and Stuart (2015) suggest that the absence of residents' active involvement in place branding can be attributed to the local leadership practices and residents' perceptions of the place governance. Should residents feel that they lack the agency to contribute, their inclination to engage with and participate in place activities diminishes. Conversely, when residents perceive themselves as empowered and heard, they tend to exhibit heightened dedication and accountability towards the place brand, becoming more inclined to engage in the brand development or communication process. Effective leadership is considered crucial in place branding, as it enables leaders to implement a coherent marketing strategy, oversee the brand structure and identity, and manage relationships with various stakeholders (Casais and Monteiro, 2019). This approach fosters an environment of trust and collaboration, encouraging active participation from all involved parties, enhancing residents' sense of responsibility and accountability towards the place brand and creating a more cohesive and impactful brand narrative.

## Future Research Directions

The chapter discussed how resident entrepreneurial acts lead to successful forms of internal branding and promotion of the destination for global tourists. By emphasising the active participation of residents and the community, the chapter underscores the critical role played by local entrepreneurs in shaping the destination's image and identity. Some ideas were mapped based on previous literature and most focused on highlighting the opportunities of resident entrepreneurship acts, without providing too much context on the challenges associated with residents' involvement or negative place meaning emerges from resident entrepreneurship acts, an areas which can be further developed.

Moreover, the seven areas covered in the resident entrepreneurial acts focused on resident and community's involvement and while the primary contributors were the residents, other stakeholders may have played an important role in these initiatives, however their engagement was not clarified by in this chapter. Thus, the chapter suggests the need for comprehensive research to delve into the broader scope of stakeholders' influence on internal branding initiatives and entrepreneurship acts. Understanding the dynamics of the contributions made by various stakeholders could offer a comprehensive perspective of the internal branding process and its overall impact on the destination. Considering the digital trends, more importance should be given to how residents entrepreneurial acts are affected by digitalisation and how resident digital footprint impact on the destination global image and reputation (Feng et al., 2023).

The chapter's discussions demonstrate the need for continued research and analysis to comprehensively understand the multifaceted nature of resident entrepreneurial acts, their impact on internal branding, and the role of various stakeholders and digitalisation in shaping the destination's global image and reputation. By addressing these aspects, future studies can be conducted to aim the understanding of the complex dynamics underlying successful internal branding strategies in the context of resident entrepreneurship acts and place branding.

## Conclusion

This chapter provides a comprehensive review that sheds light on the significant role of residents in shaping the branding and identity of their places through diverse entrepreneurial initiatives. Through a systematic investigation, the chapter has highlighted the essential role of residents' entrepreneurial acts on internal place branding, emphasising how their activities contribute to the overall identity, culture, and reputation of a place. The discussions showed that residents' entrepreneurial acts can help destination create strong images and worldwide reputations and internal branding was used as a tactic to empower residents to discover, shape, and empower the local resources and place offerings. The seven themes discussed look at how residents' entrepreneurial acts can be used in internal branding to shape a destination's identity and reputation and problematised some of the resident participatory issues in place branding. The seven themes elucidated the various opportunities that arise from residents' entrepreneurship acts, emphasising the potential for fostering community engagement, promoting cultural heritage, and generating sustainable economic development. By recognising the diverse opportunities associated with residents' entrepreneurial engagements, the chapter emphasises the significance of empowering local communities and fostering a conducive environment for entrepreneurial growth and innovation.

Nevertheless, several favourable results have arisen from residents' entrepreneurship acts and engagement in place-related endeavours. This underscores the significance of integrating residents into the decision-making processes of the place brand and supporting and empowering residents to take ownership of their place and use local resources to develop their communities. This chapter serves as a resource for students, researchers, and practitioners interested in understanding the intricate relationship between residents' entrepreneurial activities and internal place branding. By providing numerous examples of the positive impacts and opportunities associated with these acts, the chapter encourages further exploration and analysis, underscoring the need for continued research to develop more effective strategies that leverage the potential of resident entrepreneurship for holistic place development and branding. Thus, researchers and practitioners should not neglect residents' power and internal branding potential, rather it is crucial to diligently monitor and comprehensively assess the entire spectrum of activities involved in the creation of the brand meaning and examine the full picture of how place brand meaning emerges, and how it is communicated in the eyes of the locals who should be supported to shape the place reputation from within. This includes not only the promotional aspects but also the intricate interplay of local culture, community dynamics, and the residents' genuine experiences, all of which significantly contribute to the establishment of a well-rounded and authentic brand identity.

### Case Study

***Promoting Destination through Resident Entrepreneurship: A Fictional Case Study of Mountain Haven***

Introduction: Places and destinations are increasingly looking for enhancing their competitive image worldwide and internal branding is recognised as way of achieving this competitive advantage. This case presents a fictional scenario of Mountain Haven, a mid-sized mountain town nestled along the banks of a serene river. The town used internal branding approaches based on diverse resident entrepreneurship acts to shape its global reputation. The town has gained world-wide reputation due to the transformation made in the last decade in which a strategic blend of resident entrepreneurship and community involvement were used to offer a plethora of interactive activities and experiences for tourists and locals. The case investigates the strategies and outcomes of residents' entrepreneurial skills to brand and promote the destination while also posing some thought-provoking questions to stimulate deeper analysis.

***Mountain Haven's Entrepreneurial Transformation***

Mountain Haven was once a quiet town struggling to attract locals and tourists due to a poor economic environment. The place struggled to attract visitors but also residents due to being in a remote location and the absence of a cohesive branding strategy that would differentiate the place from their surroundings neighbourhood towns. However, once the local people decided to improve their community, they gather together to create a vision and a plan. Although this was not an easy task and required long-time commitments from the residents, local authorities and local businesses, with a collective vision and determination, residents harnessed their entrepreneurial skills to re-brand the place and add life into the destination. The following section summaries some of the strategies underpinning Mountain's Haven's success:

*Local Community Experiences*

Residents recognise that authenticity is a key driver for tourism experiences and embarked on creating unique tourist experiences that dwelled on local offerings. Being located in a remote place, residents have learned to create their personalised style of living, including the creation of artisanal goods. Local artists opened boutiques stores offering handcrafted personalised souvenirs that portray the town's cultural heritage. Local artisanal goods were them branded with the name of the place and sold both online and offline. This not only provided tourists with memorable experiences but also contributed to creating awareness of the place globally and preserving its identity.

*Culinary Experiences*

Taking advantage of being a remote location, locals had also developed a taste for unique foods and beverages. Residents started to provide mouth-watering gastronomic adventure to tourists using the local farms and serving traditional dishes with locally sourced ingredients. Embracing the abundance of nearby farms, residents curated a tantalising culinary journey and established the renowned 'Mountain Haven Farm-to-Table Café' that introduced a menu highlighting indigenous ingredients, such as farm-fresh vegetables, artisanal cheeses, and locally sourced meats. The local community is also considering to further commercialise its local foods through selling some of their artisanal cheeses online and partner with global brands to export their local products worldwide. Mountain Haven leveraged its rich culinary heritage by encouraging residents to open cafés and restaurants that outsource their products from local farms, enhancing not only the gastronomic tourism market but also supported the local economy.

*Community-Development Experiences*

Recognising authenticity as a key driver for global tourists, residents from Mountain Haven embarked on locally rooted activities and experiences by opening local artisan shops and boutique stores, making handcrafted souvenirs and goods that promote their local traditions and cultural heritage. One popular product that they started to sell is their handcrafted personalised wooden dolls created in local schools by young teenagers. These dolls are handcrafted from local woods and dressed in left-out re-cycled clothes collected through donations from the local communities. The creation of handcrafted personalised wooden dolls by young teenagers in local schools epitomised the community's investment in nurturing the next generation while fostering a culture of sustainability. Through the use of locally sourced woods and recycled fabrics, the residents not only celebrated the town's artisanal heritage but also instilled an appreciation for resourcefulness and environmental consciousness among the youth. This emphasis on sustainable practices within the community's cultural and educational initiatives.

*Activity-Based Tourism*

Residents tapped into their passions and skills to come up with ideas of promoting local traditions and locals' hobbies. Residents are offering a diverse range of activities, form Kayaking clubs to guided nature walks, and photography workshops led by locals who were intimately familiar with the area's natural beauty. For example, the local cultural centre organises interactive workshops where visitors could actively participate in traditional craft-making sessions, including pottery, weaving, and woodworking, under the guidance of skilled local artisans. In these activities, participants gain insights into the region's cultural practices, fostering a deeper appreciation for the town's artistic heritage. Moreover, for adrenaline-pumping enthusiasts, the local community is organising excursions in nature, catering to thrill-seekers and nature enthusiasts alike. Activities ranged from exhilarating rock-climbing expeditions to invigorating mountain biking trails, all guided by seasoned local experts well-versed in the terrain's intricacies, thus ensuring a safe yet exhilarating adventure for all participants. These experiences enable tourists to immerse themselves in the town's breathtaking natural landscapes. These activities not only engaged tourists but also fostered a sense of camaraderie between visitors and the community that generated word-of-mouth and digital content for the destination.

Conclusion: The Mountain Haven case study illustrates how resident entrepreneurship can be a transformative force in destination promotion. The town successfully integrated the passions, skills, and creativity of its residents to create a unique and appealing tourist destination. By fostering a profound sense of connection and attachment to their community, the residents of Mountain Haven effectively cultivated a robust and enduring bond with the place and its brand. Through their concerted efforts to create a welcoming and immersive destination, the residents not only strengthened their emotional ties to the place but also established a nurturing environment that facilitated the growth and prosperity of the town. This case study provides valuable insights into the power of community-driven initiatives and serves as an inspiration for future destination marketers and planners seeking sustainable and authentic ways to attract and engage tourists.

**Case Questions (Three Questions)**

1. How does Place Attachment Theory explain the role of resident entrepreneurship in destination promotion? Provide examples from Riverside Haven's case.
2. In what ways can resident entrepreneurship contribute to sustainable tourism development while maintaining the cultural integrity of a destination?
3. Beyond the activities mentioned in the case study, brainstorm three innovative resident-led initiatives that could be employed to promote a mountainous destination. How might these initiatives cater to the interests and desires of modern tourists while benefiting the local community?

**Key Terms**

- Place Branding: Is a process used to express 'the visual, verbal and behavioural expression of a place, which is embodied through the aims, communication, values and the general culture of the place's stakeholders and the overall place design' (Zenker and Braun, 2010: 5); in this chapter branding is an umbrella term for various place activities aimed to 'identify common ideas and directions for the future of the community and to produce collectively generated stories and visions' (Ashworth et al., 2015: 6).
- Internal Branding: Is a process of involving internal stakeholders, often employees in the creation and communication of the brand values and brand culture (Barros-Arrieta and García-Cali, 2021); in this chapter internal branding was discussed in relation to place internal branding that is a process of involving internal stakeholders, such as residents, in the creation and communication of a brand's values, culture, and identity, with the aim of creating a consistent internal and external image of the place brand.
- Stakeholders: Individuals or groups who have a vested interest in the success and development of a place, often including residents, businesses, policymakers, and community organisations.
- Cultural and Artistic Hubs: In this chapter, cultural and artistic hubs represent venues within a place that promote cultural and artistic activities, including art galleries, studios, theatres, and festivals, contributing to the enrichment of local culture and the enhancement of a place's cultural identity.
- Community-Led Tourism Initiatives: In this chapter, community-led tourism initiatives and the tourism initiatives that are often started bottom-up, driven by local residents to promote the place by showcasing the authentic cultural, social, and natural attributes of a place, often involving community participation in tourism management and promotion.
- Social Impact: In this chapter, social impact refers to the effect of residents' actions and initiatives on the community's well-being and development, encompassing efforts to address local or global issues, promote social enterprises, and support marginalised groups.
- Eco-Friendly Initiatives and Sustainability: In this chapter, eco-friendly initiatives and sustainability relates to the residents' efforts to promote sustainable practices and environmental consciousness within a place, including the adoption of renewable energy projects, green infrastructure, and eco-friendly policies to reduce the ecological footprint.
- Gastronomic Excellence: In this chapter, gastronomic excellence relates to the promotion and preservation of culinary traditions and local food movements within a place, highlighting residents' roles in advocating

for local food, sustainable agriculture, and supporting initiatives such as farmers' markets and farm-to-table programmes.

- Heritage Conservation and Restoration: In this chapter, heritage conservation and restoration relate to residents' involvement in activities that preserve and promote a place's cultural heritage, historical buildings, and landmarks.
- Power Imbalance: In this chapter, power imbalance relates to the uneven distribution of influence and decision-making authority among different stakeholders within a place, potentially leading to conflicts, exclusionary practices, and a lack of meaningful engagement of residents in place branding and development activities.

## References

Altman, I., & Low, S. M. (1992). *Place attachment.* New York and London: Plenum Press.

Ashworth, G. J., & Kavaratzis, M. (2011). Why brand the future with the past? The roles of heritage in the construction and promotion of place brand reputation. In M. F. Go, & R. Govers (Eds.), *International place branding yearbook 2011: Managing reputational risk* (pp. 25–38). Hampshire: Palgrave Macmillan.

Ashworth, G. J., Kavaratzis, M., & Warnaby, G. (2015). The need to rethink place branding. In M. Kavaratzis, G. Warnaby, & G. J. Ashworth (Eds.), *Rethinking place branding: Comprehensive brand development for cities and regions* (pp. 1–11). Cham, Switzerland: Springer International Publishing.

Barros-Arrieta, D., & García-Cali, E. (2021). Internal branding: Conceptualization from a literature review and opportunities for future research. *Journal of Brand Management, 28*(2), 133–151.

Baxter, J., Kerr, G. M., & Clarke, R. J. (2013). Brand orientation and the voices from within. *Journal of Marketing Management, 29*(9–10), 1079–1098.

Boisen, M., Terlouw, K., Groote, P., & Couwenberg, O. (2018). Reframing place promotion, place marketing, and place branding - Moving beyond conceptual confusion. *Cities, 80,* 4–11.

Braun, E. (2011). History matters: The path dependency of place brands. In M. F. Go, & R. Govers (Eds.), *International place branding yearbook 2011: Managing reputational risk* (pp. 39–46). Hampshire: Palgrave Macmillan.

Braun, E., Eshuis, J., & Klijn, E. H. (2014). The effectiveness of place brand communication. *Cities, 41,* 64–70.

Braun, E., Eshuis, J., Klijn, E., & Zenker, S. (2018). Improving place reputation: Do an open place brand process and an identity-image match pay off? *Cities, 80,* 22–28.

Campelo, A., Aitken, R., Thyne, M., & Gnoth, J. (2014). Sense of place: The importance for destination branding. *Journal of Travel Research, 53*(2), 154–166.

Casais, B., & Monteiro, P. (2019). Residents' involvement in city brand co-creation and their perceptions of city brand identity: A case study in Porto. *Place Branding and Public Diplomacy, 15*(4), 229–237.

Cassinger, C., & Thelander, Å. (2018). Spaces of identity in the city: embracing the contradictions. In M. Kavaratzis, M. Giovanardi & M. Lichrou (Eds.), *Inclusive Place Branding - Critical Perspectives on Theory and Practice,* (1st ed., pp. 70–81). Abingdon: Routledge.

Chang, K., Chen, H., & Hsieh, C. (2022). Effects of relational capital on relationship between place attachment and resident participation. *Journal of Community & Applied Social Psychology, 32*(1), 19–41.

Chen, N., Dwyer, L., & Firth, T. (2018). Residents' place attachment and word-of-mouth behaviours: A tale of two cities. *Journal of Hospitality and Tourism Management, 36*, 1–11.

Chen, N., & Šegota, T. (2015). Resident attitudes, place attachment and destination branding: A research framework. *Tourism and Hospitality Management, 21*(2), 145–158.

Choo, H., Park, S., & Petrick, J. F. (2011). The influence of the resident's identification with a tourism destination brand on their behavior. *Journal of Hospitality Marketing & Management, 20*(2), 198–216.

Davies, C. (2009). *Bristol public given right to decide whether graffiti is art or eyesore.* Retrieved 3 August 2023, from https://www.theguardian.com/artanddesign/2009/aug/31/graffiti-art-bristol-public-vote.

El Banna, A., & Stoica, I. S. (2021). From participation to transformation: The multiple roles of residents in the place brand creation process. In N. Papadopoulos, & M. Cleveland (Eds.), *Marketing countries, places, and place-associated brands* (pp. 97–113). Cheltenham: Edward Elgar Publishing Ltd.

Emmendoerfer, M., Almeida, T. C. d., Richards, G., & Marques, L. (2023). Co-creation of local gastronomy for regional development in a slow city. *Tourism & Management Studies, 19*(2), 61–60.

Fan, D. X. F., Buhalis, D., & Lin, B. (2019). A tourist typology of online and face-to-face social contact: Destination immersion and tourism encapsulation/decapsulation. *Annals of Tourism Research, 78*, 102757.

Feng, S., Berndt, A., & Ots, M. (2023). Residents and the place branding process: Socio-spatial construction of a locked-down city's brand identity. *Journal of Place Management and Development, 16*(3), 440–462.

Florek, M., & Insch, A. (2020). Learning to co-create the city brand experience. *Journal of International Studies, 13*(2), 163–177.

Freire, J. R. (2009). 'Local people' a critical dimension for place brands. *Journal of Brand Management, 16*(7), 420–438.

Gajdošík, T., Gajdošíková, Z., & Stražanová, R. (2018). Residents' perception of sustainable tourism destination development - A destination governance issue. *Global Business & Finance Review, 23*(1), 24–35.

Gao, H., Wang, T., Gu, S. (2022). A study of resident satisfaction and factors that influence old community renewal based on community governance in Hangzhou: An empirical analysis. *Land, 11*(9), 1421.

Go, F. M., & Govers, R. (2013). *International place branding yearbook 2012 - Managing smart growth and sustainability* (1st ed.). Hampshire: Palgrave Macmillan.

Govers, R. (2011). From place marketing to place branding and back. *Place Branding and Public Diplomacy, 7*(4), 227–231.

Govers, R. (2018). *Imaginative communities: Admired cities, regions and countries* (1st ed.). Antwerp: Reputo Press.

Heather Skinner. (2018). Who really creates the place brand? Considering the role of user generated content in creating and communicating a place identity. *Communication & Society, 31*(4), 9–25.

Hudson, S., Càrdenas, D., & Meng, F. (2017). Building a place brand from a bottom-up: A case study from the united states. *Journal of Vacation Marketing, 23*(4), 365–377.

Ind, N., & Holm, E. D. (2013). Beyond place branding. In F. M. Go, & R. Gover (Eds.), *International place branding yearbook 2012 - Managing smart growth and sustainability* (pp. 45–55). Hamphire: Palgrave Macmillan.

Insch, A. (2010). Managing residents' satisfaction with city life: Application of importance-satisfaction analysis. *Journal of Town & City Management, 1*(2), 164–174.

Insch, A., & Stuart, M. (2015). Understanding resident city brand disengagement. *Journal of Place Management and Development, 8*(3), 172–186.

Jeuring, J. H. G., & Haartsen, T. (2017). Destination branding by residents: The role of perceived responsibility in positive and negative word-of-mouth. *Tourism Planning & Development, 14*(3), 240–259.

Källström, L. (2016). Rethinking the branding context for municipalities: From municipal dominance to resident dominance. *Scandinavian Journal of Public Administration, 20*(2), 77–95.

Källström, L., & Siljeklint, P. (2023). Place branding in the eyes of the place stakeholders – Paradoxes in the perceptions of the meaning and scope of place branding. *Journal of Place Management and Development.* doi:10.1108/JPMD-12-2022-0124.

Kangjuan, L., Mosoni, G., Wang, M., Zheng, X., & Sun, Y. (2017). The image of the 2010 world expo: Residents' perspective. *Engineering Economics, 28*(2), 207–214.

Kavaratzis, M. (2004). From city marketing to city branding: Towards a theoretical framework for developing city brands. *Place Branding, 1*(1), 58–73.

Kavaratzis, M., Giovanardi, M., & Lichrou, M. (2018). *Inclusive place branding - Critical perspectives on theory and practice* (1st ed.). Abingdon: Routledge.

Kemp, E., Williams, K. H., & Bordelon, B. M. (2012). The impact of marketing on internal stakeholders in destination branding: The case of a musical city. *Journal of Vacation Marketing, 18*(2), 121–133.

Klijn, E. H., Eshuis, J., & Braun, E. (2012). The influence of stakeholder involvement on the effectiveness of place branding. *Public Management Review, 14*(4), 499–519.

Kuroda, N. (2019). Conservation design for traditional agricultural villages: A case study of shirakawa-go and gokayama in japan. *Built Heritage, 3*(2), 7–23.

Lichrou, M., Kavaratzis, M., & Giovanardi, M. (2018). Introduction. In M. Kavaratzis, M. Giovanardi, & M. Lichrou (Eds.), *Inclusive place branding - Critical perspectives on theory and practice* (pp. 1–10). Oxon and New York: Routledge.

Lucarelli, A. (2019). Constructing a typology of virtual city brand co-creation practices: An ecological approach. *Journal of Place Management and Development, 12*(2), 227–247.

Magala, S. (2011). Imagined identities of existing cities: The reputation game. In F. M. Go, & R. Govers (Eds.), *International place branding yearbook 2011 - Managing reputational risk* (pp. 12–24). Hampshire: Palgrave Macmillan.

Marine-Roig, E., & Ferrer-Rosell, B. (2018). Measuring the gap between projected and perceived destination images of catalonia using compositional analysis. *Tourism Management, 68*, 236–249.

Mohammadi, S. H., Norazizan, S., & Nikkhah, H. A. (2018). Conflicting perceptions on participation between citizens and members of local government. *Quality Quantity, 52*(4), 1761–1778.

Peighambari, K., Sattari, S., Foster, T., & Wallström, Å. (2016). Two tales of one city: Image versus identity. *Place Branding and Public Diplomacy, 12*(4), 314–328.

Seo, H. (2013). Online social relations and country reputation. *International Journal of Communication, 7*, 853–870.

Sijm, J., Neuhoff, K., & Chen, Y. (2006). $CO_2$ cost pass-through and windfall profits in the power sector. *Climate Policy, 6*(1), 49–72. doi:10.1080/14693062.2006.9685588.

Stoica, I. S., Kavaratzis, M., Schwabenland, C., & Haag, M. (2021). Place brand co-creation through storytelling: Benefits, risks and preconditions. *Tourism and Hospitality, 3*(1), 15–30.

Su, L., Huang, S. S., & Pearce, J. (2018). How does destination social responsibility contribute to environmentally responsible behaviour? A destination resident perspective. *Journal of Business Research, 86*, 179–189.

Uchinaka, S., Yoganathan, V., & Osburg, V. (2019). Classifying residents' roles as online place-ambassadors. *Tourism Management, 71*, 137–150.

Warnaby, G. (2009). Towards a service-dominant place marketing logic. *Marketing Theories, 9*(4), 403–423.

Zenker, S., & Braun, S. (2010). Branding a city: A conceptual approach for place branding and place brand management. In *The 39th EMAC Annual Conference 2010: The Six Senses: The Essentials of Marketing*. Frederiksberg: Copenhagen Business School.

Zenker, S., & Petersen, S. (2014). An integrative theoretical model for resident - City identification. *Environment & Planning A, 46*(3), 715–728.

Zenker, S., & Rütter, N. (2014). Is satisfaction the key? The role of citizen satisfaction, place attachment and place brand attitude on positive citizenship behavior. *Cities, 38*, 11–17.

Vaiou, D. (2018). Rethinking participation: Lessons from a municipal market in Athens. *Journal of Place Management and Development, 11*(2), 181–191.

Višković, R. N. (2021). Gastronomy as a social catalyst in the creative place-making process. *Acta Geographica Slovenica, 61*(1), 185–199.

# 10 Transport

## Digital Trends

*Pavlos Arvanitis*

### Introduction

This chapter explores the ways in which digital advancements in 'smart' travel aim to put people at the fore and centre of the design, development maintenance, and operation of digital assets and services by adapting transportation technologies more precisely to a diversity of human needs. The aim of this chapter is to familiarize the reader with the latest digital trends in transport.

### Background

There are different approaches, definitions, and interpretations of transport, although in general transport is nothing but the movement of goods or people from origin to destination (Sahu, 2017). While we would agree that there are different transport modes, depending on the geographical characteristics, the actual distance that needs to be travelled, and the actual transport mode used or available, there are distinct differences between long haul travel for instance, and door-to-door travel which include more than one transport mode to achieve the desired outcome. For example, when you commute to work you can either use your personal car, parked outside your home, drive to your place of work, and vice versa. Technically, there are two modes of transport used for this journey, walking and driving, since somehow you have to reach your car and to your place of work from the place you parked your car. Similarly, you could walk or cycle to a nearby bus stop or station, get on a bus, then on an underground metro, then on a shorter or longer distance train and continue on foot/bus/bicycle/tram/metro to your final destination (Rodrigue, 2020).

Similarly, everyday goods that are purchased from a local convenience store might have travelled a very long journey from the place of manufacture to a distribution or logistics centre and then to a regional or local one until they are delivered to the local convenience store before they are purchased (Rodrigue, 2020). The different volumes and sizes of products transported dictate different needs and different approaches. Increased congestion and mobility awareness have dictated different levels of digitalization, innovation, and entrepreneurship depending on which stage of transport people or goods are at (Schecther, 2002).

DOI: 10.4324/9781003369967-14

Digital trends in transport are rooted in the convergence of technology, data, and the transportation industry's drive to enhance efficiency, safety, and sustainability. There are early signs of computing and automation from the 1950s but the introduction of computers in the mid-20th century began to impact transportation, particularly in airline reservation systems and logistics (Molla et al., 2022). The Global Positioning System was introduced in the 1980s and became fully operational in the 1990s, enabling precise location tracking for vehicles. GPS technology revolutionized navigation and contributed to the development of route optimization systems for logistics and transportation planning. At the same time, the emergence of the internet in the same decade laid the foundation for digital connectivity in transportation. This led to the development of online booking systems for flights, trains, and buses became common, streamlining the ticketing process for passengers (Schecther, 2002). The introduction of mobile internet in conjunction with smartphones and social media has made this relationship even more complicated, direct, and personalized.

Since 2000 telematics systems, combining GPS and communication technologies, gained popularity in the commercial transport sector enabling real-time tracking of vehicles, driver monitoring, and remote diagnostics for maintenance (Mintsis et al., 2004). In the second decade of the 21st century, ride-sharing and on-demand transportation emerged when companies like Uber and Lyft emerged in the 2010s, introducing the concept of ride-sharing and on-demand transportation (Brodeur and Nield, 2018). These platforms leveraged mobile apps to connect passengers with drivers, disrupting the traditional taxi industry. The 2010s saw a significant increase in the development and adoption of electric vehicles (EVs) as a sustainable alternative to internal combustion engine vehicles. Digital innovations in battery technology, charging infrastructure, and range estimation improved the EV user experience. The development of autonomous or self-driving vehicles gained momentum in the 2010s, with companies like Tesla, Waymo, and traditional automakers investing in the technology (Kihm and Trommer, 2014). Autonomous vehicles rely heavily on digital sensors, machine learning, and real-time data processing to navigate and make decisions.

The rollout of 5G networks in the 2020s promised ultra-fast, low-latency connectivity, crucial for autonomous vehicles and real-time data sharing in transportation. The latter, in combination with existing technologies like GPS navigation, resulted in the emergence of smart cities and infrastructure (Santa et al., 2022). Cities around the world began adopting smart technologies to optimize traffic management, reduce congestion, and improve transportation efficiency. In addition, the increased pressure on, climate crisis and environmental sustainability, led to increased efforts to promote electric vehicles and other eco-friendly transportation options, Finally, the COVID-19 pandemic accelerated the adoption of digital trends in transport, as people sought contactless and online options for commuting and transportation services (Ahmedova, 2022; Quiros, 2020, WEF, 2021).

The background of digital trends in transport is marked by a continuous evolution driven by technological advancements, changing consumer preferences, environmental concerns, and the quest for more efficient and sustainable transportation

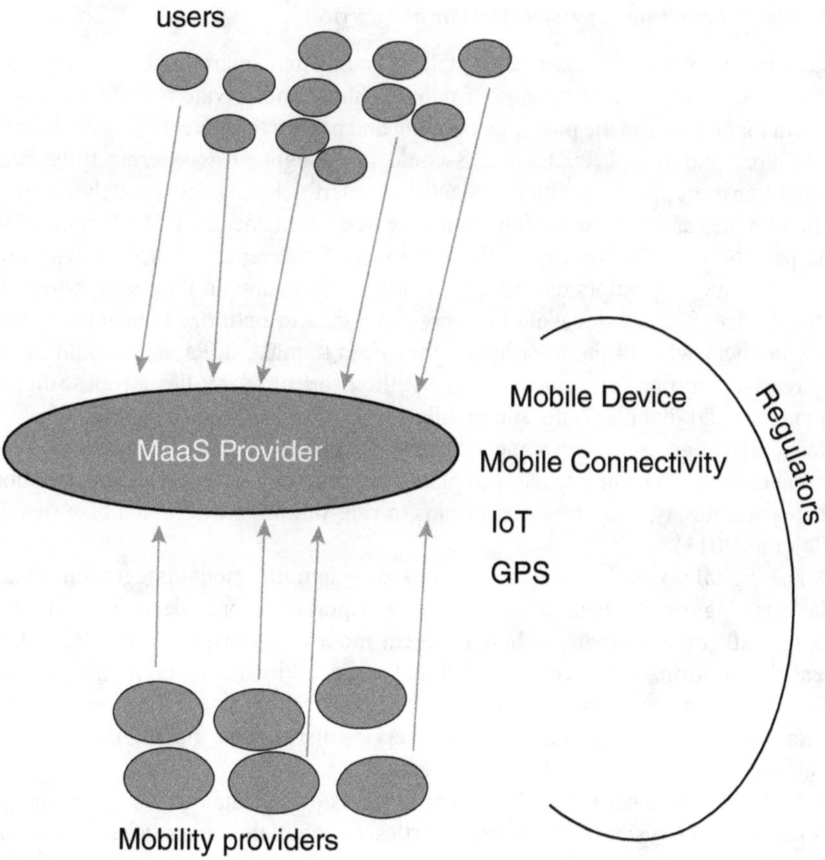

*Figure 10.1* Conceptual framework.

solutions (The Guardian, 2016). The future of transportation will likely continue to be shaped by digital innovation and the integration of emerging technologies (Figure 10.1).

## Mobility as a Service (MaaS) and Its Integration with Digital Trends in Transportation

The world of transportation is undergoing a profound transformation, driven by the rapid evolution of digital technologies. Among the most transformative concepts to emerge in recent years is Mobility as a Service (MaaS), a paradigm that seeks to revolutionize the way we access and utilize transportation services (Aries-Molinares and Garcia-Palomares, 2020). MaaS combines digital innovations with an integrated approach to transportation, promising seamless, convenient, and sustainable mobility for individuals and communities. In this section, an exploration of the intersection of MaaS and digital trends in transportation will be presented, showcasing how these developments are reshaping the way we move.

***The Digital Foundation of MaaS (Mobility as a Service)***

MaaS is a term used to describe digital services, often smartphone apps, through which people can access a range of public, shared and private transport, using a system that integrates the planning, booking and paying for travel (Aries-Molinares and Garcia-Palomares, 2020). MaaS could be thought of as offering truly integrated transport planning with the benefits of smart ticketing and through ticketing rolled-up into an easily accessible online service. Someone using MaaS could plan and pay for an entire journey, without the need for several transactions with different transport operators or multiple tickets using an app on their smartphone or other device. The service would use real-time data to optimize their journey and provide them with all the information they need to make it. Payment could be on a pay-as-you-go or subscription basis. At the heart of MaaS lies a robust digital foundation. Digital platforms and mobile applications serve as the cornerstones of MaaS, providing users with a one-stop-shop for all their transportation needs. This digital convergence enables users to plan, book, and pay for various transportation modes seamlessly, from buses and trains to ride-sharing services and bike rentals (Deloitte, 2017).

The digital revolution enables MaaS to excel in multi-modal integration. MaaS platforms aggregate data from diverse transportation providers, enabling users to craft journeys that combine different modes of transport (Deloitte, 2017). Real-time information, a product of digital connectivity, is pivotal in this context. Users can access up-to-date information on transit schedules, availability of transportation options, traffic conditions, and service disruptions, making informed decisions on the go.

Cashless payments within MaaS applications are a testament to the power of digital payment systems. Users can effortlessly complete transactions for various transportation services without the hassle of cash or multiple tickets. Furthermore, data analytics underpin MaaS platforms, analysing user behaviour, preferences, and traffic patterns (Aries-Molinares and Garcia-Palomares, 2020). This data-driven approach allows for better service optimization, route recommendations, and personalized travel experiences.

***Personalization, Smart Cities and Environmental Considerations***

Personalization is another hallmark of MaaS. By analysing user data, MaaS platforms can offer tailor-made transportation recommendations based on individual preferences and travel history. Moreover, MaaS applications incorporate environmental considerations, promoting eco-friendly transportation options when possible, contributing to sustainability in urban mobility. MaaS applications facilitate feedback and rating systems, allowing users to voice their opinions on transportation services (Aries-Molinares and Garcia-Palomares, 2020). This feedback loop is invaluable in enhancing the quality and reliability of transportation providers. Innovative pricing models are also a hallmark of digital MaaS. Subscription-based services, pay-as-you-go options, and bundled packages are all made possible through digital payment systems, making transportation more cost-effective for users.

In the context of smart cities, MaaS can be seamlessly integrated into urban planning and transportation management systems. This integration enables efficient traffic flow, reduced congestion, and improved mobility within cities (Lopez-Carreiro et al., 2023). Additionally, digital MaaS platforms enhance accessibility for individuals with disabilities by providing information on accessible routes and transportation options, thus fostering inclusivity.

The term “smart” has been adapted by several industries including tourism and transport. Smart tourism and transport refer to the dynamic fusion of human experiences with intelligent technologies. It is intricately tied to the advancement of Smart Cities and progresses in technological domains such as Artificial Intelligence, the Internet of Things (IoT), Big Data, and 5G. The primary aim of smart tourism is to enhance resource management efficiency, boost competitiveness, and promote sustainability by leveraging innovative technologies. Consequently, an increasing number of destinations are embracing this modernization in their operations, encompassing everything from payment systems to a diverse array of interactive activities (Lopez-Carreiro et al., 2023).

The surge in Smart Cities is exerting a notable influence on various sectors, including tourism, which is progressively transitioning towards a smart destination model. As previously mentioned, the growth of smart tourism is closely intertwined with the expansion of Smart Cities. These urban centres aspire to enhance the well-being of their residents while also creating more sustainable environments (Lopez-Carreiro et al., 2023). Consequently, smart tourism aligns with these objectives by offering more immersive and environmentally conscious experiences.

Given the strategic economic significance of tourism in numerous countries, this novel approach to city travel is gaining prominence. Several pivotal factors for the establishment of smart tourism within a Smart City include:

Cutting-edge infrastructure that ensures sustainable development and equitable accessibility.
Widespread availability of free Wi-Fi in public spaces and on the streets.
Adoption of electric mobility as a sustainable alternative to traditional transportation.
Promotion of eco-friendly tourism practices.
Provision of real-time information, such as traffic updates and public transport incidents.
Diverse cultural and interactive activities for visitors.

### *Regulatory Considerations*

The rise of MaaS presents challenges to existing regulatory frameworks. Issues such as data privacy, fair competition, and safety need to be addressed in the digital era. As MaaS disrupts traditional transportation models, governments and authorities are working to strike a balance between innovation and regulation to ensure that these transformative services benefit society as a whole. The case of sharing economy and/or circular economy can be experienced in the forms of Uber, AirBnB and food delivery services where real time tracking data enable users to make informed decisions on their choices (Metro-magazine, 2015). At the same time, public transport timetables are available in real time on several applications allowing passengers to make similar choices. The cases of AirBnB and food delivery services do not fully lie within the context of transport of people, however they are both anthropocentric, designed around the needs of people to be served in real time.

There are numerous countries around the world where regulatory constraints limit or even ban Uber operations following reactions from established taxi (or similar) companies. Countries like Switzerland, Denmark, Italy, Thailand, Hong Kong, Canada and Australia, do not allow partial or full operation of Uber services with all the repercussions that come with it. Reasons for these prohibitions range from alleged unfair competition to a lack of safety measures and problems with illicit dispatcher services. In Germany, for instance, a court in Cologne ruled that the app breaches a German law stipulating that taxis and similar services should be administered via a central dispatch office. However, Uber still operates in other German major cities, including Berlin and Munich.

### *Priorities and Trends in Digital Transport*

This section summarizes the key trends shaping the future of travel, since the pandemic forced individuals and businesses alike to re-evaluate their travel purposes. The ubiquitous adoption of remote work and online meetings has led to a significant reduction in business travel. According to McKinsey, there is a projected 20% long-term decrease in business travel, as the lines between business and leisure blur in the age of "work from anywhere" and "bleisure" travel. People are now seeking fewer, but longer trips, often driven by a desire for more meaningful experiences, less crowded destinations, and reuniting with loved ones. Furthermore, the crisis has prompted 53% of global travellers to commit to travelling more sustainably, recognizing the environmental impact of their journeys (Bushell et al., 2022).

The travel industry is making substantial strides towards carbon neutrality (Forbes, 2022). Airlines worldwide are committing to increased fuel efficiency, alternative fuels, new technologies, carbon capture, and carbon offset initiatives. Notably, half of the world's airlines have announced net-zero emissions goals in the past year. Companies like easyJet, British Airways' parent company IAG, and Carnival Cruises have made sustainability a core part of their mission, with initiatives ranging from carbon-neutral flights to substantial emissions reductions. Fleet restructuring, retiring older, fuel-inefficient aircraft, and adopting next-generation, more eco-friendly planes like the A350 and B787 are also contributing to reduced emissions per passenger.

The travel industry faces a post-pandemic reality where capacity shortages, rising ticket prices, and increasing awareness of emissions' impact on the environment drive travellers to make more environmentally responsible choices. Travel search engines, such as Sweden's Flygresor and Google, are beginning to display carbon emissions data during flight searches, allowing travellers to make informed decisions. Meanwhile, startups like Lilium, Vertical Aerospace, and Joby are poised to revolutionize urban mobility with all-electric vertical take-off and landing (eVTOL) aircraft, and Virgin Hyperloop aims to introduce levitation technology by 2027. Hydrogen-powered aircraft are also on the horizon, with companies like ZeroAvia and Universal Hydrogen leading the way.

Collaboration within the travel industry is vital for achieving sustainable goals. The Global Future Council for Sustainable Tourism is mobilizing industry partners to develop sustainable business models and establish industry-wide metrics to measure progress. Government support in Europe, tied to emissions reduction and domestic flying, is driving industry change. In the United States, rejoining the Paris Climate Accord and subsidizing biofuel production are signals of commitment to sustainability. Airlines like Lufthansa are working with partners to create green fuel supply chains, while Emirates has joined the Forum's Clean Skies for Tomorrow Coalition, charting a path towards carbon-neutral flying. System collaboration is advancing at an unprecedented pace (plainconcepts, 2022).

The pandemic has necessitated a pivot in careers and business strategies within the travel sector. Companies are reinventing themselves to adapt to new realities. Surf Air, originally a private aviation membership club in California, has transitioned into a provider of regional aircraft hybrid electric propulsion conversions. Lime, known for e-bikes, has transformed into a one-stop app for all electric vehicles. Frequent traveller programmes must also evolve to remain relevant, potentially incorporating eco-friendly decisions such as purchasing carbon offsets.

As the travel industry rebounds from the pandemic, it is undergoing a profound shift towards sustainability, driven by changing travel purposes, technological innovations, and collaborative efforts. The future of travel is poised to be more eco-conscious, with travellers and industry stakeholders alike embracing environmentally responsible choices (Noussan et al., 2020). This transformation reflects not only a response to current challenges but also a commitment to a more sustainable and resilient future for global travel.

In the ever-evolving landscape of transportation management, several technological innovations are poised to reshape the industry's future. This essay explores five key trends that are driving transformation in transportation: Artificial Intelligence (AI), Real-Time Visibility Solutions, Internet of Things (IoT), Time Slot Management, and Autonomous Vehicles, with a particular focus on autonomous trucks and last-mile deliveries (SAE; Forbes, 2022; International Finance, 2023).

Artificial Intelligence (AI) stands as a cornerstone of modern transportation management systems. It plays a crucial role in optimizing shipment planning by learning and adapting to various constraints, including capacity, regulations, and hours of service. AI's ability to analyse data results in more accurate Estimated Time of Arrival (ETA) predictions for shipments to warehouses, stores, and customers. Beyond ETAs, AI assists shippers in identifying carriers that consistently meet service level expectations and routes that are prone to delays. This technology empowers shippers to enhance efficiency while maintaining service quality.

Real-time visibility solutions have become indispensable for ensuring a positive customer experience. These tools enable tracking and monitoring of product movements at various stages, from warehouses to stores and end customers. While they are particularly effective for over-the-road (OTR) shipments, their importance extends to other modes of transportation as well. OTR visibility relies on integration with truck carriers' systems, while ocean shipments face unique challenges such as currents and wind speeds. Air cargo, with its predictability, benefits from real-time visibility tools that provide more accurate Estimated Time of Arrival (ETA) predictions.

IoT-enabled fleet management solutions offer versatility and visibility across diverse industries. These solutions enhance asset visibility, vehicle utilization, reduce wait times, and enable proactive maintenance cost savings. IoT technology is pivotal in addressing specific requirements within fleet management and integrated logistics, catering to the unique demands of different industries.

Time Slot Management is gaining prominence as an essential tool in organizing warehouse operations. It begins with accurate ETA predictions and extends to providing real-time updates on truck arrival, dock assignments, loading status, and documentation requirements. These applications are invaluable in mitigating the impacts of unforeseen disruptions, such as traffic delays or missed appointments, by efficiently rescheduling loadings and unloadings.

Last-mile delivery, a challenging and costly aspect of the supply chain, is poised for transformation through autonomous solutions. Drones have garnered attention for their potential to reduce delivery costs and enhance customer service. Companies like Wing, Amazon, UPS, Matternet, Flytrex, and Zipline are pioneering drone deliveries, with a particular focus on medical supplies and prescriptions.

Autonomous mobile robots are also gaining ground for last-mile delivery. While they are not entirely autonomous, they are remotely monitored, allowing for intervention in case of issues. Companies like Starship Technologies, Nuro, and FedEx are conducting pilot programmes and deploying delivery bots in various settings.

The transportation industry is at the cusp of a significant transformation, driven by technological innovations that promise to enhance efficiency, reduce costs, and improve sustainability. These innovations, ranging from AI-driven planning to autonomous last-mile deliveries, are poised to reshape how goods are transported, managed, and delivered. As these trends continue to evolve and gain acceptance, they hold the potential to usher in a more resilient and efficient future for the transportation sector.

In the modern business landscape, digital technologies have become an integral component of every organization. This fundamental shift, often referred to as "digital transformation," involves the conversion of previously analogue machine operations, service processes, organizational tasks, and managerial functions into digital formats (Pagani and Pardo, 2017, p. 185). This transformation is fundamentally altering how companies function, the products and services they provide, and their value-creation methods (Matarazzo et al., 2021; North et al., 2019).

Digital transformation serves as a catalyst for business enhancements (Fitzgerald et al., 2014) and empowers companies to optimize their operations, leading to improved operational efficiency and the collaborative generation of value (Taylor et al., 2020). Value co-creation refers to the outcome of both direct and indirect interactions, as well as the exchange and amalgamation of resources among various stakeholders (Jaakkola and Hakanen, 2013; Prenkert et al., 2019; Sandberg et al., 2018). It is recognized as an interactive process wherein different network participants actively contribute to value generation (Sjödin et al., 2016). This process extends beyond dyadic relationships since resources are distributed among diverse network participants, both directly and indirectly.

Direct and indirect interactions are thus essential elements of value co-creation, as they facilitate the exchange and combination of resources dispersed among network participants. Nevertheless, in certain cases, these interactions may lead to value co-destruction, a concept that denotes situations where the exchange of resources and activities among collaborating entities results in adverse value experiences (Järvi et al., 2018, p. 77).

Even traditional businesses that primarily produce analogue physical products are now confronted with the necessity of integrating digital services or software into their core offerings. This transformation necessitates changes in internal processes and business methodologies (North et al., 2019; Vial, 2019). Nevertheless, it's crucial to acknowledge that not every company can execute a comprehensive digital transformation. For traditional enterprises heavily reliant on manual labour and analogue procedures, digital transformation frequently engenders ongoing tensions and challenges. The coexistence of traditional analogue processes and newly introduced digital processes presents a significant and persistent challenge that impacts value outcomes over time (Vial, 2019). While prior research has delved into disruptive changes occurring during digital transformation (Nambisan et al., 2019; Vial, 2019), the topic of value co-creation, where traditional and digital activities are intricately intertwined, remains largely unexplored. Existing studies have predominantly focused on discussing the characteristics and distinctions between traditional and digital companies (Bosch and Olsson, 2021; Tekic and Koroteev, 2019), the alterations to traditional business models brought about by digitalization (Øiestad and Bugge, 2014), and the dynamic capabilities necessary for digital transformation within traditionally operating industries (Matarazzo et al., 2021; Warner and Wäger, 2019).

Current research seldom addresses the challenge of simultaneous coexistence of traditional and digital processes and resources, nor does it examine how this coexistence influences a company's ongoing operations and the resultant value creation (with exceptions including Matarazzo et al., 2021; Øiestad and Bugge, 2014; Tekic and Koroteev, 2019; Warner and Wäger, 2019).

User-Centric Innovation: Smart travel focuses on user-centric innovation, which means placing the user at the centre of the design process. This involves conducting user studies, usability testing, and incorporating user feedback to iterate and improve digital assets and services continually. By understanding user wants and needs, travel technologies can adapt and evolve to provide a better user experience.

Sustainable transportation systems are intended to facilitate the movement of people and goods while also fostering social inclusion and promoting balanced urban development (Elias and Shiftan, 2012; Gudmundsson, 2004; Miranda and Rodrigues da Silva, 2012). Public transportation is often seen as a solution to address urban mobility challenges, and an effective and user-friendly public transportation system should consider factors such as the accessibility to public transportation stations, the efficiency of the public transportation system, and its seamless integration with other transportation modes, such as railways, subways, light rail transit, and buses (Mishra et al., 2012). However, despite the potential of public transportation to contribute to transportation sustainability, many urban travellers opt not to use it (Gabrielli et al., 2014). This is often due to inadequate planning in terms of accessibility, the effectiveness of public transportation, and the convenience of transferring between different modes of transportation, which can lead to reduced ridership and increased reliance on private vehicles (Welch and Mishra, 2013).

Accessibility can be quantified by measuring the distance between residences and public transport stops or by assessing the duration of a journey from one's home to work using public transportation (Handy and Niemeier, 1997). However, research on the accessibility of public transportation remains limited (Mavoa et al., 2012). Individuals with limited mobility, such as youth and the elderly, require a reasonable level of accessibility to reach their destinations. Urban travellers are also concerned with the effectiveness of public transportation services, where effectiveness is defined as the ability to travel conveniently. The frequency of service at specific locations is the most common way to assess effectiveness (Sanchez et al., 2004). While intermodal transfers are often necessary in multimodal transportation systems (Vuchic, 2006), inconvenient transfers can decrease user satisfaction, discourage potential riders, and reduce a system's competitiveness (Wardman et al., 2001). Seamless connectivity is essential to enhance the performance of multimodal transportation systems to meet passengers' demands (Hadas and Ranjitkar, 2012). However, the application of connectivity measures to public transportation is infrequent (Mishra et al., 2012).

Previous studies that measured accessibility, effectiveness, and connectivity primarily focused on Geographic Information System (GIS)-based public transit networks (O'Sullivan et al., 2000; Mavoa et al., 2012; Tribby and Zandbergen, 2012). Comprehensive measurement indicators that assess accessibility, effectiveness, and connectivity throughout an entire public transportation service chain from the perspective of urban travellers are seldom found in the literature. The measurement approach proposed in this chapter serves as a tool that urban planners and policymakers can utilize to assess urban travellers' perceptions of accessibility, effectiveness, and connectivity. This approach allows for the identification and prioritization of underperforming scenarios. Additionally, this study takes into consideration the diversity in passenger behaviour (Bushell et al., 2022; Celic, 2020).

The literature evaluates the latent characteristics of urban travellers, including their perceptions of accessibility, effectiveness, and connectivity, and employs the Rasch method, a psychometric technique, to mitigate the biases introduced by ordinal scales through logistic linear transformations. The Rasch model is advantageous as it enables the comparison of individual parameters with item parameters, which can be transformed logarithmically along a logit scale to highlight challenges in service scenarios that certain urban travellers may find difficult to overcome (Lyons et al., 2020). Furthermore, variations in the socio-economic characteristics of urban travellers influence their perceptions of accessibility, effectiveness, and connectivity. An effective strategy can be developed to alleviate perceived travel challenges for different segments of urban travellers. The analytical outcomes generated by the Rasch model offer several practical implications for transportation agencies, urban planners, and policymakers seeking to enhance public transportation services to meet the needs of diverse passenger groups.

Overall, the goal of digital advancements in smart travel is to create a more user-focused, inclusive, and sustainable transportation system by adapting transportation technologies to meet a diverse range of human wants and needs.

## Solutions and Recommendations

Digital transformation in transport has been in line with technological developments since operational excellence has always been in the forefront of several transport applications and modes. The implications of implementing MaaS suggest that a re-organization of government structures (House of Commons Transport Committee, 2018), in regional and local levels in particular, need to be introduced. Such changes have to be introduced in the long term and short term solutions might not be applicable or well received. At the same time, it seems that the end users tend to use a combination of public and private mobility services which needs to be considered when planning for future mobility systems and networks. The mobility trends suggest though that handheld mobile devices and real time data are crucial for the selection of the mobility service (Shaheen & Lipman, 2023). Low-emission vehicles, including public transport ones like hybrid, electric and hydrogen will be introduced even further in all transport modes, passenger and cargo, despite criticism regarding the electricity-generating sources and their carbon footprint. Autonomous vehicles at the same time might take more time to be introduced commercially due to regulatory implications primarily and safety concerns secondarily. The pace of the regulatory reform differs significantly from country to country and region to region, despite genuine efforts to introduce and regulate these transport modes. Finally, it is evident that the demand for personalized services will increase further which might lead to the further expansion of on demand taxi services similar to Uber, resulting in increased competition and more choices for passengers.

## Future Research Directions

There are several future research directions that can be addressed, however the main ones, linked to the concept and theme of the current book are related to cybersecurity threats, data privacy, and infrastructure costs, the regulatory and policy changes necessary to support the adoption of emerging technologies, including but not limited to the interaction with people, like the implementation and use of autonomous vehicles and the impact on jobs in the transportation sector. Insufficient automation design often leads to conflicts, mistakes and accidents, and human factors in most cases play a crucial role in the development of these events. The role of human factors needs to be measured and assessed as essential for safety and efficiency, but in a supervisory role rather than a constant decision making one.

The latter raises another direction for future research linked to increased safety through automation in general which can potentially lead to fewer human generated failure.

At the same time, the emergence and adaptation of the regulatory and policy framework over these developments tends to be significantly more challenging and time consuming than originally anticipated. There is a lot of scepticism towards the implementation and regulation of these developments in several transport modes, despite positive steps that have already taken place. Similarly, data privacy, data use and storage has significantly increased threats linked or related to cybersecurity.

## Conclusion

Mobility as a Service represents a pivotal milestone in the evolution of transportation, enabled by the digital trends that have defined the 21st century. Through digital platforms, real-time information, payment integration, and data analytics, MaaS reimagines mobility, offering a user-centric, efficient, and sustainable approach to transportation. As digital technologies continue to evolve, MaaS is poised to play an even more prominent role in shaping the future of urban mobility, making our cities and lives more accessible, efficient, and environmentally responsible. Drivers and barriers need to be re-evaluated taking into consideration success stories and factors in order to achieve the desired outcomes.

### Case Study

***Case Study: The Gig Economy Dilemma – Legal Implications and Tax Windfalls***

The gig economy, marked by its dynamic and transient workforce, has become a transformative force in the modern labour market. Gig workers, those engaged in short-term jobs or 'gigs,' often straddle the line between traditional full-time employment and freelance work. This comprehensive case study explores the intricate legal landscape surrounding the gig economy, examining the evolving employment status of gig workers and its profound implications for companies, government, and the broader economy. Additionally, it delves into the workforce trends shaping this unique labour market and highlights the economic impact of the gig economy.

***Legal Challenges in the Gig Economy***

The legal status of gig workers has emerged as a focal point of debate and regulation within employment law. In recent years, several landmark rulings have redefined their classification, significantly affecting their entitlement to employment rights. One of the most notable cases involved Uber, the ride-sharing behemoth. The decision handed down in this case declared that Uber drivers should be considered workers rather than self-employed contractors. This groundbreaking verdict extended essential rights such as the national living wage and holiday pay to these drivers, challenging the prevailing notion that gig work exists as an entirely separate category of employment.

This shift in classification poses multifaceted operational challenges for gig economy companies like Uber. Traditionally, these platforms have operated under a business model where workers assumed the financial burden and responsibilities associated with their work. For instance, Uber drivers are currently responsible for their vehicles, fuel expenses, insurance, and maintenance, all while paying a 25% service fee to the platform. If the trend of reclassification continues, these companies may need to reevaluate their roles in supporting equipment maintenance, worker benefits, and overall well-being.

Furthermore, the gig economy's appeal lies partly in the flexibility it offers to workers. However, recent court decisions indicating that gig workers may be classified as "workers" or "employees" could result in a substantial 10% increase in payroll costs for companies. This shift may necessitate significant alterations in business strategies to maintain competitiveness while adhering to evolving labour regulations.

### *Tax Windfall for the UK Government*

The reclassification of gig workers also has far-reaching implications for government revenue. If more gig workers are classified as employees or workers, rather than self-employed individuals, employer national insurance contributions will rise. This shift could serve as a much-needed counterbalance to the UK government's projected £6 billion shortfall in national insurance revenue by the next decade. This impending financial gap, driven in part by the exponential growth of self-employment, notably accelerated by gig economy platforms like Uber, underscores the significant role the gig economy plays in the national economy.

### *Workforce Trends and the Future of Gig Work*

Beyond legal and financial aspects, the gig economy's evolution is profoundly influenced by broader workforce trends. An increasingly digitized landscape has facilitated the rapid growth of gig work, allowing workers to connect with opportunities more seamlessly than ever before. The advent of user-friendly platforms and apps has democratized gig work, making it accessible to individuals from diverse backgrounds.

Moreover, the COVID-19 pandemic accelerated the adoption of remote work and gig work, with many individuals seeking additional income sources through online platforms. This shift brought about a transformation in the gig economy, attracting a more diverse array of workers, including professionals in fields such as information technology, marketing, and consulting. These skilled individuals are now turning to gig work as a means to leverage their expertise independently. This trend is transforming the gig economy into a multi-faceted marketplace, offering opportunities for a broader range of professionals and blurring the lines between traditional employment and gig work.

Companies are also adapting to this changing landscape. Rather than relying solely on traditional employment models, they are increasingly incorporating gig workers into their workforce strategies. Gig workers provide flexibility, scalability, and access to specialized skills when needed, allowing businesses to adapt swiftly to changing market conditions.

### *Economic Impact of the Gig Economy*

The gig economy is not only reshaping the labour market but also leaving a significant economic footprint. It contributes to economic growth by providing opportunities for individuals to earn income and stimulating consumer spending. In the United Kingdom alone, gig economy workers contribute billions of pounds to the economy annually.

Additionally, the gig economy fosters entrepreneurship and innovation. It enables individuals to test business ideas, launch start-ups, and access a global customer base through platforms and digital tools. Many successful businesses today began as small gigs on platforms like Etsy, TaskRabbit, or Upwork. This entrepreneurial spirit has far-reaching implications for economic dynamism and job creation.

However, the gig economy also poses challenges to traditional labour markets. The rise of gig work has led to concerns about job security, income volatility, and the erosion of traditional employment benefits such as healthcare and retirement plans. Policymakers face the complex task of balancing the benefits of gig work with the need to protect workers' rights and financial stability.

## Conclusion

The gig economy's intricate legal landscape, fiscal implications, evolving workforce trends, and economic impact highlight the need for a comprehensive approach to address the complex challenges and opportunities it presents. As the gig economy continues to evolve, it will remain essential to strike a balance between regulatory measures that protect workers and provide clarity for companies while harnessing the economic potential and innovation it offers. The gig economy's impact on the future of work and the broader economy is undeniable, making it a critical area of focus for policymakers, businesses, and workers alike.

**Case Study Questions**

1. How have recent legal rulings, such as the Uber case, redefined the employment status of gig workers, and what are the implications for their rights and benefits?
2. What operational challenges do gig economy companies face as a result of these legal reclassifications, and how might this impact their business strategies and competitiveness?
3. What are the economic impacts of the gig economy, both in terms of its contributions to the economy and the challenges it poses to traditional labour markets, and how can policymakers address these challenges while fostering innovation and entrepreneurship?

**Key Terms and Definitions**

**Autonomous Vehicle**: There are two levels of driving automation according to the Society of Automotive Engineers (SAE) depending on the level of human/driver interaction with the vehicle. In the first level, the human monitors the driving environment, whereas in the second, the automated system monitors the driving environment. Within each level there are different degrees of automation, ranging from driver assistance to full automation (SAE.org, n.d.).

**Digital Trends**: The identification of new techniques, including trends, which use the digital ecosystem, including the internet, to execute several tasks that were previously done without or limited use of internet and automation (IGI Global, 2020).

**Digitalization**: Integration of digital technologies into everyday life by the digitization of everything that can be digitized (IGI Global, 2018).

**Electric Vehicle**: According to the British Houses of Parliament, an Electric Vehicle (EV), is more efficient compared to internal combustion-powered vehicles and produces less emissions. However, the total emissions of the vehicle depend on how the electricity is generated (Houses of Parliament, 2010).

**Maas**: Mobility as a Service refers to the integration of various forms of transport services into a single mobility service, accessible on demand (World Bank, 2022).

## References

Ahmedova, S. (2022). Covid-19 impact upon the digitalization of the transport sector in Bulgaria. *Transportation Research Procedia*, 63, 809–816. Doi:10.1016/j.trpro.2022.06.077.

Aries Molinares, D., Garcia-Palomares, J. C. (2020). The Ws of MaaS: Understanding mobility as a service fromaliterature review. *IATSS Research*, 44(3), 253–263. Doi:10.1016/j.iatssr.2020.02.001.

Bosch, J., Olsson, H.H. (2021). Digital for real: a multicase study on the digital transformation of companies in the embedded systems domain, *Journal of Software: Evolution and Process*, 33(5), 1–25.

Brodeur, A., Nield, K. (2018). An empirical analysis of taxi, Lyft and Uber rides: Evidence from weather shocks in NYC. *Journal of Economic Behavior & Organization*, 152, 1–16. Doi:10.1016/j.jebo.2018.06.004.

Bushell, J., Merkert, R., Beck, M. (2022). Consumer preferences for operator collaboration in intra- and intercity transport ecosystems: Institutionalising platforms to facilitate MaaS 2.0. *Transportation Research Part A*, 160, 160–178.

Celic, L., Magjarevic, R. (2020). Seamless connectivity architecture and methods for IoT and wearable devices. *Automatika*, 61(1), 21–34.

Deloitte Review (2017). The rise of mobility as a service. Issue 20.

Elias, W., Shiftan, Y. (2012). The influence of individual's risk perception and attitudes on travel behavior. *Transportation Research Part A: Policy Practice,* 46(8), 1241–1251.

Fitzgerald, M., Kruschwitz, N., Bonnet, D., Welch, M. (2014). Embracing Digital Technology; a new strategic imperative. Available online at: https://emergenceweb.com/blog/wp-content/uploads/2013/10/embracing-digital-technology.pdf

Forbes (2022). Top 5 transportation technology trends for 2023. Available online at: https://www.forbes.com/sites/stevebanker/2022/12/16/top-5-transportation-technology-trends-for-2023/.

Gabrielli, S., Forbes, P., Jylhä, A., Wells, S., Sirén, M., Hemminki, S., Jacucci, G., et al. (2014). Design challenges in motivating change for sustainable urban mobility. *Computers in Human Behavior*, 41(1), 416–423.

Gudmundsson, H. (2004). Sustainable transport and performance indicators. In: Hester, R.E., Harrison, R.M. (Eds.), *Transport and the Environment—Issues in Environmental Science and Technology* (pp. 35–63), vol. 20. Royal Society of Chemistry, Cambridge.

Hadas, Y., Ranjitkar, P. (2012). Modeling public-transit connectivity with spatial quality-of-transfer measurements. *Journal of Transport Geography*, 22, 137–147.

Handy, S.L., Niemeier, D.A. (1997). Measuring accessibility: an exploration of issues. In: Harrison, R.M. (Ed.), *Transport and the Environment—Issues in Environmental Science and Technology* (pp. 35–63), vol. 20. Royal Society of Chemistry, Cambridge.

House of Commons Transport Committee (2018). Mobility as a service.

Houses of Parliament (2010). Electric vehicles. Available online at: https://www.parliament.uk/globalassets/documents/post/postpn365_electricvehicles.pdf.

IGI Global (2018). What is digitalisation. Available online at: https://www.igi-global.com/dictionary/it-strategy-follows-digitalization/7748.

IGI Global (2020). What is digital trends. Available online at: https://www.igi-global.com/dictionary/digital-trends/76474.

International Finance (2023). Transportation sector 2023: Five trends to look out for. Available online at: https://internationalfinance.com/transportation-sector-2023-five-trends-look-out-for/#:~:text=The%20smaller%20trends%20include%20infrastructure,mile%20connectivity%20and%20micro%2Dmobility.

Jaakkola, E., Hakanen, T. (2013), Value co-creation in solution networks, *Industrial Marketing Management*, 42(1), 47–58.

Järvi, H., Kähkönen, A.-K., Torvinen, H. (2018). When value co-creation fails: reasons that lead to value co-destruction. *Scandinavian Journal of Management*, 34(1), 63–77.

Kihm, A., Trommer, S. (2014). The new car market for electric vehicles and the potential for fuel substitution. *Energy Policy*, 73, 147–157. Doi:10.1016/j.enpol.2014.05.021.

Lopez-Carreiro, I., Monzon, A., Lopez, E. (2023). MaaS implications in the smart city: A multi-stakeholder approach. *Sustainability*, 15(14). Doi:10.3390/su151410832.

Lyons, G., Hammond, P., Mackay, K. (2020). Reprint of: The importance of user perspective in the evolution of MaaS. *Transportation Research Part A*, 131, 20–34.

Matarazzo, M., Penco, L., Profumo, G., Quaglia, R. (2021). Digital transformation and customer value creation in Made in Italy SMEs: A dynamic capabilities perspective. *Journal of Business Research*, 123, 642–656. Doi: 10.1016/j.jbusres.2020.10.033

Mavoa, S., Witten, K., McCreanor, T., O'Sullivan, D. (2012). GIS based destination accessibility via public transit and walking in Auckland, New Zealand. *Journal of Transport Geography*, 20(1), 15–22.

Metro-magazine (2015). The 'Uber Effect': Will new ride services reinvent transit. Available online at: https://www.metro-magazine.com/mobility/article/410225/the-uber-effect-will-new-rideservices-reinvent-transit.

Mintsis, G., Basbas, S., Papaioannou, P., Taxiltaris, C., Tziavos, I. N. (2004). Applications of GPS technology in the land transportation system. *European Journal of Operational Research*, 152(2), 399–409. Doi:10.1016/S0377-2217(03)00032-8.

Miranda, H.D.F., Rodrigues da Silva, A.N., (2012). Benchmarking sustainable urban mobility: the case of Curitiba, Brazil. *Transporation Policy*, 21, 141–151.

Mishra, S., Welch, T.F., Jha, M.K. (2012). Performance indicators for public transit connectivity in multi-modal transportation networks. *Transportation Research Part A: Policy Practice,* 46(7), 1066–1085.

Molla, A., Duan, S. X., Deng, H., Tay, R. (2022). The effects of digital platform expectations, information schema congruity and behavioural factors on mobility as a service (MaaS) adoption. *Information Technology & People*. Doi:10.1108/ITP-03-2022-0226.

Nambisan, S., Wright, M., Feldman, M. (2019). The digital transformation of innovation and entrepreneurship: progress, challenges and key themes. *Research Policy*, 48(8), 1–1.

North, K., Aramburu, N., Lorenzo, O.J. (2019), Promoting digitally enabled growth in SMEs: a framework proposal. *Journal of Enterprise Information Management*, 33(1), 238–262. Doi: 10.1108/JEIM-04-2019-0103

Noussan, M., Hafner, M., Tagliapietra, S. (2020). *The future of transport digitalization and decorbanization; trends, strategies and effects on energy consumption*. Springer Open.

O'Sullivan, D., Morrison, A., Shearer, J. (2000). Using desktop GIS for the investigation of accessibility by public transport: an isochrone approach. *International Journal of Geographical Information Science*, 14(1), 85–104. Doi: 10.1080/136588100240976

Øiestad, S., Bugge, M.M. (2014). Digitisation of publishing: exploration based on existing business models, *Technological Forecasting and Social Change*, 83(1), 54–65.

Pagani, M., Pardo, C. (2017). The impact of digital technology on relationships in a business network. *Industrial Marketing Management*, 67, 185–192.

Plainconcepts (2022). Smart tourism: The future of the sector is technological. Available online at: https://www.plainconcepts.com/smart-tourism/.

Prenkert, F., Hasche, N., Linton, G. (2019). Towards a systematic analytical framework of resource interfaces, *Journal of Business Research*, 100, 139–149.

Quiros, T. (2020). COVID-19 brought urban transport to its knees. Digital technology will put it back on its feet. World Bank Blogs. Available online at: https://blogs.worldbank.org/transport/covid-19-brought-urban-transport-its-knees-digital-technology-will-put-it-back-its-feet.

Rodrigue, J. P. (2020). *The geography of transport systems*. Routledge. ISBN 978-0-367-36463-2. Doi:10.4324/9780429346323.

SAE.org (n.d.). What vehicle automation means to SAE International. Available online at: https://www.sae.org/what-is-automated-and-unmanned/.

Sahu, K. K. (2017). *Handbook of research on economic, financial, and industrial impacts on infrastructure development*. IGI Global. Doi:10.4018/978-1-5225-2361-1.ch010.

Sanchez, T.W., Shen, Q., Peng, Z.-R. (2004). Transit mobility, jobs access and low income labour participation in US metropolitan areas. *Urban Studies*, 41(7), 1313–1331.

Sandberg, E., Pal, R., Hemilä, J. (2018). Exploring value creation and appropriation in the reverse clothing supply chain, *The International Journal of Logistics Management*, 29(1), 90–109.

Santa, J., Katsaros, K., Bernal-Escobedo, L., Zougari, S., Miranda, M., Castaneda, O., Dalet, B., Amditis, A. (2022). Evaluation platform for 5G vehicular communications. *Vehicular Communications*, 3. Doi:10.1016/j.vehcom.2022.100537.

Schecther, D. (2002). *Delivering the goods: The art of managing your supply chain*. Wiley. ISBN 978-0471211143.

Shaheen, S., & Lipman, T. (2023). MaaS and the future of mobility: A review of the literature. *Transport Reviews*, 43(2), 272–293.

Sjödin, D.R., Parida, V., Wincent, J. (2016). Value cocreation process of integrated product-services: effect of role ambiguities and relational coping strategies, *Industrial Marketing Management*, 56, 108–119.

Taylor, S.A., Hunter, G.L., Zadeh, A.H., Delpechitre, D., Lim, J.H. (2020), Value propositions in a digitally transformed world, *Industrial Marketing Management*, 87, 256–263.

Tekic, Z., Koroteev, D. (2019). From disruptively digital to proudly analog: a holistic typology of digital transformation strategies, *Business Horizons*, Elsevier Ltd., 62(6), 683–693.

The Guardian (2016). Four of world's biggest cities to ban diesel cars from their centres. Available online at: https://www.theguardian.com/environment/2016/dec/02/four-of-worlds-biggest-cities-to-bandiesel-cars-from-their-centres.

Tribby, G., Zandbergen, P. (2012), High-resolution spatio-temporal modelling of public transit accessibility, *Applied Geography*, 34, 345–355. Doi: 10.1016/j.apgeog.2011.12.008

Vial, G. (2019). Understanding digital transformation: a review and a research agenda, *Journal of Strategic Information Systems*, 1, 13–66.

Vuchic, V.R. (2006). *Urban Metro Systems and Technology*. John Wiley & Sons, Hoboken, NJ.

Wardman, M., Hine, J., Stradling, S. (2001). *Interchange and Travel Choice*, Volume 1. Scottish Executive Central Research Unit, Edinburgh, Scotland.

Warner, K.S.R., Wäger, M. (2019). Building dynamic capabilities for digital transformation: an ongoing process of strategic renewal. *Long Range Planning*, 52(3), 326–349.

Welch, T.F., Mishra, S. (2013). A measure of equity for public transit connectivity. *Journal of Transport Geography*, 33, 29–41.

World Bank (2022). Mobility-as-a-Service (MaaS) can help developing cities make the most of complex urban transport systems—If they implement it right. Available online at: https://blogs.worldbank.org/transport/mobility-as-a-service-can-help-developing-cities-make-most-complex-urban-transport-systems-if-they-implement-it-right.

World Economic Forum (2021). A new era of sustainable travel prepares for take-off. Available online at: https://www.weforum.org/agenda/2021/06/new-era-sustainable-travel/.

Section 4

# Management, Policy and Research Insights in Tourism

# 11 Managing Cultural Diversity and Communication Inside the Tourism Industry

*Leszek Wypych and Ijaz Ahmad*

## Introduction

The very essence of the tourist experience is based within new encounters and immersing oneself within a new location and its sounds, smells, language, and culture, all of which are also potential sources of dissatisfaction. By understanding culture and the role of communication, we are able to improve satisfaction for visitors, and the quality of interaction for teams and their guests. The objectives of this chapter are to consider the make-up of a multicultural identity, to understand the tools that are available to improve Intercultural Competence, and to identify how to apply these techniques to create effective intercultural teams within the tourism industry.

## Background

Cross cultural contact within a tourism setting has the potential to build an appreciation of cultures between the host and visitor, and allows the traveller and the resident to build a respect and liking for one another (Allport, 1979). However, the opposite can also be evidenced, where interactions with locals may increase negative perceptions towards a destination, and generate panic, worry, and uncertainty (Fan et al., 2023). Exacerbated by communication gaps which can lead to isolation of the tourist from the host society, Cohen (1972) argues that not knowing the local language makes forming acquaintances so hard, that so few tourists try to build any relationships.

In considering those who visit a new country, there are different interacting partners (Fan, 2023); those that are pertinent for this chapter are tourist-resident, tourist-service personnel, and tourist-tourist (be they the same culture or not). Then there are those who may stay for slightly longer, such as sojourners and immigrants; in the travel industry this may look like a manager on an overseas contract to open a new hotel, or a student seeking a new life in a new country, who may decide to work as junior staff for a number of years. For these last two categories there is an expectation for these newcomers to adjust culturally to the mainstream society, the latter more than the former. For instance, Berry's (1997) model, that can be applied to a number of different acculturating groups, indicates that it is an

DOI: 10.4324/9781003369967-16

individual who decides how much interaction and participation one wants to have with the dominant culture, and adapts one of four acculturation strategies. These include attempting to assimilate, to integrate, or may result in separation if the individual wants to hold onto their culture, or finally marginalisation occurs if there is no interest in cultural groups having relations with one another (Sam and Berry, 2006). The process can only take place if one has a freedom to decide and the mainstream society welcomes newcomers. The more effective adjustment results in more successful outcomes in professional and private life. This is not a linear process, because culture per se creates additional obstacles.

There are certain key areas why we need to have an appreciation of interactions across cultures. Regarding culture and the importance of an intercultural competence, Helen Spencer-Oatey describes culture as "a fuzzy set of basic assumptions and values…" (Spencer-Oatey and Franklin, 2009) which starts to give us an indication of just how complex the subject area is; each one of us has an understanding of our own culture, yet how we explain what our culture is, what links us to others, or what causes us to create a sense of "otherness" are all understood individually, and are often difficult to explain to others. We are living in an era not just of increased global travel, but also of global communities, where an average metropolitan city has multiple races and ethnicities within. Lustig and Koester (2010) note the *demographic imperative* for intercultural competence is important, as working and socialising across cultures has an importance for us at home, let alone overseas.

Fan et al. (2023) argue that culture is a key component to understand intergroup relationships, and also note that there is a lack of research related to the dynamic cultural effects on tourist behaviour and perceptions. In light of this, as we journey through this chapter, we aim to introduce some core ideas related to culture and intercultural communication and the needs to communicate with others, focussed within the tourism setting. Understanding these areas becomes vital to appreciate the impact of a tourist on destination and the destination's context including the locals, and then what can be done to improve the situation when we encounter others and manage interactions across cultures. Research Questions: What is the significance of communication in cross-cultural encounters? What factors related to perception impact cross-cultural experiences? What can be done to mediate cross-cultural interactions and improve effectiveness of teams and leaders in the tourism space? This research aims to ensure the reader understands themselves and the constituents of a multicultural identity, explores the tools to that individuals and leaders can use to become more effective across cultures, all of which then supports an international tourism business to become more sustainable (Figure 11.1).

## Communication, Culture, and Cultural Universals

The very act of communication is the core of living in societies and communities, and it is by understanding the nature of communication itself that allows us to consider where misunderstanding can potentially occur in an intercultural setting; intercultural communication by its essence involves "culture" and "communication" (Alkelani, 2023).

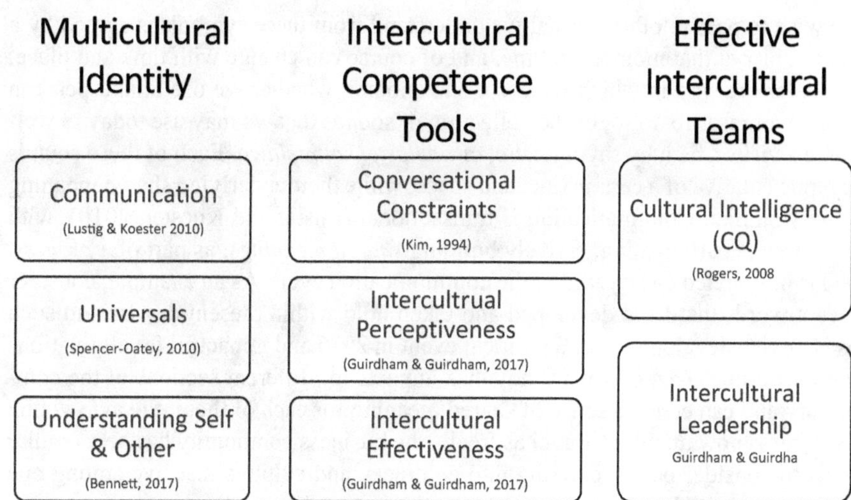


*Figure 11.1* Model of research.

### *Communication*

Before considering the basics of communication, it is important to note the nature of communication as a negotiation tool between the parties involved. In a cross-cultural setting this can add a certain level of complexity to the communication due to the parties having both common and conflicting interests, essentially bringing a different set of expectations and assumptions to the conversation (Moran and Stripp, 1991). We may find that due to cultural preferences, someone may be entering into a communication in order to build relations and to get to know the other person, while the other party may simply be present for the purpose of a task, essentially it is a *transactional* interaction. Further, the idea of pragmatic transfer (Žegarac and Pennington, 2000) within the communication can be an issue of concern where an interlocutor applies their main values and culturally governed behaviours to a new setting; if the cultural schema of an individual is applied to a situation which merely appears to be the same, there is a potential to misread the reality of the situation. As an example, a young person in a new country may be being friendly, warm and open, however this may be misunderstood as a romantic intention by someone from a conservative culture. Where cultural assumptions are applied to new situations without thought, these may lead to confusion and miscommunication. In focussing on the basics of communication, Lustig and Koester (2010) define it as:

"…symbolic, interpretive, transactional, contextual process in which people create shared meanings" which will be explored in brief below.

*Symbols* can be considered to be a word (a collection of sounds), an action (such as hand gestures), or objects (such as a wedding band) that represent a unit of meaning (Lustig and Koester, 2010). The collection of these symbols into a package, allow us to communicate with one another as we formulate *messages* that we wish to pass to others. What we understand from these symbols is agreed by a community at that moment in time, and of course can change with time and place. Consider the way in which we greet one another, whether we use Shakespearean English phrases to do so, or the collection of sounds that we may use today to welcome a visitor, be it by saying *hello*, or *namaste, konnichiwa*. Each of these sounds is representative of a certain time and space, while their underlying shared meaning is very similar. Communication is transactional (Lustig and Koester, 2010), with both parties participating, actively building *shared meanings*, as part of a *process*, and is interpreted clearly within the communication cycle. As an example, consider recent words that have developed and taken hold within present day Britain such as "Brexit", developed after a political event in 2016 and impactful for all relations within Europe. This new word now has impacts for different sections of the community and can create a sense of shared meaning for each of them; the avid young explorer is now unable to travel as freely, the business community has new regulations to consider before entering into contracts, and cultures that live among one another now are deciding whether or not they are welcome.

All communication occurs within a *context*, or more simply put, the communication happens in different situations or settings. This is discussed more fully with reference to the work of Hall (1976) below.

### *Culture*

Defining culture is complex. It is something that we all have and hold onto dearly, yet is something that is challenging to explain to another exactly what our own culture is. This is reflected in the academic field related to studies of culture, with Kroeber and Kluckhohn collating 164 different definitions just after the Second World War (Spencer-Oatey, 2012). Here are just two definitions that will allow us some direction as we progress through this chapter:

> [Culture] is the collective programming of the mind which distinguishes the members of one group or category of people from another.
>
> (Hofstede, 1994)

> …the set of attitudes, values, beliefs, and behaviors shared by a group of people, but different for each individual, communicated from one generation to the next.
>
> (Matsumoto, 1996)

These definitions give us a general direction of ideas related to what culture may be, and they hint at ideas such as groups of people that belong together, and then by default, otherness – or those who are not like us. We also get ideas that a culture must be shared, so there must be a collective of people, and that there is a mechanism to pass these shared ideas to the next generation. To add to the complexity of explaining our culture to an outsider, cultures are not only continually changing through both internal and external forces, but they borrow elements from other cultures in the process of *cultural diffusion*. Anthropologists present that as much as 90% of artefacts, ideas, and behaviours had their origins elsewhere, and this borrowing is not only in an "advanced" cultural to "inferior" cultural sense, but as a two-way process between cultures in contact (Spencer-Oatey, 2012), and can lead to cultural convergence (Kazi, 2022).

***Cultural Universals***

There are two ways in which we can consider the universalities of culture; at a simpler, higher order, all humans have overarching systems by which we all function by virtue of being human. We all have family systems by which we live together as part of a community and arrange the manner in which we ensure the continuation of our communities. We all partake in some sort of economic system in order to barter or pay for our varied needs, either as individuals, family units, or as a community. There are education systems to further the family and economic systems, there are systems linked to supernatural beliefs, and an organisation of these beliefs, and finally we are all managed by social control systems, be they formal legislative, or informal "gossip" systems. A detailing understanding of what these may look like in a culture-by-culture sense would require the anthropological study of each, giving us a culture specific "emic" viewpoint (Pike, 1967). The second manner in which we can consider the cultural universals are those that are driven by a culture general, "etic" approach (Pike, 1967), where pre-determined categories can be applied to each culture, in search of those comparable, analysable parts of culture that can be determined to exist in a more or lesser extent. It is this approach, the *etic* approach, that will be focussed upon for the rest of this chapter and explored below.

One of the first large scale comparative studies that took place had an impact upon the research in the field as it used many respondents, across many countries, and the data was easily accessible and comparable. Geert Hofstede in his 1980's work proposed that cultures were made up of dimensions that were all present to a more or lesser degree, and that these (etic) comparisons were a useful tool in understanding cultural behaviour. In considering just two dimensions out of the six presented by Hofstede (see Hofstede Insights, 2023 for an exploration of all six) that can readily impact a tourism setting and the tourist experience, the first dimension of Power Distance is explained as "the degree to which less powerful members of society expect that power is distributed unequally" (Hofstede Insights, 2023). This may manifest in a member of staff from a country with a high-power distance score feeling obliged to listen to the requests of those who have power over them, regardless of their opinion. This could be a manager making requests that may not be accepted in a low-power distance country, or tourists making requests that may be viewed as unethical. The second dimension of Individualism is explained as "a preference for a loosely-knit social framework in which individuals are expected to take care of only themselves and their immediate families" (Hofstede Insights, 2023), with those from a Collectivist culture being more used to a more closely knit social framework. Here, a manager from an individualist nation may be unsuccessful in managing a team that is collectivist in nature, unless adaptations are made to management styles. In the next section, the discussion is considered from an alternative perspective, that of culture as an intertwined, integrated part of the communication process. Culture as communication

> Culture is communication, and communication is culture.
>
> (Hall, 1959)

Edward Hall theorised that communication was a very large notion, not limited just to verbal but also included non-verbal communication, notions of time, the importance of context, and the speed of information, and he noted how these variables were culture specific (Mahadevan, 2023). Hall (1976) identified two perspectives of communication linked to the "content versus context" of communication. He argues that there were two major differentiators; were the key messages of the communication to be found within the *content* of the communication (the words, the speech, the written text), or in the *context* of the communication (who is speaking, where are they speaking, is it a formal or informal event, and what words are being missed out of the communication?). He labelled the first a *Low-Context Communication* and the latter a *High-Context Communication.*

An area of importance from the work of Hall is the identification of "ingroups" and "outgroups". Simply, an ingroup is a social group with which you identify strongly, and an outgroup is one with which you don't (Giles and Giles, 2013); anyone who is not from the same culture and fails to understand the artefacts, heroes, rituals, and values (Hofstede, 1980) in the same way is a part of the outgroup. To add a little more complexity to the discussion, Avruch (cited in Spencer-Oatey, 2012) argues that no population can be characterised by a single cultural descriptor and that as we identify with a variety of criteria, so there is a need for a variety of subcultures; these may be differentiated by a range of socially constructed ideas, such as social classes, or political interest groups, or criteria linked to nationality. As an example, a British person living in Spain may not identify as a tourist, nor a Spaniard, nor a resident of the UK, and as such a new subculture may be appropriate, and a new ingroup would be created among the British expatriate population.

In trying to understand the *Context* orientation of another, it has to be deduced from the situation. Through the readings of his work (and that with his wife), Hall starts to name countries along the spectrum of context, noting for example that "Low context people include Americans, Germans, Swiss, Scandinavians, and other Northern Europeans… The French are much higher on the context scale than either the Germans or the Americans…" while noting "Japanese, Arabs, and Mediterranean peoples… are high-context…" (Hall and Hall, 2001). As such, while it is tempting to place people within categories of low or high context, the more realistic approach is to understand that cultures present upon a spectrum, and it is only by the interaction and the interpretations of those interactions that we can decide where these individuals may sit on the scale. Further, as a result of globalisation, some of these differences may be less pronounced (Guffey and Loewy, 2013). In a tourism setting, we may find that a guest from a high context background may not voice dissatisfaction or unhappiness, but this may be communicated through silence, or body language; displeasure may only be verbalised with people at the appropriate power level in a private setting. This starts to give us an indicator of just a few areas that can lead to miscommunication across cultures, where, as we will discuss in the next section, cultures have a different *reliance* in communication using them.

## Understanding Self and the Other

There are two areas of reflection when considering the potential to be effective in an intercultural interaction. One of the areas that is important to consider is our awareness of our own cultural identity (Wiersma-Mosley and Garrison, 2022) and interacting with others, and the second is to consider how we view others and therefore should we accept the possibility of adapting and integrating new aspects of culture into our own. It is through consideration of these areas that allows us to monitor our own outlooks to other cultures; if we view culture from a perspective of our own cultural superiority, the chances of effective communication across cultures decrease (Belaid, 2022). This section considers the self, and the views we have upon the cultures of others in this process.

As discussed already, culture is universally present, and a manner in which we can start to understand not only our own culture, but to appreciate other cultures is by thinking about cultural metaphors. As a result of this it can become quite normal to compare aspects of our culture against aspects of other cultures; the culture-specific *emic* research allows the observer to gain an understanding of the way an insider views the world through their lens, how they create meaning systems, belief systems, and values (Wypych, 2020). It is important to note that *etic* perspectives that consider the culture-general, bring in the views, opinions, and beliefs of outsiders (Hennink et al., 2020).

In an *emic* sense, comparing the use of cutlery to eat, versus the use of chopsticks to eat, versus the use of one's (right) hand is entirely normal, but entirely useless. If we were to ask which one of these modes of eating is correct, the answer simply is "all of them", none is more correct than the other and rather they are correct for the society in which they function. In essence, it becomes important that we understand that the way we do things, may not be the same as others (Jones and Quach, 2007).

### *Intercultural Sensitivity*

The emotional, affective response to intercultural difference has been called *Intercultural Sensitivity* (Straffon, 2003) where the individual at the heart of the interaction has an active desire to understand, appreciate, and accept differences among cultures (Chen and Starosta, 1998). Chen and Starosta argue that this must remain within the affective, but an increase in intercultural sensitivity leads to an ability to deal with (behavioural) cross cultural interactions more effectively (Hammer et al., 2003). It is important to note that the areas measured to indicate levels of Intercultural Sensitivity include interaction enjoyment (whether the individual enjoys meeting others or finds it a source of stress and unhappiness), respect for cultural differences, interaction confidence, and interactive attentiveness (Chen and Starosta, 2000). When looking at the experiential aspects of cultural difference, Milton Bennett's Developmental Model of Intercultural Sensitivity (DMIS) explains how people view cultural difference, and notes how as a person's experience becomes more complex, their potential for intercultural competence increases (Hammer et al., 2003).

The model was initially developed to explain language communication adjustment, but it is often used to explain cultural adjustments. Bennett (1993, 2013) linked the process of acculturation with the individual personal experiences; he believes that the perception of reality that we hold and interact with has an impact on the way in which we experience our realities. Elaborating further, the barriers we create between us and others also governs the way in which we experience intercultural interactions (Wypych, 2020). The first three of the stages presented by Bennett are rooted in Ethnocentrism and demonstrate a lower level of sensitivity, while the latter three stages are moving towards the ethnorelative. These demonstrate not only a higher level of sensitivity (Perry and Southwell, 2011), but also allows a single individual identity to embody multicultural relations and ethnicity (Wypych, 2020).

In brief, the first stage of *Denial* is at one end of the continuum, where "others" may be seen as foreigners, minorities, or not perceived at all; Bennett (2017) argues that in this stage, people perceive themselves as more "real" than others. This may manifest in a tourism setting where an allocentric tourist will claim cultural superiority. It can further manifest in (tourist) organisations as an outcome of not having policies and procedures in place to identify and respond to cultural differences (Wypych, 2020). In the *Defense* stage others are perceived more fully, but this may manifest as a more clear "us and them". This may be more readily witnessed by the host nation imploring the perceived socio-cultural impacts that tourism is having on a location, as witnessed by acts and policies against tourists in Barcelona in 2016, Venice in 2017, and the new approach towards tourists in Amsterdam with the "Stay Away" campaign of 2023. As an example, see Sabaghi (2023). It is possible that instead of defence, *reversal* would take place, where an individual may note that "us" are inferior, and "them" are superior through romanticisation or exoticisation of the new culture. This can be more common in a tourism setting where travellers may "awaken" their cultural self, or embark on a journey of self discovery by going "native" (Wypych, 2020). By focussing upon the commonalities of humans and the shared values, the "us and them" elements can be reduced, allowing for a *Minimization of cultural difference* and the focus of the intercultural event becomes the similarities between self and others. In a tourism setting, we may find that as tourists explore a new location, they will engage with fairness and equality, purchase trinkets and souvenirs, but never really question the deeper meanings of what the artefacts symbolise, what the historical inferences are, or what rituals represent. Bennett (2017) argues that exploring these deeper differences through deeper questioning, reflection, and deliberation actually has the potential to cause people to regress to the *defense* stage.

The first of the stages on the ethnorelative scale is that of *Acceptance*. It does not mean agreement, however. The lack of cultural flexibility at this stage prevents people from adapting themselves to different cultural settings (Wypych, 2020). It is argued that the shift from *Acceptance* to the next stage requires a paradigm shift to occur related to ethicality of issues, and that to move to the *Adaption* stage, people need to understand the (ethical) validity of choices in different cultures and contexts. This may mean that one needs to consider what is good for them in a particular situation, before they make a decision about what behaviour change is acceptable from an ethical perspective (Wypych, 2020). Perry (1999) presents the Ethical Scheme which demands that leaders and managers engage in contextual relativism, allowing decisions to be made based on "goodness in context" or "feelings of appropriateness" to take place, and allows us to experience the world as if we are participating in a different culture, ultimately leading to us operate in a bi-cultural manner. In an international business sense, we may find that staff are able to operate effectively in two (or more) cultures, and that the organisation has policies that are flexible enough to allow unhindered operation in multiple locations, resulting in both global and domestic approaches to be absorbed into the business. Individuals in this situation may develop a set of behaviours for two different settings (biculturalism), or multiple settings (multiculturalism); the question that arises at this stage is in which cultural setting does their true identity belong to (Wypych, 2020), or are we, as Bennett (2017) suggests, a living and dynamic container that is constantly evolving.

The last stage along the continuum is *Integration* where we find the sustainable incorporation of cultural difference into communication. This supports the individual to move in and out of different worldviews, allowing the creation of cultural bridges, and the potential to mediate across cultures with sophistication. Here we may find international managers are able to negotiate effectively across cultures and bring about the resolution of complex roadblocks to communication.

## Multicultural Communication Competence

This section builds upon the constructs of language and considers the changes in outlook that individuals need to have in order to communicate effectively across cultures, having taken a look at oneself, and an appreciation of the etic nature of the cultures they are interacting with. To be able to effectively communicate, the correct symbols for the given culture must be utilised, they must be interpreted correctly (so the correct choice of symbols must be utilised for that given context). Alkelani (2023) noted in his research how it wasn't just the actual words that needed to be correct, but the accents became important in the communication process. Using low-context communication pattern in a High Context culture is ineffective, and in cultures where relations take precedence over tasks, an individual maybe seen as brash and lacking in refinement (Guirdham and Guirdham, 2017).

As an individual moves between cultures, so the challenges in selecting the correct balance of communication increase. One such manner in which an international manager or worker in a tourist setting may balance the need for clarity and to avoid hurting others is to consider the idea of *Conversational Constraints* (Kim, 1994). The theory argues that an interlocutor in a setting will choose a strategy for the conversation, balancing between clear (imperative) words and phrases, or indirect (hinting) phrases with the aim of balancing clarity, minimising the possibility of offence, and minimising imposition. If the concern in the communication is for clarity (described here also as getting one's own way), the interlocutor will make a judgement about how to adapt their communication towards a more direct form, and away from a hinting form, being explicit in speech (Kim and Aune, 1997). This may happen in a communication that needs to define travel itineraries or detailing task responsibilities to staff. If the concern is low, then the communication may shift to indirect hinting, and may occur in a situation such as receiving and welcoming guests, where you may need to detail some small information such as times for meals, but the overall communication is based around relationship building. Someone who aims to have a high concern for the feelings of the other person may find indirect language utilised instead, while if there is no concern for the other, then more direct language will be used, potentially damaging the honour or "face" of others in the communication. Brown and Levinson (1978) described this as "the want to maintain the hearer's positive face" and may occur in situations where tourists are endangering themselves, others, or disrespecting cultural norms. Finally, the concern for effectiveness will change the communication. A tour guide in a bustling city centre will need to conduct the tour with efficiency and effectiveness, ensuring that the timetable is adhered to, that all the guests are kept safe, and there is minimum disruption caused to the local community; this will lead to an increase in direct language. Conversational Constraints are based within the context of speech, and Guirdham and Guirdham (2017) argue that this can be extrapolated to the written word.

There are clear interrelated concepts from the work of conversational constraints (Kim, 1994), and the foundational work of Hall (1976) and the context of communication noted above. By understanding the adaptations that are required in communication in a culture that is different from one's birth culture, having an appreciation of the levels of context that vary from culture to culture, and then by supporting this with aspects of *concern*s within communication, the interlocutor engaging in a cross-cultural exchange is more likely to be successful. The next section builds upon this, and considers further steps in effective communication, related to aspects of perceptiveness of cross-cultural encounters.

## Intercultural Effectiveness

A variety of theorists provide advice and guidance on determinants of effective communication in an Intercultural setting. As an example, Schwarz (1993) notes that when articulating a message the sender of the communication must be coherent and comprehensible, give neither too little nor too much information, be relevant, and convey the message appropriate to the context; Alkelani (2023) even noted the appropriateness of "loudness" in communication. There are certain key areas that may support effective communication and presentation of oneself in a cross-cultural setting. These include speaking with clarity, using person centred messages, showing empathy in discussion, communicating a relationship orientation, sharing the correct level of information, and communicating an appropriate level of assertiveness for the context (Guirdham and Guirdham, 2017). First, achieving *clarity* is important to ensure that the message has been understood; as a key requirement within a work setting, communicating effectively for task completion is often vital. The Maxim of Manner (Grice, 1975) supports this discussion, as he notes that brevity, clarity, and orderliness avoids ambiguity. As noted earlier, due to conversational constraints (Kim, 1994), a person may choose indirect hinting forms of communication if the need for "getting one's own way" is reduced, thus impacting clarity in communication. This can be further complicated when power-distance (Hofstede Insights, 2023) complexities are at play, where a member of staff in a tourist setting may feel obliged to change their communication based upon the position of power that tourists may hold. Using *person-centred* messages recognises the perspectives of others within the communication, and may start with using open questioning to obtain information related to the values, beliefs and attitudes of others. This then necessitates the communicator to internalise and process this information, incorporating it into the exchange, ultimately requiring greater effort from those involved. Building *empathy* simply means seeing things through another's eyes and being non-judgemental (Guffey and Loewy, 2013); it does not require agreeing with them, but simply considering the interaction from the perspectives of others. This may be achieved through seeking verbal and non-verbal clues, and can be demonstrated through the mirroring of some of the non-verbal clues such as body-language, speed of discussion, and tone. By saying things that are pertinent, relevant, and valuable to the conversation, the Maxim of Relation (Grice, 1975) also supports the building of empathy. In contrast, Moriizumi and McDermott (2017) note that in collectivist cultures, the high-context of communication may mean that the mere presence of a person is highly valued and enough to demonstrate empathy, and verbalised communication is of a lesser importance.

As noted earlier in the discussion related to work-life balance, Hofstede Insights (2023) noted the Masculinity dimension, indicating the balance of a task orientation, or a relationship orientation within a culture. By communicating a *relationship orientation*, someone from an individualist culture is more likely to build effective relations in a collectivist culture. This can be achieved through remembering details about people that are met, names or nicknames that get shared (as an example, a *Kunya* is given to most Arabs, and they may be named after their first child), or details about families or health. Thanking others, reciprocating gifts are all other steps that can support in this regard. Linked to the masculinity dimension noted here, communicating the *appropriate level of assertiveness* for the culture is a key skill to possess. An individual may communicate with a certain level of authority in a home country, yet in another this will be seen as too aggressive, and in a third this may be viewed as not authoritative enough. The key skill is to understand the suitable level of assertiveness that is appropriate for the person being interacted with. Alkelani's (2023) research was based on Malaysian students and noted an absence of assertive communication, indicated by not asking questions (as this may also damage "face" by showing that they don't know), or interrupting others. Finally, *sharing information* or being open assists in relationship building, and creating a positive impression of individuals. The Maxim of Quantity notes that it is important to be informative, giving relevant information as needed, and no more, while also embracing the Maxim of Quality, by being truthful and supporting discussion with evidence (Grice, 1975). The step breaks down pre-conceptions, and can help with slowly reducing negative stereotypes, and building higher levels of social integration.

## Intercultural Leadership

The very nature of the globalised world that we live in has created a demand for leaders who are effective in leadership in many cultures simultaneously; Bratianu and Paiuc (2022) summarised their work on multicultural leadership with "the future is already here, and it is multicultural" (p. 335). Culturally adapted leadership results in positive outcomes for all involved including the leaders, their followers, and indeed the organisation itself (Guirdham and Guirdham, 2017). The following section considers just a few of the areas in which leaders can improve intercultural teamworking, for the betterment of all and include areas such as working with strengths, developing these strengths, promoting natural forces of harmony, and undertaking managerial interventions where necessary (Guirdham and Guirdham, 2017). Each of these is explored further below.

By recruiting team members that have positive attitudes to diversity, an organisation is already in a position to start to *work with strengths* of the team, as the natural mental outlook of the team to varied opinions, assorted outlooks and different communication approaches are readily accepted. The positive outlooks should foster brainstorming performance, measured by the quality of ideas generated (Nakui et al., 2011). Team members who have an experience of working across cultures and import that knowledge and experience help to promote a knowledge sharing culture (Zakaria et al., 2004); this may start from sharing credit and showing benevolence, supporting the development of social structures within the organisation and an action orientation. Adults who spent a period of their youth living across cultures are more likely to be effective in international assignments (Westropp et al., 2016) and have the potential to become especially competent global leaders (Lam and Selmer, 2004). *Developing strengths* that have been brought into the business can occur through training, or helping team members develop interpersonal congruence. Training is key in the early stages of a group's formation, with leaders looking for potential failures early on in the establishment of the team (Hambrick et al., 1998).

The training may centre around how one reacts to others in terms of beliefs, how others react to situations, and job performance expectations (Knouse and Dansby, 1999). Guirdham and Guirdham (2017) note that in *promoting forces of harmony*, the naturally occurring energies that can be found in a diverse team can be supported by the channelling of interactions through a central, often leadership, role. This creates a natural "check mechanism" for the team, where every request is noted in some capacity, be it a formal or informal one, and a natural monitoring opportunity takes place within the team – although this could be problematic in countries with a low Masculinity score (Hofstede Insights, 2023) as this may be seen as micromanaging the team. Increasing time spent together is also a driving factor in reducing perceived differences. In global teams, cross site visitation is a key step in building rapport and intercultural understanding. Co-workers become more accustomed to communication styles, personalities, and roles within work and outside of work (Hinds and Cramton, 2014), while techniques such as radical relocation furthers the general idea of getting teams to spend time together by placing them in their own large rooms. Radical relocation brings about a shortening of time to deliver, performance to increase, productivity to improve, and higher satisfaction both within teams, and for customers (Teasley et al., 2000). In considering potential failures, *managerial interventions* are key to managing intercultural team differences. Resolving problems through brainstorming, and managing conflict through persuasion and negotiation are key areas to consider.

**Solutions and Recommendations**

In order to become more effective in a cross-cultural setting, building cultural awareness, understanding of context, and considering one's own perceptions are all key steps in this journey and allow for the perceptions of the intercultural event to be more accurately understood (Guirdham and Guirdham, 2017). They argue that the first step towards a better intercultural understanding is to build an *awareness of your own and others' cultural sensitivities*, potentially by seeking out exposure to other cultures, be it in a work setting or otherwise. The greater the number of interactions there are with other cultures, the more likely it is that real differences can be confronted, however, a process would need to be established to logically discuss and reason the similarities and differences that exist between the groups (Wypych, 2020). Trying to approach the cross-cultural experience with an ethnorelativistic approach in communication would allow the process to be more successful, and learning from these interactions can be sped up by reflective observation, with the aim of focussing upon one's own ethnocentric tendencies (Guirdham and Guirdham, 2017). The second step is to build an *awareness of contexts*, through considering a few key elements. It is vital to consider the *Power & Status* of the individuals that exert influence over the discussion, The sheer wealth disparity from western tourists visiting developing nations is likely to be the greatest dynamic which impacts much of the interaction. This lack of balance has an impact within culture, with the local people feeling inferior or subservient to the tourists (Telfer and Sharpley, 2008); this power imbalance may further impact the language strategies chosen by locals, as presented by Kim (1994) earlier. The third and final step is the need to *overcome perceptual barriers*, by trying to ensure that you appreciate the uniqueness of each culture, becoming more sensitive to the verbal and non-verbal language of others, by becoming aware of stereotypes, and to avoid evaluating another culture as inferior. Tour guides in particular may need to act as mediators who interpret local culture from a living or a heritage perspective, or a cultural identity perspective, for customers who expect the tour guide to show a sensibility towards the culture of the tourists too (Rabotić, 2010). On occasion, they may need to control the tourist with more direct language in order to ensure that tourists remain within cultural boundaries (as discussed above with conversational constraints, Kim, 1994).

Once the perceptual factors have been considered, certain cognitive and affective stages can be undertaken. The employee that wishes to increase in cross cultural effectiveness will need to *unlearn* and learn certain attitudes, attempt to *predict* the responses of others accurately, and be *mindful*, and non-judgemental. "Unlearning" can be explored as a necessary step to free oneself from preconceptions, prejudices, and expectations, to allow new information to be absorbed and "learnt". This process is concerned with behavioural changes, that are part of acculturation processes Berry (1997, 2017) mentioned at the very beginning of this chapter. Berry refers to the unlearning process as "culture shedding". This however is a challenging stage to undertake, as it often involves looking at oneself deeply, and asks a person to consider their ethnocentric outlooks that they have held. Bennett argues that at the *minimization* stage, if an individual is confronted with deep dissimilarities, he or she may regress on the scale back further towards the *defense* stage (Wypych, 2020). Finally, being mindful involves a mental paradigm shift, turning from a "mindless" reliance on old categorisations, into the "mindful" continual creation of new ones, (therefore reducing implicit biases (Ngnoumen, 2019). There are important areas to note here. The first is that the process is constant, as new information arrives, the categorisations are continually updated. The second is that this mental state becomes necessary to prevent misinterpreting the behaviour of others. There is a link here to being non-judgemental, as judging behaviours and actions are often reliant on stereotypes and can lead to negative attributions. In order to change one's outlook and minimising judgement of others, the colleague working across cultures may listen openly to others without interrupting or criticism, with a view to understand, and with a view to learn and acknowledge differences. They may question and interact in a way that is open, soliciting feelings of others, and aims to understand the deeper, underlying nuances at hand.

## Future Research Directions

In an era where the global village is a reality for many, technology has increased the speed of information transfer, and in part the communication of globalised cultural messages may be impacting upon the acculturation process in developing nations that are tourist dependent. Global culture is often defined in Western terms, often defined by capitalist icons such as Coca Cola, Levi's and McDonald's (Telfer and Sharpley, 2008). Image-based apps such as TikTok and Instagram are increasing the speed of codified messages that are transferred from developing nations, influencing perceptions of expected behaviour, norms, and culture of the typical tourist. Research into the impact that app-based technology (such as TikTok) has in the acculturation process for staff in tourist destinations should be considered. In considering the communication, focussing upon service level employees in high-context cultures would allow us to determine the impacts that technology is having in communicating across cultures.

## Conclusion

In laying forth the foundations of communication, it becomes possible to understand the building blocks by which we all communicate, and it allows us to acknowledge how these building blocks are universal in their nature, yet unique for each and every culture. An individual living in a town or country that is largely heterogeneous, potentially with a single overarching culture, may consider that their language may be the only valid manner in which to communicate. As we explore deeper aspects of communication and understand that communication by its very nature includes negotiation, (Moran and Stripp, 1991) and as we negotiate from our own tacit worldviews, we start to incorporate culture into the communication process. Edward Hall (1976) identified two broad categorisations of Low Context and High Context communication, and the links to the nature of communication, and the impact that culture has on the communication process was explored in this grouping; an individual from a low-context background may experience challenges in communicating effectively with someone from a high-context background, and the same being true regarding the opposite (Guirdham and Guirdham, 2017). It is important to understand ourselves and our views to outgroups in the intercultural sensitivity process as the (metaphorical) barriers that we create between ourselves and the other has an impact in the way in which we experience intercultural interactions (Wypych, 2020). In moving from an ethnocentric to an ethnorelative position on the continuum postulated by Bennett (2017), he argues that sequential steps take place with individuals based on their need to become more competent in communicating in a social context different to their own. When the need is established, building a more detailed, intricate, nuanced understanding of the new culture (or our perceptions of it) allows us to move along the scale towards complete integration of these cultural realities into our personas (Bennett, 2017).

As the awareness of cultural variabilities increases, as our perceptions of other cultures change, and as our desire to understand these cultures intensifies, there are certain practical steps that individuals and leaders can take in order to increase in intercultural effectiveness. Guirdham and Guirdham (2017) argue that certain approaches such as using person centred messages, showing empathy, and communicating the appropriate level of assertiveness for the context allow teams to be more effective across cultures. From a leadership perspective, building a diverse team allows the natural cross-cultural strengths of a team to be established, and it is through developing these strengths, promoting the natural forces that can be found, and then by being prepared to intervene in a managerial capacity where necessary, that an interculturally competent team is able to flourish, and support the creation of an effective international business.

**Case: Tourism and the Resentment to Tourists; Japan in Focus**

Tourists behaving badly is not a new phenomenon. A quick online search of visitors behaving unacceptably by the host culture will present with a plethora of examples and range from popular Western countries to further away, long-haul destinations, where cultural expectations and values are very diverse. In many of these countries, the local community has resented the overtourism and the socioeconomic changes that this has brought. Japan may be more sensitive to the concerns of overtourism; as a mono-ethnic nation, reserved, private culture, there is a concern that tourists represent a sort of "contamination" (Duignan et al., 2022, p. 9) which is a concern for some of the residents. Japan had almost 32 million tourism arrivals in 2019 and analysts predict that tourism levels will return to pre-pandemic levels at some point; in May 2019 there were just under 2.8 million tourist arrivals, while in May 2023, just seven months after the relaxation of post-covid travel restrictions, the arrivals had reached just under 1.9 million arrivals (JTB Tourism Research & Consulting Co., 2023), and it is steadily rising. Concerns have been voiced by the local population related to overtourism as a key concern (Duignan et al., 2023), so much so that "Kankō Kōgai" became a popular local phrase coined by the Japanese to describe "tourism pollution" (Alyse, 2023).

Guided tours of Japan can become an important way in which to communicate in Japan as not only do they mediate the understanding of the language, they ensure that the tourist is safe and comfortable throughout the experience (Tokyo Localized, 2023). However, despite the tour-guide responsibility as an "ambassador" (Yu et al., cited in Rabotić, 2010), there are instances where the tour guide is unable to control the behaviour of the tourists. As an example, tour guides reported poor behaviour when taking tourists to local baths, with the guests failing to adhere to local norms and practices; they would not wash before entering, or enter with footwear, or with clothing when the custom was to be naked, resulting in the owners very angry at the tour guides (Duignan et al., 2022). An example of how some businesses have now dealt with the situation, a pub in the centre of Tokyo has taken the approach of telling groups of more than five tourists that the restaurant is fully booked when it isn't (Duignan et al., 2022).

**Questions**

1 Some residents refer to tourists as a "contamination". What are some of the factors that explain this outlook?
2 What role does a tour guide have in a setting such as Japan? Consider the answer to this question from both a resident and tourist perspective.

3 Why would pubs and restaurants claim to be "full"? Why would business owners behave like this? What advice and recommendations could you offer local business owners?

**Key Terms and Definitions**

Attribution Bias: Tourists use internal attributions for positive tourism outcomes and external attributions for negative outcomes (Jackson et al., 1996).

Attribution Theory: *How* and *why* things occur; attribution theory explains how people arrive at causal explanations for events (Fiske and Taylor, 1991).

Allocentrism: The individual equivalent of (societal) collectivism. An individual can be an allocentric in an individualist culture, or indeed a collectivist society. The opposite, idiocentrism is the individual equivalent of (societal) individualism, and indeed, an individual can be an idiocentric in a collectivist culture, or indeed an individual one (Mio, 2020).

Cultural Home: A sense of identity that revolves around particular cultural constructs such as history, ethnicity, or even similar physical features. This can link to geographic ideas where an area is densely populated by people migrants who have a shared heritage and create an ethnic enclave, or symbolic, where those with a nomadic lifestyle take their language, traditions, clothing and customs with them (Vivero and Jenkins, 1999).

Ethnocentrism: The view that one's own culture is superior to the culture of others.

Face: Concerned with identity, dignity, and one's sense of worth (Wypych, 2020), Positive face links to one's desire to be accepted and valued by some selected others (friends, work colleagues, the community), while negative face is linked to the want to be free, autonomous, and not impeded by others (Brown and Levinson, 1987).

Social Distance: The degree to which an individual perceives a lack of intimacy with individuals different to themselves, driven by cultural factors such as race, religion, ethnicity, or other variables.

Sojourner: Originally developed by Park (1924) as the concept of the Marginal Man, a sojourner holds onto his or her original culture as they visit another culture, with an intention to return home. As with the Marginal Man, the sojourner is on the edge of two cultures, and has a unique perspective on viewing both host and home cultures.

Stereotyping: Is a valuable cognitive shortcut that allows attributes to be placed upon a particular social group, and it allows a reduction in the processing time and mental strain placed on an individual to make decisions about others.

Subculture: A group of people that aims to differentiate itself from the larger culture within which they exist, categorised by ethnicity, lifestyle, socioeconomic status, sexual orientation, or another social construction of the era.

## References


Alkelani, W. (2023). Intercultural miscommunications faced by UTM students and the strategies they use to overcome them. *Eximia, 11*, 225–235.

Allport, G. W. (1979). *The Nature of Prejudice*. Addison-Wesley Pub.

Alyse. (2023). Overtourism, did Japan become a victim of its own success? [online]. Available at: https://www.theinvisibletourist.com/overtourism-in-japan-tourist-pollution/.

Belaid, L. (2022). Exploring the prevalence of ethnocentric traits at university: An initiative to perpetuate ethno-relativism as a cultural mindset. In Saricoban, A., Yazgi, C., Sari, S., and Metin Tekin, B., (Eds.), *AELTE 2022 21st Century Challenges in English Language Teaching,* Ankara Haci Bayram Veli University, Ankara, (pp. 273–285).

Bennett, M. J. (1993). Towards ethnorelativism: A developmental model of intercultural sensitivity. *Education for the Intercultural Experience, 2*, 21–71.

Bennett, M. J. (2013). *Basic Concepts of Intercultural Communication: Paradigms, Principles, and Practices*. Hachette UK.

Bennett, M. J. (2017). Developmental model of intercultural sensitivity. In Kim, Y., (Ed), *The International Encyclopedia of Intercultural Communication,* Wiley (pp. 1–10).

Berry, J. W. (1997). Immigration, acculturation, and adaptation. *Applied Psychology: An International Review, 46*(1), 5–68. Available at: https://iaap-journals.onlinelibrary.wiley.com/doi/epdf/10.1111/j.1464-0597.1997.tb01087.x?saml_referrer.

Berry, J. W. (2017). Theories and models of acculturation. In Nathan, P.E., Schwartz, S.J., and Unger, J.B., (Eds.), *Oxford Handbook of Acculturation and Health,* Oxford University Press, Oxford (pp. 15–27).

Bratianu, C., & Paiuc, D. (2022). A bibliometric analysis of cultural intelligence and multicultural leadership. *Revista de Management Comparat International, 23*(3), 319–337.

Brown, P., & Levinson, S. C. (1978). Universals in language usage: Politeness phenomena. In Goody, E.N., (Ed.), *Questions and Politeness: Strategies in Social Interaction* (pp. 56–311). Cambridge University Press.

Brown, P., & Levinson, S. C. (1987). *Politeness: Some Universals in Language Usage.* Cambridge University Press.

Chen, G. M., & Starosta, W. J. (1998). *Foundations of Intercultural Communication.* Allyn & Bacon.

Chen, G. M., & Starosta, W. J. (2000). The development and validation of the intercultural sensitivity scale. Paper presented at the Annual Meeting of the National Communication Association, Seattle, WA.

Cohen, E. (1972). Toward a sociology of international tourism. *Social Research, 39*(1), 164–182.

Duignan, M. B., Everett, S., & McCabe, S. (2022). Events as catalysts for communal resistance to overtourism. *Annals of Tourism Research, 96*, 103438.

Fan, D. X. (2023). Understanding the tourist-resident relationship through social contact: Progressing the development of social contact in tourism. *Journal of Sustainable Tourism*, *31*(2), 406–424.

Fan, D. X. F., Qiu, H., Jenkins, C. L., & Lau, C. (2023). Towards a better tourist-host relationship: The role of social contact between tourists' perceived cultural distance and travel attitude, *Journal of Sustainable Tourism*, *31*(2), 204–228. https://doi.org/10.1080/09669582.2020.1783275.

Fiske, S. T., & Taylor, S. E. (1991). *Social Cognition* (2nd ed.). McGraw-Hill.

Giles, H., & Giles, J. (2013). *Ingroups and Outgroups*. SAGE Publications, Inc. https://doi.org/10.4135/9781544304106.

Grice, H. P. (1975). Logic and conversation. In P. Cole, & J. Morgan (eds.) *Studies in Syntax and Semantics III: Speech Acts* (pp. 183–198). Academic Press.

Guffey, M. E., & Loewy, D. (2013). *Essentials of Business Communication*. Cengage Learning.

Guirdham, M., & Guirdham, O. (2017). *Communicating across Cultures at Work* (4th ed.). Palgrave.

Hall, E. T. (1959). *The Silent Language*. Doubleday.

Hall, E. T. (1976). *Beyond Culture*. Doubleday.

Hall, E., & Hall, M. (2001). Key concepts: Underlying structures of culture. In M. H. Albrecht (Ed.), International HRM: Managing diversity in the workplace (pp. 24–40). Blackwell Publishers.

Hambrick, D. C., Davison, S. C., Snell, S. A., & Snow, C. C. (1998). When groups consist of multiple nationalities: Towards a new understanding of the implications. *Organization Studies*, *19*(2), 181–205.

Hammer, M. R., Bennett, M. J., & Wiseman R. (2003). Measuring intercultural sensitivity: The intercultural development inventory. *International Journal of Intercultural Relations*, *27*(4). 421–443.

Hennink, M., Hutter, I., & Bailey, A. (2020). *Qualitative Research Methods*. Sage.

Hinds, P. J., & Cramton, C. D. (2014). Situated coworker familiarity: How site visits transform relationships among distributed workers. *Organization Science*, *25*(3), 794–814.

Hofstede, G. (1980). Motivation, leadership, and organization: Do American theories apply abroad? *Organizational Dynamics*, *9*(1), 42–63.

Hofstede, G. (1994). *Cultures and Organizations: Software of the Mind*. Harper Collins Business.

Hofstede Insights. (2023). Intercultural management: What you need to know [online]. *Hofstede Insights*. Available at: https://www.hofstede-insights.com/intercultural-management#whatisinterculturalmanagement.

Jackson, M. S., White, G. N., & Schmierer, C. L. (1996). Tourism experiences within an attributional framework. *Annals of Tourism Research*, *23*(4), 798–810.

Jones, A., & Quach, X. (2007). *Intercultural Communication*. The University of Melbourne.

JTB Tourism Research & Consulting Co. (2023). Japan-bound statistics [online]. Available at: https://www.tourism.jp/en/tourism-database/stats/inbound/#:~:text=According%20to%20JNTO%2C%20the%20estimated,to%20Japan%20in%20October%202022.

Kazi, M. d. M. R. (2022). Globalization and folk culture of North-East India: An analytical review. *International Journal of Health Sciences*, *6*(S1), 6744–6751. https://doi.org/10.53730/ijhs.v6nS1.6773.

Kim, M. S. (1994). Cross-cultural comparisons of the perceived importance of conversational constraints. *Human Communication Research*, *21*(1), 128–151.

Kim, M. S., & Aune, K. S. (1997). The effects of psychological gender orientations on the perceived salience of conversational constraints. *Sex Roles*, *37*, 935–953.

Knouse, S. B., & Dansby, M. R. (1999). Percentage of work-group diversity and work-group effectiveness. *The Journal of Psychology*, *133*(5), 486–494.

Lam, H., & Selmer, J. (2004). Are former "third-culture kids" the ideal business expatriates? *The Career Development International*, *9*(2), 109–122. https://doi.org/10.1108/13620430410526166.

Lustig, M. W., & Koester, J. (2010). *Intercultural Competence: Interpersonal Communication across Cultures*. Pearson.

Mahadevan, J. (2023). *Cross Cultural Management: A Contemporary Approach*. Sage.

Matsumoto, D. (1996). *Culture and Psychology*. Brooks/Cole.

Mio, J. S. (2020). Allocentrism vs. idiocentrism. In Carducci, B.J., Nave, S.N., Jeffrey, S.M., Riggio, R.E., *The Wiley Encyclopedia of Personality and Individual Differences: Clinical, Applied, and Cross-Cultural Research*, Wiley (pp. 205–208).

Moran, R. T., & Stripp, W. G. (1991). *Dynamics of Successful International Business Negotiations*. Gulf Pub. Co. The managing cultural differences series 0872011968.

Moriizumi, S., & Mcdermott, V. (2017). The role of narcissism and face concerns in providing comforting messages: A cross-cultural comparison between Japan and the United States. *Japanese Journal of Communication Studies*, *46*(1), 23–41.

Ngnoumen, C. T. (2019). The use of socio-cognitive mindfulness in mitigating implicit bias and stereotype-activated behaviors (Doctoral dissertation, Harvard University).

Nakui, T., Paulus, P. B., & Van der Zee, K. I. (2011). The role of attitudes in reactions toward diversity in Workgroups 1. *Journal of Applied Social Psychology*, *41*(10), 2327–2351.

Park, R. E. (1924). The concept of social distance. *Journal of Applied Sociology*, *33*(6), 881–893.

Perry, L. B., & Southwell, L. (2011). Developing intercultural understanding and skills: Models and approaches. *Intercultural Education*, *22*(6), 453–466. https://doi.org/10.1080/14675986.2011.644948.

Perry, W. (1999). *Forms of Cognitive and Ethical Development in the College Years (With a New Introduction by Lee Knefelkamp)*. Josey Bass.

Pike, K. L. (1967). *Language in Relation to a Unified Theory of Structure of Human Behavior*. Mouton.

Rabotić, B. (2010). Tourist guides in contemporary tourism. In: Proceedings on international conference on tourism and environment. Philip Noel-Baker University, Sarajevo, Bosnia & Herzegovina, 4–5 March, 2010, pp. 353–364.

Sabaghi, D. (2023). Amsterdam launches 'stay away' campaign targeting wild party behavior of young British tourists [online]. *Forbes*. Available at: https://www.forbes.com/sites/dariosabaghi/2023/03/31/amsterdam-launches-stay-away-campaign-targeting-wild-party-behavior-of-young-british-tourists/?sh=ecf33f764f04.

Sam, D. L., & Berry, J. W. (2006). *The Cambridge Handbook of Acculturation Psychology*. Cambridge University Press.

Schwarz, N. (1993). *Judgement in a Social Context: Biases, Shortcomings and the Logic of Conversation*. ZUMAArbeitsbericht, 1993/03. Zentrum für Umfragen, Methoden und Analysen – ZUMA. https://nbnresolving.org/urn:nbn:de:0168-ssoar-69961.

Spencer-Oatey, H. (2012). What is culture? A compilation of quotations. *GlobalPAD Core Concepts*. Available at: GlobalPAD Open House, http://www.warwick.ac.uk/globalpadintercultural.

Spencer-Oatey, H., & Franklin, P. (2009). *Intercultural Interaction: A Multidisciplinary Approach to Intercultural Communication*. Springer.

Straffon, D. A. (2003). Assessing the intercultural sensitivity of high school students attending an international school. *International Journal of Intercultural Relations*, *27*(4), 487–501.

Teasley, S., Covi, L., Krishnan, M. S., & Olson, J. S. (2000). How does radical collocation help a team succeed? In Kellogg, W. and Whittaker, S., *Proceedings of the 2000 ACM Conference on Computer Supported Cooperative Work*, Association for Computing Machinery, New York (pp. 339–346).

Telfer, D. J., & Sharpley, R. (2008). *Tourism and Development in the Developing World.* Routledge.

Tokyo Localized. (2023). Japan guided tours 2023: Everything you need to know about guided tours in japan [online]. Available at: https://www.tokyolocalized.com/post/japan-guided-tours-everything-you-need-to-know-about-guided-tours-in-japan.

Vivero, V. N., & Jenkins, S. R. (1999). Existential hazards of the Multicultural individual: Defining and understanding cultural homelessness. *Cultural Diversity and Ethnic Minority Psychology*, *5*, 6–26.

Westropp, S., Cathro, V., & Everett, A. M. (2016). Adult third culture kids' suitability as expatriates. *Review of International Business and Strategy*, *26*(3), 334–348.

Wiersma-Mosley, J. D., & Garrison, M. B. (2022). Developing intercultural competence among students in family science: The importance of service learning experiences. *Family Relations*, *71*(5), 2070–2083.

Wypych, L. (2020). Changes in motivation and acculturation strategies: A case of Indian students at the UoB. Available at: https://research.ebsco.com/linkprocessor/plink?id=46122ce4-c8ba-309e-b3b5-bf835feac8e1.

Zakaria, N., Amelinckx, A., & Wilemon, D. (2004). Working together apart? Building a knowledge-sharing culture for global virtual teams. *Creativity and Innovation Management*, *13*(1), 15–29.

Žegarac, V., & Pennington, M. C. (2000). Pragmatic transfer in intercultural communication. In Spencer-Oatey, H., *Culturally Speaking: Managing Rapport through Talk across Cultures,* Continuum (pp. 165–190).

# 12 Changing Government Attitudes

## Development and Management of New Knowledge on Security and Safety in Tourism

*María Jesús Jerez-Jerez*

### Introduction

The primary aim of this chapter is to determine the central and emerging challenges for asset protection and procedural safety given contemporary and ongoing social/political/economic threats to stability. These challenges are compounded by internal difficulties with respect to the workforce and prerequisite skills. The further intention of this chapter therefore, is to provide recommendations for industry and management in addressing and overcoming these challenges while maintaining competitive operations.

### Research Background

Experience in operations management over the last 20 to 30 years will almost certainly involve familiarity with making difficult or unpopular decisions that "keep you up at night" in an effort to ensure the continued survival and growth of a given organisation (Hartz & Quansah, 2021; Malinao & Ebi, 2022). This presents a particularly challenging set of circumstances for multinational companies (MNCs) operating in various states globally (Meyer et al., 2011). Moreover, ongoing political unrest and polarisation in key markets as well as its coincidence with significant social and working adjustments present a risk of an unprecedented global recession in terms of scale and impact (Behie et al., 2023). This complex set of risks, challenges and possibilities require that management identify innovative and otherwise underdeveloped prospects. To that end, the research presented herein seeks to first identify the industrial adjustments required by climate change and energy transitions, as well as an environmental crisis demanding changes to production and processes including accelerated digitalisation, while adhering to the restrictions imposed by a global pandemic. The research presented will then also identify the extent and impact of the socio-political fallout of Covid-19 including conspiracy theories and risks to cyber security, as well as Russian aggression in Ukraine and the inherent risk of significant economic and political divisions of global proportions. In light of these multifaceted and unprecedented challenges, the survival of corporations and particularly MNCs is acutely dependent on their ability to adapt, adjust and adopt new policies, procedures and sources of expertise.

DOI: 10.4324/9781003369967-17

The research presented here will therefore provide guidance with respect to just such new approaches from the perspective of both risk management and human resources.

The second section, Research Background, of this research presents an account of the background and previously successful fundamental principles that can therefore be maintained. The third section, Developments Risks and Responses, will then present an analysis of the various and complex underlying issues, controversies and problems requiring MNC and organisational adjustment. This will lay the groundwork for the fourth section, Solutions and Recommendations, which will seek to establish prospective solutions and recommendations for management as well as MNC operations and procedures, before addressing Future Research Directions in the fifth section and presenting a Conclusion in the sixth section.

In the interests of clarity and accuracy, this paper is not intended to reflect the standard research article model including a problem analysis, methodology, findings, etc. Rather, the paper is constructed as an account of the aforementioned changes and developments in terms of industry operations, human resources, economic impacts, and socio-political tremors, all of which threaten a significant risk to organisational survival if they are underestimated or if responses are inaccurately planned. The paper is based on a key-note address by the first author to the 17th Symposium on Loss Prevention and Safety Promotion, at Prague, in June 2022.

### *The Foundations of Success*

Given the unprecedented challenges that loom for MNC's and major organisations, persistent commitment to such principles is more critical and valuable. Moreover, continued financial success is reinforced and buttressed by a foundation of confirmed ideals, beliefs, values, and principles expressed and maintained by all staff and across all operations (Szyszka, 2017). These form the seven pillars of operational traits, encouraging successful operations and including, accountability, being results oriented, maintaining a skilled staff, collaboration, prioritising a work life balance, taking proactive measures, and encouraging creativity. To that end, the aim is to challenge staff to maintain these principles or pillars in terms of being creative, accountable for their actions, collaborative, proactive, focusing on results and adopting a measured but risk-taking approach. Moreover, executing operational traits based on these pillars provides a significantly improved chance of success for process safety and asset integrity, among other cross organisational programmes, as well having a positive impact on operations (Behie et al., 2023). It is in this respect therefore that key to ongoing financial success is the commitment to what can be cultural traits, as well as foundational principles (Kwak & Anbari, 2006).

Commitment to such principle-based operations, at all levels of an organisation, will result in producing a learning organisation since a prerequisite for maintaining these principles is effective communication, self-examination programmes and ongoing training. Furthermore, this tends to reinforce resilience to the inherent difficulties of daily operations, particularly with respect to MNCs, and improved coherence in their navigation such that they become characteristic of the corporate

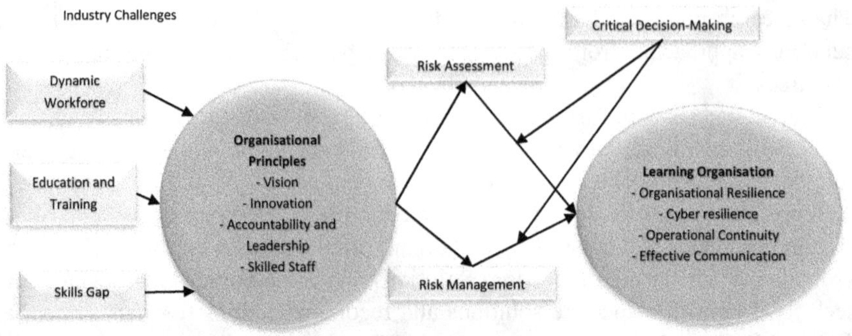


*Figure 12.1* Learning organisation conceptual framework.

culture (Cai et al., 2023). These standards must then be extended to all jurisdictions, towards a "world-wide standard of care" in order that these foundational principles function as such. The cost of failure to apply a sincere and equal standard across all operations in all jurisdictions is evidenced by the examples of companies who consequently experienced significant downturns or collapse (Behie et al., 2023) (Figure 12.1).

### *The Importance of Integrity*

The most important principle among those identified is integrity since commitment to this is foundational to the demonstration of organisational values, to maintaining the other principles across all operations, as well as indicating the sincerity and trustworthiness of the organisation (inside-out paradigm) (Cashman, 2017). This also has the effect of reinforcing trust in leadership of an organisation since their commitment and maintenance of integrity indicates the sincerity of the approach at all levels. Moreover, the consequent trustworthiness of the organisation provides a pathway to develop and encourage the outside-in paradigm of third-party trust (industry groups, suppliers, government, etc.) (Habib et al., 2012).

### *The Balance of Innovation and Exploitation*

Organisational success tends to be preceded by maintaining balance between investment for short- or near-term growth (exploitation), as well as evaluating, anticipating, and investing in future industry trends (innovation). In other words, an exploratory attitude is a near prerequisite for success since companies and organisations must combine an understanding of the industry with the inclination to identify opportunities in change to increase efficiencies, efficacy and improve their financial position (Che-Ha et al., 2014).

In fact, there are countless cases of organisations whose failure is predictable by their exclusive focus on existing production at the cost of adjusting in time to new industry-wide changes and trends (Behie et al., 2023). Conversely, the same is also true of those organisations for whom failure is preceded by a total focus of financial resource and effort on anticipated trends (the green shadowed part). An underlying characteristic of these organisations is the persistent scepticism of current success, buttressed by a willingness to invite and accommodate challenges and change.

***Key and Critical Corporate Decision-Making***

Decisions taken by operations or corporate management that have the potential for large-scale or organisation-wide consequences, particularly negative consequences, are "critical". Such decisions are premised on and necessitate a risk assessment which projects potential consequences taking into consideration the possibility of inaccurate evaluation of current conditions (Aven, 2016).

There are two categories of critical decisions: planned and unplanned. Planned decisions typically require that senior management source internal, expert advice. For example, the decision to complete a major acquisition may require detailed evaluation and assessment with respect to legalities, market reactions and any operational consequences, as a significant prerequisite for such critical decisions (Choi et al., 2018). Maintaining aforementioned fundamental, organisational principles is also central concern (Sawalha, 2015). Consequently, executive management must be involved in every stage such that final decisions and execution is a result of weighing all the relevant factors.

Conversely, unplanned decisions typically occur at the level of plant operations and often as a result of a warning or significant incident. Such events and the resulting actions can also have long-term negative impacts for an organisation (Jain et al., 2020). That risk is exacerbated by the fact that the nature of unplanned decisions are such that they do not afford management time to conduct an in-depth assessment considering impacts, otherwise exclusively overseen by C-Suite executives.

In light of that and following a Chevron Refinery fire in California, recommendations were made by a 2015 CSB investigation report that individuals submitting critical decisions be required to seek review from their manager, and then further up the chain of responsibility, in an auditable process. These processes, in accordance with the CSB document recommendation, would therefore intersect with all stages of decision making. These 2015 recommendations echo the conclusions of previous CSB investigation reports (Kalantarnia, Khan, & Hawboldt (2010); Clarke (2017)) wherein it was determined that Process Safety is insufficiently incorporated into all levels of managerial decision-making (CSB, 2007, 2014). It is therefore in the best interests of organisations that risk assessment processes are ingrained at all levels of decision making to mitigate the potential negative impacts that particularly arise from unplanned critical decisions (Islam, 2023).

In order to ensure efficacy, risk assessments must be directed by a senior risk engineer who is sufficiently familiar with the operations and facilities involved, and must also include all the relevant representatives, safety professionals, management and operators. That assessment will hypothesise and assess multiple scenarios and potential consequences or outcomes of each decision option. These are then cross referenced against the prevailing conditions or circumstances at the time and the organisations existing risk matrix, producing a probability of occurrence for each consequence of each decision scenario. The resulting decision is therefore taken only with an understanding of the determined risk level; the greater the risk, the higher the position of management in the organisation required to take responsibility. In line with this approach therefore, all senior and executive management will be aware of all critical decisions since the final point of responsibility will include them and require their approval. A thorough-going knowledge of risk-based decision management is a crucial requirement for executives and management. To that end, a "risk register document" outlining processes for critical decision-making, can provide a basis for consistency in critical decisions with respect to every level of business operations (Smalley & Chebotar, 2017).

Of course, rigorous risk assessment processes alone will not be sufficient to pre-empt, anticipate or avoid all incidents but may present a substantive means of mitigation and moreover, significantly reduces the associated risk of poor decision-making exacerbating negative incidents. Openness and the willingness to engage with this approach by leadership will therefore be necessary to improving organisational operations and risk mitigation if only to reinforce resilience practices. Constitutive of such resilience capability is not only improved anticipation and reception to unexpected threats but also to improved lines of communication with leadership. That leadership is of course charged with the responsibility to take contingency measures to mitigate and streamline recovery but efficient communication, particularly in the case of warnings, will permit an appropriate and adequate mobilisation of emergency responses.

### *Managing Incidents, Responding to Emergencies and Disaster Recovery*

Of course, most operating facilities are required to have emergency plans and contingencies in place in order to permit speedy tiered response to unexpected incidents. Such Emergency Response Plans describe and define incidents, emergencies and their management from initial response at the level of plant personnel and operators, to escalation requiring further resources to ensure an appropriate response. The Crisis Management Plan defines such escalated responses and triggering decision points and includes plans for mutual aid as well as coordination with local emergency response, as and when necessary. Of course, incidents that significantly damage crucial facility systems or components, and are therefore disruptive to business operations, can also undermine the long-term viability of an organisation which depends, in part, on its capacity for quick recovery from such significant incidents. Such recovery and the resumption of regular operations is determined by the development, preparation, and implementation of recovery plans, amplifying the importance of having a Business Continuity Management Plan (BCMP) in anticipation of all reasonable cases (Razzetti, 2022).

There are eight steps in a BCMP, requiring that Senior Management both actively participate in and review the process in order to identify possible adjustments for increased efficiency. Following initial analysis and identification, risk assessments can be conducted as a precursor to developing plans and preparing measures that can be implemented efficiently. The steps in a BCMP complete a closed loop ensuring audits, feedback, and reporting.

In a practical sense, therefore, a BCMP underlines the responsibility of C-suite management to ensure that operations are sustainable and capable of continuity and quick recovery, i.e., resilient. Therefore, management are required to assess risk management and implement plans in order to meet their responsibilities and deliver the goals agreed with shareholders. Moreover, the Board of Directors carry a judiciary responsibility to ensure that senior management are accountable for the success, continuity and resilience of their operations.

## Developments, Risks and Responses

### *Issues, Controversies, Problems*

#### *Changes, Threats and Responses*

The combined effect of high retirement numbers over the last several years and consequent shortage in skills, as well as changes in attitude among the workforce, present an immediate threat (Stewart, 2010). Success in such an environment will require that companies and organisations meet these challenges, developing new approaches to attracting and retaining technically skilled operators within the workforce. Such approaches might include promotion of schemes and programmes for virtual work; extending employment options for women and older employees; encouraging and nurturing talent development and mobility. Of course, such approaches are not without their own challenges in light of the trends away from globalisation and towards nationalisation. To date, globalisation has had an ongoing and significant impact on varying nations' ability to reinforce their economies and improve general quality of life, if by no other means than supporting a global connection between states and peoples. Nevertheless, that same connectivity means that instability in any one region can have a significant global impact. The Covid-19 pandemic presents one such contemporary and large-scale example, still effecting global markets, organisational operations, and governmental policy two years later. Yet, the greatest global risks likely to be confronted are indicated by climate change and cyber-attacks of increasing sophistication and scale (Naveenan & Suresh, 2023). This is exacerbated by the more familiar but no less destabilising risk of war and aggression between states such as Russian aggression in the Ukraine, considered a significant contributor to the danger of global recession (Saliu & Memaj, 2023).

These changes in the profile, qualities, and commitments of the workforce, as reflected by the significant difference between contemporary workers and those of past decades, will persist such that the workforce of tomorrow will have different expectations in terms of work/leisure balance, experience and expectations. This is an issue of particular pertinence for companies and organisations within the energy sector over the next two decades, in terms of attracting and retaining workers with the required technical knowledge (Sumbal et al., 2017). To that end therefore, adaptation and the development of new approaches become critical organisational qualities as competition for technical and managerial talent becomes increasingly acute.

Three primary generations were active participants in the workforce of the 1900s, generally organised hierarchically in respect of seniority (North, 2019). Conversely, the contemporary workplace is increasingly complex in respect of that "traditional" division of labour (Calk & Patrick, 2017). To that end, it is projected that by 2025, 75% of the workforce will be millennials epitomising these new attitudes to work and society (Pathy, 2021). Chemical process industry (CPI) companies depend on their ability to attract and retain technical staff to maintain success. However, the demographic gap means that such companies must navigate a significant lack of talent for mid-career professionals and must identify senior talent tasks while developing an ongoing resolution to address this issue and retaining workers with technical skillsets. This strategy is further complicated as the interests, demands and focus of new generations continue to change (Bencsik et al., 2016).

In order to meet these challenges, companies must not only understand the nuances and dynamics of changes among the workforce but also establish structures and programmes that anticipate and adjust to meet those changes in the future. By way of contextualising this necessity, the oil and gas industry are expected to see a departure of 22,000 petro-technical professionals over the coming five years, and will attract only 17,000 such professionals, according to the US Census Bureau (Behie et al., 2023). Ultimately, this will result in a 20% deficit of technical staff even if accounting for improved technological and operational efficiencies, and increased productivity. Resolving this issue will require more than an acceleration of skill development but also taking steps to ensure the provision of viable and attractive career propositions for new talent.

Similarly changes in roles will not perpetuate the norms preceding them. The scale and force of the generational changes among the workforce are exerting pressure on the workplace itself with no signs of abating. The effect of these forces and pressures can be seen across various levels of the workplace such as HR recruitment processes and remuneration; approaches to work; internal and external communications; training and development; employee motivation and advancement; organisational structure. This means, among other things, that flexibility across many of these areas including working times, places and structures are vital requirements. To that end, some organisations have already begun efforts in redesigning their workplace to accommodate open office concepts or expand remote working policies and practices.

*Training and development.* Similarly, employee development and its associated programmes also must adapt and adjust such that they offer high-quality technical and professional training, while providing substantive assessments, reports, and feedback. Such adjustments will include appealing to new generations' technological and communicative preferences. Simulation tools, for example, as well as on-the-job training can be implemented to address core issues and with greater efficiency than has historically been the case. To that end, the oil and gas sector includes multiple organisations who have developed training centres, employing a wide range of techniques which already include simulation and the use of props to demonstrate the functionality and operation of key components (Morrison et al., 2019).

*Global workforce and technical skills shortages.* Compounding the difficulties, challenges and risks involved in navigating contemporary dynamics inherent to a multi-generational workforce, a further challenge is already being identified. With specific regard to the oil and gas industry, this is the challenge of diminishment in the availability of workers who meet the skill requirements to adopt roles without significant disruption to operational efficiencies.

Among the published reports of the World Economic Forum, they state,

> The global economy is approaching a demographic shock of a scale not observed since the Middle Ages. There are several factors that exacerbate this development. The energy transition to fight climate change with all its new energy sources and processes, is demanding for its development and construction a growing number of technically skilled people, whereas a technical profession in the decades before was not valued highly. The working-age populations of many developed economies will start to decline shortly. Numerous organisations will be unable to find enough employees in their home markets to sustain profitability and growth. By 2050, the global population of those 60 years old and older will exceed the number of young (less than 15 years old) people for the first time in history.
>
> (WEF, 2010)

Projections of skilled worker availability from 2020 to 2030 for multiple states across the Americas, Europe, Asia, and Oceania are indicated by Figure 12.2.

Therein, grey colours indicate attractiveness for migrant workers ranging from high represented by dark grey, and low, indicated by light grey. Skill gaps are also colour coded: very high skills gaps (black); high skills gaps (medium grey); low skills gaps (light grey); no skills gaps (white).

Following these projections, 2020 to 2030 will see high and very high skills gaps appear in the North American continent in construction and engineering, as well as IT and education, among others. Moreover, among the difficulties and challenges these conditions present is the risk to undermining economic growth globally, further impacting changes in the workforce. The acute nature of that risk is indicated by the fact that 80% of projected workforce shortages reflect diminishment of skills rather than availability of workers. Consequently, organisations which are particularly effective in talent attraction and retention will find themselves better positioned and with significant advantage within their industries.

Talent attraction and retention will therefore be dependent on accurately identifying and satisfying the needs and preferences of workers, critical to which is the provision of opportunities for development and training at work. Learning and Management Systems must therefore incorporate parameters anticipating these shortages so they may be appropriately pre-empted and in order to reflect the differing demands of the dynamic generational mix among the workforce of tomorrow. Satisfying the expectations associated with millennials, for instance, will require implementation of development plans that advance requisite technical skill and existing talent within the workforce. Moreover, such plans must also be replicable,

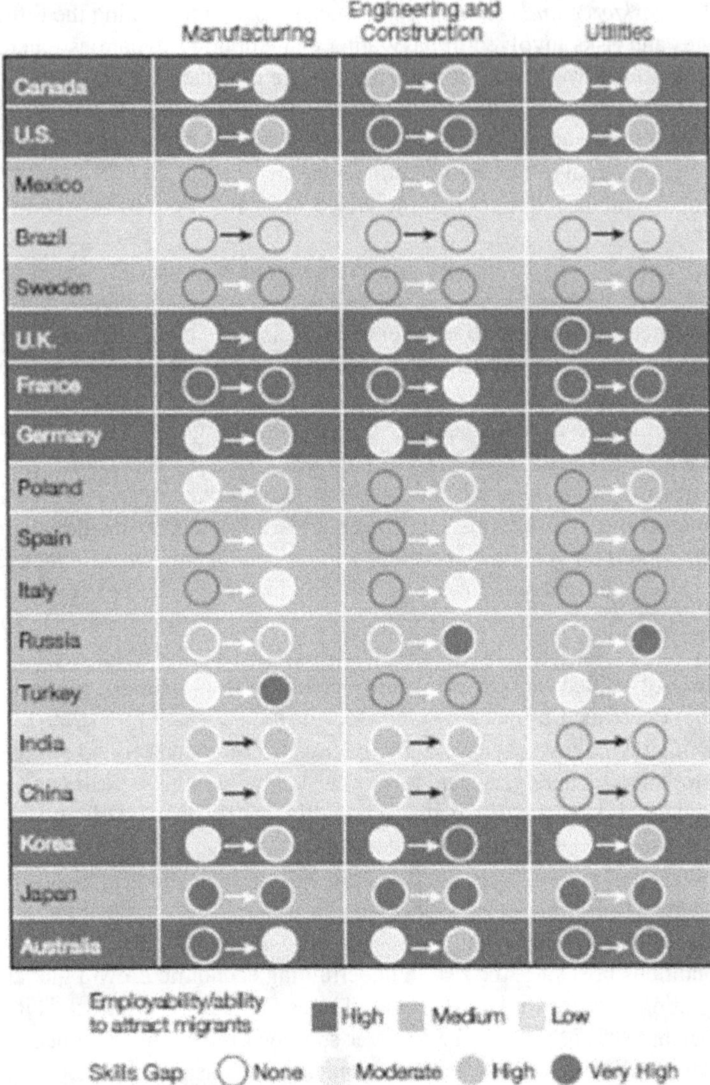


*Figure 12.2* Skills gaps and ability to attract migrant workers (WEF, 2010).

sustainable and subject to appraisal, while equipping workers with an accelerated path to progress.

*Closing the skills gap.* Addressing seven particular factors will present a means by which to begin reducing the gaps indicated by these upcoming crises, at least with respect to the CPI: introducing strategic workforce planning, easing migration, fostering knowledge-sharing, developing a talent trellis to create more growth opportunities, encouraging temporary and virtual mobility, extending the talent pool by becoming an employer of choice, and increasing employability through community outreach and vocational programmes (Behie et al., 2023).

*Strategic workforce planning.* A key step in the process of addressing these concerns will require that the functions of Human Resources are incorporated into strategic planning. To that end, anticipations, projections and profiles of workforce supply and demand can and should be conducted with a view to identifying potential gaps five to ten years in advance. These adjustments will also have to be echoed by potential workers who must be constantly aware of changes in skill requirements, reinforcing the importance of government roles in fostering opportunities and developing relations. Of course, this aspect of workplace adjustments can be streamlined by increasing automation and digitalisation, producing greater efficiencies in training especially if enhanced with interactivity and a degree of entertainment among teams of workers.

### *More Issues, Controversies, Problems*

Two virtual conferences involving panel discussions were conducted by the Mary Kay O'Connor Process Safety Centre (MKOPSC) during Covid lockdowns. Over the course of the first panel discussion entitled, "Integrating Pandemic Preparedness and Response into Business Continuity and Risk Management Planning", the difficulties and options available for maintenance of operations within the CPI were discussed by senior management representatives of major chemical companies. The second panel entitled, "The challenges of supporting critical infrastructure operations during the Coronavirus pandemic", focused on suppliers and contractors required to buttress industry operations during the pandemic such that they could navigate restrictions. Both discussions identified concurrent principles based on an understanding of day-to-day obstacles to operations. Among those principles, assuring the involvement of senior management in routine communications with workers was considered critical. Moreover, conducting such discussions with sufficient sincerity and efficacy required that they involve and include workers relatives, as well as experts of relevant fields who can disabuse stakeholders of rumour and uncertainty. Furthermore, the general recognition of a lack of preparation on the part of all major social institutions including industry and government was a critical lesson for both panels. Key lessons identified included: Planning insufficiency for adverse conditions and the inadequate scope and detail of existing plans, as well as inadequacies of existing risk assessments; Worker resilience and their commitment in the face of unforeseen challenges cannot be underestimated; Limitations on the capacity to maintain operating under crisis conditions; The necessity of developing sufficient indication mechanisms permitting early response at varying organisational levels; The necessity of benchmarks and indicators defining conditions of recovery; Reliance on scientifically verifiable evidence; Plans for returning to work must be constructed such that they are appropriate for and relevant to all workers; Improving operational efficiencies requires relevant training with respect to new technologies.

### *Even More Issues, Controversies, Problems*

Despite the advantages that came with an enforced move away from the traditional working environment into an increasingly digital space, consequent and ongoing cybersecurity breaches are becoming proportionally catastrophic. By way of example, there has been a 435% increase in ransomware cases, and 95% of cybersecurity concerns originating in human error, according to a report by the cybersecurity company, Deep Instinct (Alford, 2021). Importantly, large organisations are not the exclusive focus of ransomware attacks. Rather, they target any individual, group or body possessing valuable data. To that end, healthcare and public health were the primary ransomware targets in the US in 2021, as indicated by the IC3 report from the FBI. Among their threats, such cyber-attacks hinder infrastructural continuity by seeking to destabilise performance, undermining or deteriorating products, impeding services, and may even lead to loss of life (Behie et al., 2023). Given the susceptibility of centralised data storage characteristic of modern systems, such attacks carry an implicit, reverberating risk to an entire system (Cazorla et al., 2016). Ensuring the minimal possible degree of risk will require that organisations implement both cyber security and cyber resilience measures.

Therein, cybersecurity refers to the impediments to various cyber-attacks, and cyber resilience refers to protocols for damage incurred by attacks, intended to mitigate impacts and permit efficient resumption of normal functions.

Previously, cyber-security was sufficient for data breach prevention and general technology infrastructural protections, but the combination of technological sophistication, access and reliance has rendered these approaches largely superficial. The sheer depth of dependency on internet and cloud-based applications requires a proportional degree of emergency response planning to mitigate and restrict any impact on operations. This is exacerbated by the increasing complexity and interdependence of systems (Lewis, 2019). For instance, traditional systems tend to function with the same operating systems and data formats but contemporary systems such as Cyber-Physical Systems, function with application-specific operating systems and data formats. Thus, organisational leadership have a duty to ensure provision of the requisite tools for cyber-attack prevention and mitigation, permitting a path by which to maintain operations and limit negative financial or media impacts. Nevertheless, no recognised cyber resilience framework is currently in practice and there is no model by which organisations can measure advancements in base-level cyber resilience requirements.

Typically, reference to “conspiracy theories” inspires thoughts of political actions and national security concerns (Georgiou et al., 2020). Recent history, however, has witnessed the expansion of conspiracy theories extending across multiple nations, organisations, as well as religious, social and political institutions. A contemporaneous example is amply provided by the plethora of Covid vaccination conspiracy theories. Such conspiracies and attitudes of institutional scepticism threaten exacerbating social divisions at precisely the time that confrontation with and responding to major global threats requires good faith, mutual engagement. The ostensible democratisation of information that has developed with the rise of accessible technology has coincided with a disintegration of the divisions between fact and fiction. The reach, influence and impact of news platforms, for instance, multiply the implicit threat of editorialising and publication of disinformation in order to reinforce an ideological position; truth has become difficult to identify.

Furthermore, ongoing Russian aggression in the Ukraine is not only having a destructive impact on that nation and its infrastructure but also the global economy particularly in respect of supply chains (IMF, 2022). Moreover, the consequences for other nations threatens the stability of peace in Europe as Russian actions escalate. These risks are, at least in part, amplified by the economic effect for European dependence on Russian Energy. The combination of these economic impacts in the wake of a global pandemic will invariably exacerbate inflation and may tip the world's major economies into a recession. Such a complex and dangerous set of conditions demand proportionally dynamic and unprecedented responses from global leaders not only in order to address the impending economic challenges and mitigate their impact, but also to engage in a transformation of energy supply towards improved sustainability and integration, ensuring long term resilience.

**Solutions and Recommendations**

Integration of critical decision-making into management processes and derived from substantive risk analysis is a prerequisite to ensure those procedures are included in the general approach of management. Moreover, it is necessary that such risk assessment processes be replicated throughout a given organisation, at all levels, including corporate guidance documents with the knowledge and support of management and Directors. This will mean that, upon implementation, both C-Suite and relevant management will have input and involvement as a matter of procedure, ensuring that such decisions are appropriately assessed, and that determining decisions are executed at appropriate organisational levels.

*Corporate Risk Register Development*: A substantive risk register protocol, compiling an account of high-level risks and reported to the Board bi-annually, must be constructed and maintained by organisational risk groups. The focus of such a register would be exclusive to actions corporate risk matrixes designate as "high" and should therefore be promoted and endorsed by senior executives. Concurrently, BU related risk registers would accumulate an account of changes, decisions and actions over several years, particularly reflecting changes following the previous reporting period.

*Directors' Enhanced Roles*: For most large scale or international organisations, Boards of Directors are composed of a wealth of experience and expertise that can provide greater input at appropriate levels. This may, in certain instances, require expansion in board member numbers to reinforce that expertise, supporting executives, and crucially, reassuring shareholders that oversight and advice are routinely available.

*Technical Staff Mentoring / Coaching*: Retired staff and relevant experts represent an opportunity for part-time mentoring and coaching of entry level technical workers, maximising the output and impact of their knowledgebase for continued operations. Moreover, recruitment of such experienced staff and subject experts will address the challenges presented by skills shortages, present compelling opportunities for retired workers, and reduce the learning curve of entry level workers.

*Senior Management and C-Suite Executive Coaching*: Multiple organisations have already found success in providing coaching for their executives. Moreover, these benefits extend beyond developing technical skills but also improved training in human factors and providing the requisite skills and tools to address contemporary challenges to operational efficiency.

*Skilled Staff Attraction and Retention*: The projected manpower crisis must be addressed, in part, by organisations' adjustments to Learning and Management Systems in order that they can be appropriately prepared and responsive to changes in needs and preferences of both new and existing intergenerational workers. These concerns are amplified by restrictions and shortages on skilled worker migration in the wake of the Covid pandemic, exacerbated further still by the combination of a military aggression in the Ukraine inclining markets and economies towards a global recession and an unprecedented combination of underlying factors.

*Talent Pool Development*: There are some substantive possibilities for organisational actions that may mitigate the skills shortage, as well as developing and retaining an existing skilled workforce:

- Developing a leadership structure reflecting the diversity of the workforce, adopting HR functions as central to functional corporate strategy, and extending these principles to a general understanding of the requirements for being "an employer of choice".
- Progressing understanding of the preferences and needs of an intergenerational workforce, as well as the provision of relevant and appropriate training opportunities including virtual internships and work schemes.
- Construction, development and promotion of innovations in employment opportunities of senior staff, as well as considering provisions to improve day care opportunities for parental engagement.
- Innovation and development in training programme construction and promotion, aligned with the attributes across an intergenerational workforce, providing opportunities for ongoing refinement through feedback, as well as skills and competency assessments.
- Governmental support for such organisational programmes and projects may also be a significant contributor by means of investment in education, and revisions in immigration and retirement policies.

*Championing Cultural Honesty*: Executives, Senior Management and leadership must adopt, encourage and champion a "just culture" approach wherein errors, failures and shortcomings are reported openly, honestly facilitating improved recovery, prevention and resilience. To that end, all levels of leadership must champion a just culture and can do so in the first instance, by ensuring that review discussions openly addressing success as well as failures, weaknesses, and shortcomings. In the case of being confronted by such failures, leadership must adopt the responsibility to provide encouragement and support throughout the organisation in order to adapt and improve. Concealing weaknesses carries the risk of embedding poor operational practice, compounded by failing to respond at the most opportune moment for improvement.

*Reinforcing Commitments to Safety*: Developing a "Safety Motto" programme that involves an intra-organisational poster campaign as well as formalised ceremonies recognising staff, management and other relevant individuals' continued commitment to safety may present one effective means by which to encourage such commitment across the entire organisation. To that end, posters must be positioned in numerous places throughout the company and in clear, elevated places to serve as permanent reminders of staff commitments.

*Celebrating All Successes*: In light of global difficulties and challenges, acknowledging and celebrating success, no matter the scale, is a crucial feature of positive reinforcement. Though opportunities for such celebrations abound, it's crucial that organisations understand the impact such acknowledgment can have on morale and therefore performance.

*Incident Recall*: A poignant reminder of incidents across numerous industries as well as the importance of anticipatory and preventative measures is encapsulated in, *What Went Wrong?: Case Histories of Process Plant Disasters and How They Could Have Been Avoided* (2009), by Dr. Trevor Kletz. Nevertheless, repeated incidents of identified and avoidable root causes suggest that industry is yet to learn from past disasters. A presentation entitled, *Guidance for Effective Management of Risk Associated with Highly Hazardous Chemicals* before the Safety Advisory Board of the OPCW in November of 2021, offers one such example of this failure to learn. Therein, presenters discussed over a century of ongoing ammonia nitrate incidents, indicating a similarly ongoing failure to learn from past incidents across industry.

The construction, development and maintenance of an incident database designed to record incidents and remind workers of such incidents on their anniversary, reinforcing the lessons drawn from these cases, may go some way towards addressing the failure of industry to learn. To that end, an opportunity is presented for a safety engineering programme making use of such databases, already developed but unadopted by industry.

By and large, regulators across all jurisdictions do not possess technical expertise reciprocal with industry and are further hampered by being understaffed, and consequently overworked. Greater efficacy in their work will therefore require that industry meet their obligations to work in concert with regulators, particularly in light of their mutual goals. Historical tensions between industry and regulators are largely unhelpful and a review of rules of engagement should be included.

The environments, applications and uses of technology are increasingly being interspersed throughout daily and working life (Schwab, 2019). Consequently, it will be necessary for business to engage in a fundamental reassessment of both their approach and operations, changing work and the work environment of the future. Moreover, cooperation with government will be critical for business to meet the demands of adaptation at the requisite speed while also ensuring appropriate provisions and conditions for workers. Predictably, those organisations with significant recent success will be better positioned to meet and satisfy those demands sustainably and cost-effectively. Examples can be drawn from developing automation which, despite its near ubiquity across sectors, has, perhaps consequently, revealed an increasingly problematic knowledge gap in the competency of the workforce.

Digitalisation has similarly improved operational efficiencies and resiliency, as well as encouraging creativity in the working environment but presupposes a workforce with requisite knowledge to navigate these tools. Thus, leadership and executive management are required to maintain responsiveness in the integration of the appropriate tools and in the appropriate manner relative to their workforce. This might mean, for example, that hiring criteria begin to include degrees of digital competence as pre-requisites even when employing workers outside of IT related roles. These requirements for new workers will be reflected across multiple sectors, as a general principle of hiring practices, evidencing a knowledge base in terms of digital technology in the form of experience or certification. Finally, this will also require that management and human resources have available and appropriate means for measuring digital competence and maturity such that it can be communicated to C-suite leadership towards ensuring and maintaining organisational competency levels.

## Future Research Directions

The exponential pace of global changes, whether political, social, technological, or industrial, show little to no sign of abating. In light of that and notwithstanding the tragedy, trauma and turmoil caused by the global pandemic, lockdowns and restrictions on normal business operations compelled general reconsideration, reassessment, and reinvention of general business realities. This was, at least in part, reflected by the workforce reaction in the form of the great resignation (Crowley, 2022), suggesting a social priority reassessment. Among these multiple, long-lasting impacts of the global pandemic, was the accelerated utilisation of existing tools permitting continued communications and operations such as virtual communications and distance learning. Nevertheless, such decentralisation and dependence on virtual connections also precipitated a substantial expansion of misinformation, reinforcing the importance appropriate reference frameworks in order to undermine the influence of falsehoods.

Moreover, an already deepening knowledge and skills gap in the workforce was exacerbated by the large number of early retirements encouraged as a means of avoiding and mitigating the spread of the virus. This loss of the most experienced workers also had the impact of weakening organisational understanding of cross generational needs and preferences. Yet, steps can be taken to mitigate against any such losses including encouraging mutual goals, team engagement and a shared vision across workers, improving on training and development by implementation of innovative mental models, as well as the incorporation of systems thinking to better understand interconnectivity and fractionation. Though these do not represent new concepts, their continued efficacy indicates an applicability even in dynamic and changing conditions. It is to that end, for example, that learning laboratories have been effectively employed across a wide range of industries towards refining managerial views (Haiyun et al., 2021). Implementation of such methods towards improved team harmony and leadership actions has been consistently effective in general improvements to the work environment. Moreover, confrontation with and acceptance of organisational weaknesses has improved both general safety and productivity (Gotsis & Grimani, 2016). Conversely, heightened internal competition producing increased anxiety tends to undermine creativity and will encourage a gradual discarding of virtues and values in favour of a sense of victory (Behie et al., 2023).

Recognition programmes in the US have seen significant success leading to consistency in implementation of and improvements upon current practices, born primarily as a consequence of increased and effective interaction between government and private organisations. The impact of that success extends to improvements in employee turnover, productivity and cost efficiencies (Mustard & Yanar, 2023).

## Conclusion

Ongoing success is predicated on the ability of organisation to identify, understand, and replicate underlying factors of previous success, as a foundation for future actions. Moreover, this must be coupled with an organisational commitment to develop and consistently meet expectation in respect of social responsibilities in order to maintain competitiveness. In this sense, the experiences and difficulties involved in operational maintenance in the midst of pandemic restrictions have compelled the development of both processes and tools that will permit of improved resilience in the future. Among those lessons is the value of developing and enhancing relationships to further deepen and reinforce cooperative inclinations across all departments and individuals within an organisation. Such relationships can even include, and are not limited to, closer integration with regulatory bodies; improving informational communication with particular respect to training and learning; encouraging close and operationally specific engagements with academia to develop substantive training programmes for intergenerational and diverse workers. That diverse workforce must also feel understood and to that end, organisations must be aware of and responsive to their needs, investing in innovative opportunities for extending their talent pool and existing skills. Additionally, an attitude of perpetual learning must be ingrained and embraced by management, as a necessary feature of critical decision-making. Finally, resiliency measures can be further reinforced by the systemisation of preparation including annual rehearsal of continuity and response plans, ensuring organisation-wide familiarity with all possible scenarios. Given these lessons, industry leaders and stable organisations of the future will be those that adapt, adopt and embrace the inevitabilities of fundamental change, basing their actions on Principles of a "Learning Organisation" as a permanent value.

In light of recent experience, management must maintain permanent awareness of political and socio-economic conditions in regions of operation in order to appropriately project potential impacts. This includes the prospective risks and rewards of increasing digitisation and AI-based automation.

Large organisations may increasingly turn their attention to improved human performance, as working conditions begin to improve. Moreover, the anticipated impact of the 4th industrial revolution requires proportionate preparations across all industries and major organisations. Though this may mean job elimination in a number of areas, anticipated job creation resulting from industry and global changes, demand developments in performance. A core asset of any given organisation therefore is Operator 4.0 as well as it's training and development (Peruzzini et al., 2020). Thus, new empathetic paradigms must be oriented towards human-centric designs with a focus on the support and assistance technology may offer workers (Lu et al., 2022).

Though anticipated difficulties, obstacles and challenges abound, potential rewards cannot be understated for those organisations adaptable and committed to new horizons in operations.

**Case Study**

One of the biggest challenges for continued organisational success are the ongoing threats of economic instability exacerbated by domestic politics. A new holiday and tourism agency, "XBorders", provides tours and trips based almost exclusively on influencer marketing, targeting millennials and Gen Z who represent the diverse needs and preferences, nuanced by experience of the global pandemic.

To address these challenges, XBorders invites travellers from all over the world, to contribute to the brand image by uploading images, videos and blogs documenting their experiences in return for rewards, discounts and point accumulation opening up new trip opportunities. This approach of XBorders will require close coordination with both Marketing and Cyber Resilience departments, securing against the ongoing and complex threats of malicious bots and hacks, while amplifying the of a diverse and complex consumer base. Reinforcing this approach will require a workforce that is reflective of the demographics, beliefs and principles of that consumer base, ensuring a commitment to ongoing learning as a fundamental value of the organisation. To that end, senior management engage the workforce in innovative training programmes designed to develop the existing skills base of their workforce, combining group, workaway trips and internal rewards systems that offer a clear path to professional progression. An internal function of such training and development is the construction of reports to be presented to executives to permit accurate decision-making and anticipatory business continuity planning.

Active engagement in brand development from the consumers, combined with a dynamic workforce reflective of their needs and preferences, positions XBorders to take direct advantage of the changing and reactive nature of modern socio-political conditions. Moreover, the growth and development of the brand presents a set of unique possibilities for engagement and cooperation with domestic and foreign tourism industry councils and ministers, allowing for an accelerated recovery from the restrictions imposed by the global pandemic.

**Case Questions**

- What levels of management were involved in the critical decision-making involved in this case?
- What dynamics must be considered given the complex and intergenerational workforce? This will include reference to the limitations as well as opportunities presented by diversity.
- Given the outstanding challenges of a post pandemic environment, as well as ongoing socio-economic and political difficulties, what are the innovative possibilities presented by greater governmental and industry cooperation?

**Key Terms and Definitions – Definitions for the Key Constructs**

**Critical Decision-Making**: Decisions taken by corporate leadership characterised by potential for organisation-wide consequences.

**Cyber Resilience**: Mitigating control measures implemented to reduce and respond to cyberattacks and general threats to cyber security.

**Dynamic Workforce**: Changes in the contemporary workforce characterised by an intergenerational mix and introduction of new preferences, beliefs, and requirements.

**Education and Training**: Government or industry led programmes focused on improving the knowledge and skills base of the current and future workforce.

**Learning Organisation**: Organisations with principle-based operations, championing ongoing training at all levels, communication and self-examination programmes.

**Organisational Principles**: The ethos, values and attitudes within an organisation as reflected in policy, practices and procedures.

**Organisational Resilience**: The capacity of an organisation to sustain and continue operations described by total mitigating and preventative measures characterised by accurate anticipation and effective communication.

**Risk Assessment**: Regularised evaluation of projected consequences and by-products of operations conducted by management and presented to leadership for informed decision-making.

**Risk Management**: Identification, assessment and control of operational, financial, legal and security dangers for general business continuity typically outlined in a plan developed by management.

**Skills Gap**: Discrepancy between the operational requirements of roles and the available workforce possessed of the requisite knowledge and abilities.

## References

Alford, F. (2021). Current and Emerging Threats to Data-How Digital Curation and ARM Principles Can Help Us Brave this New World: A Case Study. Master's paper. Available at https://doi.org/10.17615/9ccw-ws75.

Aven, T. (2016). Risk assessment and risk management: Review of recent advances on their foundation. *European Journal of Operational Research, 253*(1), 1–13.

Behie, S. W., Pasman, H. J., Khan, F. I., Shell, K., Alarfaj, A., El-Kady, A. H., & Hernandez, M. (2023). Leadership 4.0: The changing landscape of industry management in the smart digital era. *Process Safety and Environmental Protection, 172*, 317–328.

Bencsik, A., Horváth-Csikós, G., & Juhász, T. (2016). Y and Z generations at workplaces. *Journal of Competitiveness, 8*(3), 90–106.

Cai, Y., Rowley, C., & Xu, M. (2023). Workplaces during the COVID-19 pandemic and beyond: insights from strategic human resource management in Mainland China. *Asia Pacific Business Review, 29*(4), 1170–1191.

Calk, R., & Patrick, A. (2017). Millennials through the looking glass: Workplace motivating factors. *The Journal of Business Inquiry, 16*(2), 131–139.

Cashman, K. (2017). *Leadership from the inside out: Becoming a leader for life*. Berrett-Koehler Publishers.

Cazorla, L., Alcaraz, C., & Lopez, J. (2016). Cyber stealth attacks in critical information infrastructures. *IEEE Systems Journal, 12*(2), 1778–1792.

Che-Ha, N., Mavondo, F. T., & Mohd-Said, S. (2014). Performance or learning goal orientation: Implications for business performance. *Journal of Business Research, 67*(1), 2811–2820.

Choi, T. M., Wallace, S. W., & Wang, Y. (2018). Big data analytics in operations management. *Production and Operations Management, 27*(10), 1868–1883.

Clarke, I. (2017). Risk Based Inspection–an insurer's view of incidents in the oil and gas industry. *Loss Prevention Bulletin*, 253.

Crowley, M. C. (2022). *Lead from the heart: Transformational leadership for the 21st century*. Hay House, Inc.

Georgiou, N., Delfabbro, P., & Balzan, R. (2020). COVID-19-related conspiracy beliefs and their relationship with perceived stress and pre-existing conspiracy beliefs. *Personality and Individual Differences, 166*, 110201.

Gotsis, G., & Grimani, K. (2016). Diversity as an aspect of effective leadership: Integrating and moving forward. *Leadership & Organization Development Journal, 37*(2), 241–264.

Habib, S. M., Hauke, S., Ries, S., & Mühlhäuser, M. (2012). Trust as a facilitator in cloud computing: A survey. *Journal of Cloud Computing: Advances, Systems and Applications, 1*(1), 1–18.

Haiyun, C., Zhixiong, H., Yüksel, S., & Dinçer, H. (2021). Analysis of the innovation strategies for green supply chain management in the energy industry using the QFD-based hybrid interval valued intuitionistic fuzzy decision approach. *Renewable and Sustainable Energy Reviews, 143*, 110844.

Hartz, D. E., & Quansah, E. (2021). Strategic Adaptation: leadership lessons for small business survival and success. *American Journal of Business, 36*(3), 190–207.

Islam, S. M. (2023). Impact risk management in impact investing: How impact investing organizations adopt control mechanisms to manage their impact risk. *Journal of Management Accounting Research, 35*(2), 115–139.

Jain, P., Pasman, H. J., & Mannan, M. S. (2020). Process system resilience: from risk management to business continuity and sustainability. *International Journal of Business Continuity and Risk Management, 10*(1), 47–66.

Kalantarnia, M., Khan, F., & Hawboldt, K. (2010). Modelling of BP Texas City refinery accident using dynamic risk assessment approach. *Process Safety and Environmental Protection*, *88*(3), 191–199.

Kwak, Y. H., & Anbari, F. T. (2006). Benefits, obstacles, and future of six sigma approach. *Technovation*, *26*(5–6), 708–715.

Lewis, T. G. (2019). *Critical infrastructure protection in homeland security: Defending a networked nation*. John Wiley & Sons.

Lu, Y., Zheng, H., Chand, S., Xia, W., Liu, Z., Xu, X., … & Bao, J. (2022). Outlook on human-centric manufacturing towards Industry 5.0. *Journal of Manufacturing Systems*, *62*, 612–627.

Malinao, C. M., & Ebi, R. G. (2022). Business management competencies as the driver of small-medium enterprises' survival during COVID-19 pandemic. *Puissant*, *3*, 296–315.

Meyer, K. E., Mudambi, R., & Narula, R. (2011). Multinational enterprises and local contexts: The opportunities and challenges of multiple embeddedness. *Journal of Management Studies*, *48*(2), 235–252.

Morrison, G. R., Ross, S. J., Morrison, J. R., & Kalman, H. K. (2019). *Designing effective instruction*. John Wiley & Sons.

Mustard, C. A., & Yanar, B. (2023). Estimating the financial benefits of employers' occupational health and safety expenditures. *Safety Science*, 159, 106008.

Naveenan, R. V., & Suresh, G. (2023). Cyber risk and the cost of unpreparedness of financial institutions. In *Cyber Security and Business Intelligence* (pp. 15–36). Routledge.

North, M. S. (2019). A GATE to understanding "older" workers: Generation, age, tenure, experience. *Academy of Management Annals*, *13*(2), 414–443.

Pathy, L. T. (2021). Understanding millennial values and how they will shape the future workforce. In *Advancing Innovation and Sustainable Outcomes in International Graduate Education* (pp. 21–33). IGI Global.

Peruzzini, M., Grandi, F., & Pellicciari, M. (2020). Exploring the potential of Operator 4.0 interface and monitoring. *Computers & Industrial Engineering*, *139*, 105600.

Razzetti, E. A. (2022). *Hardening by auditing: A handbook for measurably and immediately iimrpving the security management of any organization*. AuthorHouse.

Saliu, F., & Memaj, F. (2023). Global financial economy crisis. *International Journal of Integrative Sciences*, *2*(2), 109–120.

Sawalha, I. H. S. (2015). Managing adversity: Understanding some dimensions of organizational resilience. *Management Research Review*, *38*(4), 346–366.

Schwab, K. (2019). World economic forum. Global Competitiveness Report (2018–2019).

Smalley, P. C., & Chebotar, K. (2017). Event-based risk management for subsurface risks: An approach to protect value generation from oil and gas fields. *AAPG Bulletin*, *101*(9), 1473–1486.

Stewart, J. (2010). The UK national infrastructure plan 2010. *EIB Papers*, *15*(2), 28–32.

Sumbal, M. S., Tsui, E., See-to, E., & Barendrecht, A. (2017). Knowledge retention and aging workforce in the oil and gas industry: A multi perspective study. *Journal of Knowledge Management*, *21*(4), 907–924.

Ryan, L. V, Gasparski, W. W., & Enderle, G. (Eds.). (2017). Basic ethical aspects of American companies operating in Poland. In *Business students focus on ethics* (pp. 61–72). Routledge.

# 13 Synergies to Promote Successful PMI in the Tourism Industry

*Leszek Wypych, Ijaz Ahmad and Sandhya Sastry*

## Introduction

Globalisation by its very nature exemplifies acquisition of foreign firms or mergers between organisations, yet high failure of these ventures is also endemic; both national and organisational culture are cited as main reasons for these failures. The key objectives of this chapter are to explore (i) The key frameworks that impact national and organisational culture in a M&A setting, with a focus on the tourism industry, (ii) the link that dynamic capabilities and experiential learning have on success post-merger. The Research Questions are (i) what role do dynamic capabilities play in synergising post-merger successes? and (ii) what impact does experiential learning have on the success of integration after a merger or acquisition?

## Background

The legitimacy of acquisition can be raised by the staff of an acquired organisation for a variety of reasons including disruption to resources, cost of coordination and implementation, greater process complexity, and unfavourable judgements of the process. These can build up due to differences in organisation culture, national cultural differences, or indeed the creation of 'us' and 'them' identities (Malik et al., 2023), and it is the organisational culture, and aspects of national culture that will be considered as part of this research. In recent years, foreign direct investment (FDI) was on the increase until the COVID-19 pandemic. The pandemic had a negative effect on the tourism industry (Uğur and Akbıyık, 2020). At the same time, cross-border mergers and acquisitions failed drastically across the globe. However, Quer (2021) reports that transnational investment has been increased in tourism by emerging economies such as China. That activity contributes to national, as well as cross-border tourism mergers and acquisitions (M&As).

Of course, such investment is not risk free. Any organisation involved in cross-border acquisition needs to be aware that the uncertainty and risk comes from financial markets, macroeconomics, the political and geographical environment and cultural factors (Xie et al., 2017; Chiu and Lee, 2020). In relation to risk associated with cultures, Jha et al. (2019) believes that the closer culturally firms

DOI: 10.4324/9781003369967-18

are, the more likely alliances will take place in the international arena (also Chen and Lin, 2020; Lian et al., 2021).

Regardless the advantages M&As Have to offer, there is a large proportion of companies that have failed in the endeavour. The literature reports that a large majority of mergers fail, up to the rate of 85% (depending which study one refers to). In many cases, the cultural elements played a vital role in cross-border M&As. Often companies which failed have blamed culture as the reason for not being successful. There are a number of frameworks that attempt to account for failures. Sastry (2015) designed a framework that allows different aspects to be taken into account in order to avoid such failures, and to improve intercultural synergies; this is very fruitful in post-merger environments. She believes that intercultural competencies, dynamic capabilities, experiential learning and internationalisation are interrelated and interconnected, leading to achieving synergy. In her view, experiential learning provides the essential linkage between dynamic capabilities and intercultural competence. After all, it takes place when staff of diverse cultural backgrounds interact to envisage and implement development, growth and consolidation approaches and strategies (i.e., M&As or joint ventures (JVs) with foreign businesses), all of which fundamentally depends on the context a company operates within (Figure 13.1).

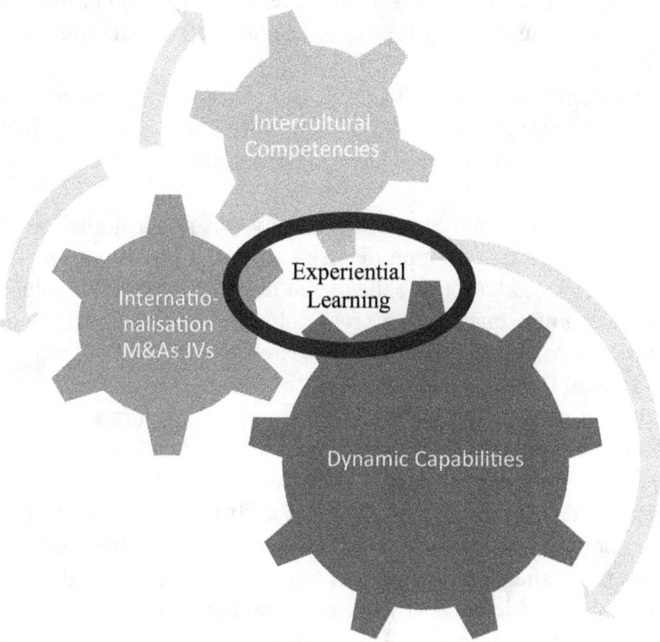


*Figure 13.1* Relationship among IC competence, dynamic capabilities, experiential learning and internalisation (printed with permission).

In order to comprehend the framework, we will focus our attention on the leading cultural theories and models.

## The Dominant Paradigms

Probably the most often frequently work on cultural dimensions is produced in Geert Hofstede's (1980) 'Culture's Consequences' where he identified cultural dimensions using country names as representations of diverse cultures. However, Kluckhohn and Strodtbeck (1961) who presented work much earlier than Hofstede were able to identify six core values (or orientations) that societies practice on a regular basis:

- Relationship with nature; should we take an approach of *Mastery* over nature, *Submission* to nature, or live in balance and ultimately in *Harmony*?
- Attitudes to time; should we primarily focus on the *Past*, *Present*, or the *Future*?
- Views of human nature; are we *Good*, *Evil*, or a *Mixture*?
- Activity; in our behaviours do we aim to express ourselves through internal motivations, for our own pleasure rather than that of others (what they defined as *Being*), or to *Achieve* (therefore relying somewhat on external motivation and external approval), or to grow and develop in abilities which we value (defined as *Being-in-Becoming*)
- Relationships between people; should we relate to one another through a hierarchy (*Lineal*), as equals (what they called *Collateral*), or according to individual merit (*Individualistic*)
- Space; what does space look like, is it *Here*, *There*, or *Far-Away*? This was not explored much further by Kluckhohn and Strodtbeck, but its influence in the work of later theorists can be evidenced (Hills, 2002).

Even though Kluckhohn and Strodtbeck's study was rather limited as it was conducted on five groups representing diverse cultures, it is still very much relevant. What is more, Fink et al. (2006) believe that the six fundamental values provided foundations for future quantitative cross-cultural studies assessing cultural values and dimensions. For instance, the research conducted by Trompenaars and Hampden-Turner (1997) supports managers in constructing and expanding cultural understanding of different cultures, as well as explaining core values and assumptions about those cultures. Their study was based on situational questionnaires resulting in the collection of responses from over 50,000 participants from more than fifty countries. What is very important with this study is that the data was collected among managers and therefore has a practical application within business. Trompenaars and Hampden-Turner believe that decision makers would benefit from reconciling dilemmas at work as it could provide competitive advantage, instead of employing their own polarised ideas, beliefs, and actions. They believe that in order to achieve positive solutions managers should be able to recognise differences, reconcile predicaments, and preserve managers' and employees' cultural styles. Covas and Pirlog (2019) took this further noting how this was not just important for managers, but for *future managers* within business schools.

The situations and dilemmas used in the study were related to interactions among people, environment, and passage of time. That allowed Hampden-Turner and Trompenaars to identify and explain seven different dimensions:

**Universalism v particularism**: Universalism can be described as a rule-based culture, applied to all, equally, at all times. As an example, crossing the street when the light is red, even when no one is around is still frowned upon in countries such as Switzerland (Trompenaars and Hampden-Tuner, 2006). Particularism allows for the exceptions that need to be made in the present circumstances.

**Neutral v affective**: details the balance between reason and emotion. An affective culture may show emotions with laughing, smiling, grimacing, scowling, and may receive emotional responses in return. Neutral cultures are more subdued and controlled in their expressions and reactions.

**Individualism v communitarianism**: Individualisation is seen as a prime orientation to the self, while the opposite, communitarianism, is defined as a prime orientation to common goals and objectives (Trompenaars and Hampden-Tuner, 2006).

**Specific v diffuse**: Specific cultures separate out work responsibilities from other dealings in life (Trompenaars and Hampden-Turner, 2006). Meeting a manager for a leisure activity is just that, and authority outside of work is curtailed. A diffuse culture may find the lines between work and personal life more blurred, with a manager seen as a source of influence even if meeting for leisure.

**Achievement v ascription**: an achieved status culture attributes status based upon the actual accomplishments of the individual, while in an ascribed status culture, it results from ideas such as age, class, gender, or education.

**Inner- v outer-directed**: should we as mankind impose our will upon nature and control it, or are we a part of nature and obey the laws of nature? The first of these is an 'inner-directed' (in that those things around us can be controlled by our inner will), while the latter is 'outer-directed', as we are a product of nature and the favours bestowed upon us.

**Sequential v synchronic**: There are two key areas to appreciate here. The first is the relative importance that a culture gives to the past, present, and the future. The second is what our view is of them; are the past, present, and future interrelated (synchronic) so that the ideas of the future and memories of the past shape what we do today or are they a series of passing events (sequential) (Trompenaars and Hampden-Turner, 2006).

As one would expect the above-described framework has its supporters as well as critics. It has to be remembered that when it comes to theoretical concepts, it is not what it is but what you do with it. A similar view to this is presented by Browaeys and Price (2008, p. 87) when they call it 'a model-to-learn-with'.

Another very influential study, even though it is not often employed in cross cultural management works, was conducted by Schwartz (1992) which investigated ten different value types:

- Power
- Achievement
- Hedonism

- Stimulation
- Self-direction
- Universalism
- Benevolence
- Tradition
- Conformity
- Security

Schwartz (2021) argues that values are cognitive representations of motives that are necessary to cope with the universal requirements for human survival, and as these requirements are indeed found in every human, so the values are likely to be recognised across cultures. Later Sagiv and Schwartz (2000) propose a framework that builds on the previous model and moves from the individual level to the societal one, allowing countries to be placed on three linear continuums:

- Hierarchy versus egalitarianism; hierarchical systems with clear roles, and unequal distribution of power are a feature. Egalitarian systems seek to induce people to recognise one another as equals, and organisations are built on cooperative negotiation (Sagiv and Schwartz, 2007).
- Embeddedness versus autonomy; in embedded cultures meaning is found in social relations and participating in their shared goals. Organisations here are more likely to function as an extension of the family. Autonomous cultures applaud uniqueness and internal attributes. Organisations here focus on staff as independent actors (Sagiv and Schwartz, 2007).
- Mastery versus harmony; mastery encourages self-assertion, and therefore organisations are likely to be competitive. Harmony focuses on the appreciation of the world, and organisations may therefore seek to undertake business in non-exploitative ways to achieve their goals (Sagiv and Schwartz, 2007).

The above-described values allow us to comprehend better the way we behave and communicate. Hall (1976) identified high- and low-context cultures, a framework which originally was designed to explain the first styles of communication. Even though high- and low-context communication styles are present in all cultures, one of them is dominant. Gudykunst et al. (1988) recognise that cultures which are more individualistic tend to use a low-context style of communication (direct), while group-oriented cultures employ high-context communication patterns (indirect) more often.

At the same time, it needs to be remembered that the communication process involves nonverbal approaches, when information is provided either by the context or behaviour. When we examine our behaviours, it becomes clear that there are similarities between high-context (collectivistic orientation) and low-context (individualistic orientation) as presented by Hall (1976) with Schwartz's values (1992). For example, high context pattern of communication would involve elements that are identified as harmony with nature, tradition, status (hierarchy), cooperation (focus on relationships), indirectness and formality. On the other hand, low context pattern of communication is an outcome of believing that one is in control over the environment, that we are in power of changing things, equality and an empowerment, competition, achievements, and competition as well as directness and openness.

Different cultural dimensions, values (or rather leaders' characteristics and attributes), and leadership styles were identified by House et al. (2004) through our study called 'Global Leadership and Organizational Behavior Effectiveness' (GLOBE). The main objective of the research was to identify how culture impacts the effectiveness of social groups, organisations, and leaders.

The research team were able to identify nine cultural dimensions across different societies:

- power distance
- humane orientation
- assertiveness
- collectivism I (institutional)
- collectivism II (In Group)
- performance orientation
- future orientation
- gender egalitarianism
- uncertainty avoidance.

These dimensions are similar in many respects to the ones we all have already discussed, principally due to GLOBE building on concepts developed by other researchers; for instance, Hofstede (1980) and Schwartz (1992, 1994). More specifically, in-group (family) collectivism and institutional collectivism are very closely linked to Hofstede's (1991) individualism/collectivism. Gender egalitarianism and assertiveness are very much alike in that they mention describing muscular and feminine societies. Uncertainty avoidance and power distance are the same in terms of descriptions and names. Long term orientation is reflected in future orientation. Humane orientation shows similarities with a dimension identified by Kluckhohn and Strodtbeck (1961) called Good versus Bad Human Nature. Lastly, elements of Need for Achievement by McClelland's (1961) are recognised in Performance Orientation. To conclude, there are strong similarities between the GLOBE and the dimensions developed by Hofstede (2006).

As an output of the empirical research, the team at GLOBE were able to develop cultural clusters (similar to Schwartz (1992) and other studies). They are presented as: Germanic Europe, Latin Europe, Anglo, Nordic Europe, Latin America, Confucian Asia, South Asia, Eastern Europe, Middle East, and Sub-Sahara Africa, creating a typology of behaviour in each cluster. The final finding of the GLOBE study is related to leaders' effectiveness. That concept is highly contextualised as what is or who is effective would differ from society to society, from one organisation and its norms to another, and beliefs and values of employees' values. To start with there were 112 characteristics identified, however, they were conceptually and statistically reduced to six:

**Charismatic/value-based style** – based on the assumption that leaders are to inspire their staff, who would become passionate about what they do. At the same time, it is expected that those involved would operate within high standards, be performance and task-oriented, be decisive, innovative, a visionary, and have high integrity.

**Team-oriented style** – based on collaboration among members of the organisation and focuses on goals, common purpose, and cohesion of the team. It is expected that the team members are benevolent, able to integrate in a team, diplomatic, collaborative and competent.

**Participative style** – the style would encourage members of the team to provide input, generate ideas and then take part in decision making processes, as well as in the internal implementation processes. The style employs equality and delegation processes, and elements of democracy and participative approach are clearly visible.

**Humane style** – this style is very supportive; focusing on compassion and understanding the psychological well-being aspects of the team members is very important for leaders, who are very much humane-oriented, patient, and modest.

**Self-protective style** – behaviours that are face-saving-oriented are visible in this style (face saving aspects are explained in Chapter 10), as managers would like to strictly obey procedures and regulations of the organisations. The focus is on the security and safety of each member of the group, including the leader. Leaders are conscious of their position, self-centred and they might even induce a conflict among members of the team.

**Autonomous style** – this is very much an individualistic and independent approach to leading others, and leaders are often self-centric.

The outcomes from research and the subsequent development of frameworks are often questioned, with each concept and theory often finding supporters and detractors alike. However, each one of them contributes to the existing knowledge. A large part of the literature on leadership and organisational behaviour is based on western theories and research, and by employing them in practice we are using rather ethnocentric views and approaches. House et al. (2004) hoped that the GLOBE could break that hegemony.

Even though theories, cultural dimensions, values and so on can explain most of cultural similarities and differences (Fink et al., 2006), our main objective shouldn't be to identify managerial and cultural differences, but our focus should be on organisational learning and adopting synergy (Holden, 2002).

## Alternative Models in Management Literature

We have looked at dominant western paradigms in cross-cultural management. This section will turn our attention towards an alternative way of thinking in management. In general, there is an agreement that there is a relationship between corporate culture and national culture. Different studies report findings based on diverse variables and values. Even though the links exist, it is difficult to establish to what extent dimensions and values overlap in both constructs. For instance, Xu and Shenkar (2002, p. 615) noticed that 'institutional distance complements, rather than replaces the cultural distance construct', when they investigated Foreign Direct Investment (FDI) behaviour of Multinational Enterprises (MNE).

Different studies identify direct links between Hofstede's dimensions and institutional elements (Parboteeah and Cullen, 2003), which supports the claim that both are central and important aspects of work. Cultural and institutional perspectives at work were investigated by Lau and Ngo (2001), who concluded that western companies where more willing and responsive to developmental and organisational interventions in comparison to those businesses based within Hong Kong. It has to be remembered, that dimensions such as cognitive, normative, and regulatory, people's beliefs, knowledge, behaviour, roles, as well as a representation of authority systems are interrelated, interconnected and overlap with national cultures (Scott, 2002). Also, that later concept is embedded in multidimensions and multiple layers of social bodies. What is more, the above examples only support the understanding that institutions and firms are cultural by nature (please see a discussion on cultural universals in Chapter 10 (Murdock, 1945; Brown, 1991)).

One element of cultural universals is the kinship system (Parkin, 1997) which explains social structure among people. The kinship perspective allows researchers to consider a sense of family that goes beyond blood or marital relationships, and in the setting of collectivist cultures, allows the inclusion of wider definitions of family, including language, place of origin, or ancestors (Manik et al., 2023).

Kinship typology can explain main forms and differences (Parkin, 1997), and that would include how one distinguishes and assigns another member of specific society to a particular kin group (Stone, 2002). It seems to be the case that many societies, especially collectivist and relation-oriented ones (such as India), pay close attention to culturally expected obligations to their kin groups in both day-to-day and professional arenas. However, individualistic, and task-oriented societies (such as the UK) appear to pay attention mainly to genetical kinships. Kinships facilitate and support providing a safe working environment, whereas working in societies and groups that are heterogeneous from a cultural point of view requires more mental effort in understanding others and processing information, producing meaningful communication, and performing tasks together. As a result, such teams deliver different levels in terms of quality and performance in general (Staples and Zhao, 2006). On the other hand, homogeneous groups were able to deliver higher levels of performance than other heterogeneous groups. In most cases the reason for such success is receiving feedback on process-related performance. Researchers reported a number of different issues heterogeneous teams experience on regular basis, such as language and communication styles differences, diverse application of power and hierarchy, application of diverse leaderships, misunderstanding of norms and values and so on (for instance, Maznevski and Chudoba, 2000; Maznevski and Di Stefano, 2000; Marquardt and Horvath, 2001; Hampden-Turner and Trompenaars, 2008).

The following are cultural differences which have been identified as influencing teams:

- Similarity attraction hypothesis: individuals representing the same culture are aware and conscious of sharing similar if not the same beliefs and values, which results in drawing and attracting them to each other (Williams and O'Reilly,

1998; hypothesis developed by Berscheid and Hatfield (1969) and then expanded, finalised, and formalised by Byrne (1971)).
- Social identity and social categorisation theory; quick and long-lasting identification of people in multinational teams and groups happens through social categories such as ethnicities, races, and nationalities (Tajfel, 1982).
- Members of multinational teams use diverse sources and ways of processing information due to cognitive frameworks, deep cultural differences, and perspectives (Stahl et al., 2010).

It is believed that training could be provided for leaders in order to allow them to understand new cultural realities. This would give them a tool that would contextualise new and diverse behaviours, along with an appreciation of the changes that occur in verbal and nonverbal communication. This tool is called Cultural Intelligence (CQ), which would help with understanding and making clearly visible links among organisations, its structures, cultures, behaviours, actions, and perceptions (Earley et al., 2006). Cultural Intelligence is 'a frame of mind' enabling individuals to be effective in multicultural contexts (French, 2015, p. 188) and is formed with the following components:

- knowledge about culture one operates within
- self-efficacy
- development of culturally acceptable behaviours by copying other people representing culture one operates within
- learning strategies.

What is more, there are behavioural, motivational, and cognitive facets which form part of CQ, and they could function as universals, idiosyncratic to the individuals, and culture specific. Not only does it help with understanding one's behaviours in an international environment, but it also allows us to predict one's behaviours more accurately. Simply speaking, cultural intelligence is one's capabilities to adjust themselves to new cultural settings. That would involve social and emotional intelligence, which are culture specific and therefore they might be incompatible with the adaptation processes. Earley and Mosakowski (2004) identify emotional, motivational, physical, and cognitive elements within CQ. In addition, Earley and Peterson (2004) include learning strategies represented by both the cognitive and metacognitive, along with behavioural as well as the previously mentioned motivational dimensions. They see these as one's abilities and skills used to collect information and then to understand and adjust their own behaviours in order to become effective in diverse multicultural contexts. CQ was defined in a similar way by Earley et al. (2006) and then Ang et al. (2007) who also added an individual aspect to these capabilities. Overall, the fundamental domains of cultural intelligence are:

- cultural metacognition – relates to consciousness of mental experience and effectiveness of motivational circumstances related to cultural settings which impacts one's choice of behaviour during intercultural interactions (Thomas et al., 2008). These processes are energy and effort consuming because they involve constant and conscious control of self, control of goal attainment, and it continues to access information and processing of the interaction and possible goal-related responses.

Ting-Toomey (1999) employed mindfulness, concepts that originate in Buddhism, in the context of cross-cultural communication. Mindfulness appears to represent higher level attention and awareness of what is being experienced at the time (Brown and Ryan, 2003). The concept of mindfulness links all aspects of cultural intelligence, knowledge and encountering different actions. Awareness, on the other hand, provides foundation for conscious awareness when the internal aspect of self and external elements are being consciously monitored. Because of this, not only can one be mindful of external and contextual pieces of information, but also all of his or her emotions, thoughts, and drives. In addition, cognitive processes are being actively supported by mindfulness resulting in measured responses instead of passing automatic responses driven by personal thoughts and emotions (Charoensukmongkol and Pandey, 2021), and developing new schemas and categories which in turn supports the development of multiple and diverse views.

In fact, Gardner and Laskin (1995) believes that mindfulness could help with many areas including:

- Developing awareness of one's emotions, ideas perceptions attributions and assumptions.
- Realising aspects of other people behaviours and synchronising with them.
- Taking in all of the contextual information (not just words) in order to comprehend situations more holistically.
- Being open-minded and being able to recognise different views of the same situation.
- Employing empathy.
- Developing understanding of others' personalities and cultural roots (mental maps), so our responses would be appropriate.
- Developing new categories of others (grouping or sometimes called labelling).
- Gathering additional information in order to prove or disprove one's previous understanding.

  - Cross-cultural skills and abilities – Thomas et al. (2008) see them as a dynamic and growing integral part of CQ. Such skills are related to a person's individual traits and differences which can be developed and improved. The researchers identify the following types:

- relational skills
- perceptual skills
- adaptive skills.

It is claimed that both, skills and knowledge, could fulfil the function of CQ providing that cultural metacognition is involved in the process and keeps both cultural elements engaged.

- cultural knowledge – or rather its content, is the foundation of decoding and understanding people's behaviours. Cultural knowledge includes understanding the logic to the underlying behaviours, as well as the processes of acquiring knowledge about new phenomena. Once the new knowledge is gathered, it has to be processed and organised in different categories before new perspectives are understood (Wade-Benzoni et al., 2002). What is more, there is a type of knowledge which is related to procedures and processes that would recognise the cultural impact on different interactions, nature and their circumstances, and related to challenges and how these are resolved. The above involves a learning process which employs observation, reflection, and conceptualisation, which in turn could result in the development of an improved cognitive system.
- Managers who wish to be successful across cultures and global environments would benefit from training in CQ, and organisations that wish to minimise negotiation challenges caused by differences in cultural values should extend this training to their employees (Caputo et al., 2019).

**Dynamic Capabilities**

Hansen et al. (2004) claim that businesses could achieve a competitive advantage providing that it would develop and use capabilities that are distinctive and unique. In this framework, attributes and processes are central in positioning a company. Decision making procedures would allow new identified opportunities to be used in practice. It also allows certain flexibility to regroup and to align with covert technology or markets if needed.

It is also believed that a part of the dynamic capabilities is value creation. This would involve resources that are already owned, to be developed further, reconfigured and deployed so the company would maximise its potential and be congruent with the others (Bowman and Ambrosini, 2003). You et al. (2023) take this further by noting that dynamic capabilities can include not just the higher-order competencies found within an organisation (that include internal resources) but can include external stakeholder relationships (You et al., 2023). In this way businesses could employ dynamic capabilities to influence its resources in a way which would promote the company's strengths to have advantage in exploring new possibilities. It seems to be the case that learning, and the skills acquisition processes are the focal strategy to improve dynamic capabilities, which encompasses sensing, seizing, and transforming activities. When the markets became more open and globalisation processes had taken place, companies had more opportunities to employ diverse forms of manufacturing and innovation. Innovation was taken even further by synthesising inter-industry in the process, which result in using platforms and developing new regulations and structures (Evans et al., 2008). That created additional pressure on smaller businesses (e.g., SMEs), which in order to survive had to develop and upgrade competencies and skills, as this process is more cost and energy effective in comparison to developing routines which fits all purposes.

It is worth pointing out that some of the companies' resources are actually people who are employed by those companies, and therefore capabilities, which are related to cultural, and intercultural aspects exist within people. Hence it is not surprising that some researchers identify synchronisation of different employees in production with diverse technologies, and collective learning as core elements of dynamic competences. In other words, competence which needs to be core is cultural intelligence. It is claimed that CQ is collective and provides essential support for short and long-term survival for businesses. CQ provides competitive advantage as it is unique on one hand, and on the other competitors are almost not able to clearly see that competence, and therefore they are not able to copy it. CQ turns into much more significant capability when treated as a collective entity, comprising processes, resources, and skills developed over a longer period of time and employed in different markets within diverse cultural settings, directly and indirectly at the same time. Such employment of CQ allows companies to find and develop sustainable sources of profit which is critical to any company.

In order to accelerate internalisation a specific skill is necessary, which is the management of PMI. What is more, the skill is a must in order to achieve synergy, which in turn is needed to obtain essential resources and knowledge from sources outside the company. At the same time internalisation of newly acquired leads is essential as merging would result in adjustment of the existing company structure or even in the development of a new one. Mathews (2006) points out that creating activities on international markets is easier and adopting a network-based structure helps developing small international businesses.

It is obvious that networking and using external sources brings direct benefits to international business, however, to maximise companies' developments on the international markets, intercultural values need to be synergised. That would require to develop abilities to be able to operate in much more complex environments, as well as to be able to adapt not only to foreign markets but also to different countries and cultures. Even though this has not been identified as dynamic capability, it is a vital one in international business and strategy. Adaptive capabilities require a holistic approach and a usage of intercultural competence to be used effectively. In this way intercultural competence is manifested through adaptive capability, which in the context of global companies becomes strategic competence for organisations (Saner et al., 2000).

## Experiential Learning

Experiential learning is the process of transforming firm experiences into applied knowledge (Karami and Tang, 2019) and we find that managers who operate only within their comfort zone will experience performance related issues. Often it is the case that in order to develop different capabilities and take people's development to the next level requires taking our learning to areas we have not yet dealt with. People who are pragmatic in nature would identify a problem in the workplace and then work around it to find out a solution. Mintzberg and Westley (2000) explain that identifying a problem and providing a diagnosis is an outcome of 'thinking' processes. Then conceptual stage has to take place to come up with a solution, which they refer to as 'seeing'. This stage is followed by putting the solution in action – 'doing' (actually this framework resembles Kolb's (1984) learning circle). This process is likely to be influenced by a person's individual traits.

A different framework dealing with changing dynamic capabilities over a period of time incorporates environmental velocity (McCarthy et al., 2010). The authors of the framework believe that management control system is vital in in the process of dynamic capabilities development, which is to leverage members of staff's organisational behaviour. The framework uses 'As – Is' and 'To – Be' styles of feedback in relation to the following dynamic capabilities processes:

- **Coordination/integration** – both processes are supposed to complement each other. They involve obtaining, assigning, organising, and integrating resources. Usually, it involves pathways leading to growth of a company such as acquisitions and mergers, employing latest and innovative technologies and so on.
- **Learning** – usually happens through drilling (repetition) and trying and experimenting in order to find a better way to perform. Learning takes place at two different levels: (1) exploration (promotes efficiency and less time-consuming tasks) and (2) experimentation (fosters creativity and novel way of thinking).
- **Reconfiguration** – changing and adjusting ways of completing usual tasks as well as resource is in order to transform or refine competencies. This process involves integrating knew learned elements and then required changes not implemented.

Each global organisation must choose their own approach and processes as a more rapid adaptation provides a more stable pathway for the company, however, with lower returns. On the other hand, organisations employing much more rapid adaptation are likely to increase their performance and rewards, conversely risk is also higher (Raisch and Hotz, 2010). Also, learning ways, practices, and modes change, and therefore younger or newer organisations are likely to have some advantages when it comes to learning processes. That is because learning new involves unlearning some of the previously acquired knowledge and behaviours.

Repetitive activities and experimenting with new ones as a learning process result in performing at higher levels, with a shortened timeframe used for the same task as previously, and additionally new opportunities are recognised as an outcome. The process of learning encompasses individual skills as well as organisational ones at the same time. Learning does not happen in separation; it is rather a process which involves other members of staff to help by contributing and providing explanations related to complex issues, and therefore learning is both collective and social at the same time. Communication processes, modes and codes are used as well as information gathering procedures which are coordinated by one. Consequently, new routines, patterns and logics are established in the organisation. What is more, creativity and innovation seem to be impacted by cultural aspects of an organisation, and in particular by beliefs and values. Literature indicates that creativity is linked with experiential learning as this is linked with an intrinsic type of motivation.

Learning styles by Kolb (1984) is probably the most significant framework, which is based on learning through experience which leads to developing new knowledge. In his view, experience is a foundation for learning which is continuous. The model requires a holistic approach to learning, taking into account a person's experience that requires adaptation – or rather the adaptation to the environment one operates in. It involves two continuously interacting continuums: process and reception. When both continuums interact, they create different processes at four steps:

- Gaining Experience
- Reflecting
- Conceptualisation
- Experimenting

Each person would have individual preferences when it comes to choosing the most suitable phase of learning for them depending on how one perceives information, feelings, how they experience, their background, and how their personality traits affect them. However, according to Kolb, the most effective learning takes place when all four steps are used in balance. Thus, when the learner moves from concrete experience to the stage when they start reflecting on them, diverging processes begin to take place. Before rushing into action and trying to amend previous behaviour and solve problems, divergers tend to observe and collect information through experience (Randolph, 2000). Often it is the case that they co-operate with others in groups, using brainstorming to see different options. They use their senses, innovation, and imagination to consider the issue from different angles, allowing them to come up with creative resolutions. The next learning stage requires assimilating previously developed ideas and solutions into a holistically developed framework. Assimilators focus on practicality and logic and therefore they are good at creating conceptual models and process these through reflective observation (Randolph, 2000). They are idea-oriented people. When ideas or concepts are created, it is time to convert them into a solution and practical use. It is often the case that decisions are based on data. Convergers have a tendency to experiment with thought and concepts before employing them in reality. The chosen solution needs to be put in practice in real life by somebody who has hands-on approach, and that's when accommodators are very effective. They will try to put the new concept into practice even if it means making mistakes. The process is often practical rather than conceptual as accommodators need to adapt and change circumstances very often, and therefore learning takes place through discovery, and indeed self-discovery (Randolph, 2000).

In the context of global organisations, the development of dynamic capabilities in relation to the experimental elements of intercultural encounters is the main aspect of learning. Even though there is a scope, experimental and/or cultural differences are not accounted for in the model. Literature indicates that partnerships and collaborations could become a platform for organisational learning. In the next sections we will discuss how such companies could effectively use internalisation strategies.

**Internalisation Strategies**

Global companies conduct very complex operations across international markets which are often results of strategic choices. Cross border activities have become vital not only to achieve sales targets, but also their sustainable existence is often secured by international expansion (Berndt et al., 2023). Wall et al. (2010) claim that it happens through a number of different modes of market entry; just to name a few:

- Foreign direct investment (FDI; highest risk and maximum return)
- Joint ventures
- Direct exporting
- Licensing
- Indirect exporting (subsidiaries; lowest risk and minimum return)

It is worthwhile to discuss acquisition and integration when looking into growth strategies. Haspeslagh and Jemison (in De Wit and Meyer, 1998, pp. 339–345) listed the following integration types:

- Absorption – as the name suggests, a full integration is to take place where boundaries between both companies will disappear. Some of the decisions might be very difficult but unavoidable. Often it limits the companies' abilities and impacts cultural sensitivity because interdependency has to be achieved. This type of acquisition is characterised by a low level of autonomy and a high level of strategic interdependence.
- Preservation – the goal is to maintain the source of the original benefits that are integral to each organisation. One way in which the businesses can achieve value creation is through fostering interactions that bring positive change, professionalism, risk taking, as well as learning, all of which are likely to help with future acquisitions as well as the existing business. This type of acquisition is characterised by high level of autonomy and a low level of strategic interdependence.
- Symbiosis – a significant quantity of capabilities and resources must be transferred and then they need to be embedded into the acquiring company's systems, and that makes the process complex and challenging. This type of acquisition is characterised by a high level of autonomy and strategic interdependence.

When we consider cultural aspects within the field of international business, these areas play a vital role in ensuring that the chosen strategy is fruitful in terms of relational aspects. Intercultural resources play significant roles as they are likely to have an impact on policymakers, help with accessing informal networking structures and lobbying systems. Some researchers notice cultural differences when it comes to relational approaches. For instance, it is believed companies from North America open evaluate relationships in the context of possible benefits and profits. In practical terms it means that cultural, human, political, and organisational aspects do not get the attention they deserve. On the opposite side, companies from Asian countries are the ones which want to develop relationships and are proficient in employing and nurturing them. Somewhere in the middle of the continuum are European companies, however it has to be remembered an approach of each company is biased by cultural differences.

During the process of internalisation there are a number of different views related to the process of globalisation that occurs at the same time. Some researchers believe that globalisation activates converging processes. On the other hand, there are some who claim the opposite, that diverging processes are present as well. If it is the case that divergence occurs then internalisation becomes even more complex as businesses need to keep the balance right between adapting to local environment and aligning standards between acquiring an acquired enterprises (De Wit and Meyer, 1998).

This is not a single view, as it is claimed that international diversity is actually encouraged by global integration (Porter, 1980). That type of international diversity would have to allow to be tapped into resources related to specialisation and innovation.

Because of the rapidly changing environments in the global arena, the length to which globalisation takes place is usually different. It creates an additional challenge for SMEs on the global market, which could be addressed by improving responsiveness, efficiency, and learning. However, due to insufficient research in the area, especially in the tourism sector, one could argue that existing frameworks might not be sufficient enough to explain behaviours and goals this type of businesses from emerging economies. One could attempt to generalise; however, would it create a clear picture?

## Future Research Directions

The total frameworks and literature discussed in this chapter just scratch the surface of issues related to intercultural management and its impact on PMIs and M&As. It seems to be the case that more research in the area of experiential learning is needed, when relationships among dynamic capabilities and different types of internalisation could be assessed.

It could be beneficial to investigate how cultural differences impact managing organisational culture. Furthermore, it would be interesting to see how data explains integration processes of both dimensions.

Usually, cultural conflict might be taken as a possible confrontation when two parties interact and mediate. It would be very interesting to investigate if a confrontation in the context of cultural conflict in M&A and PMI encounters could be used as a way to develop and adjust self-image and social identity.

Research findings so far provided us with invaluable knowledge about M&A and PMIs and different aspects impacting the process is, such as cultural impact, dynamic capabilities, using different platforms, biculturalism and so on. Yet, it seems to be the case that more work is needed especially when it comes to collaboration between the researchers representing different cultures.

## Conclusion

In concluding, the chapter aims to consider the intersection between the areas of Intercultural competences, dynamic capabilities, and the overarching role that experiential learning has within the success of post-merger integration.

This chapter noted certain theories to understand national culture dimensions, along with some theories that support the exploration of corporate cultures. It is important to reiterate here that there is a relationship between these two, although the varying studies report their findings based upon their diverse variables. In a Foreign Direct Investment setting, Xu and Shenkar (2002) noted that there was a correlation between institutional distance and cultural distance, noting them as theoretical counterparts in their respective areas.

In considering kinship groups, it was noted that working within homogeneous groups often facilitates simpler working environments that often deliver higher performance (Staples and Zhao, 2006) while heterogeneous working groups are often challenged by a number of issues ranging from the simple, such as communication issues, to the complex, that can include the misunderstanding of norms and values.

One of the key issues linked to dynamic capabilities is its role in value creation. By looking at the resources already owned, developable, reconfigurable, and utilised effectively by the organisation(s), the company would be able to maximise its potential (Bowman and Ambrosini, 2003). The process of learning and skills development are key to improving dynamic capabilities, and are often revolved around the activities of sensing, seizing, and transforming.

There are a variety of learning styles postulated by theorists, and the process of learning through experience is key within the process of successful post-merger integration. Younger organisations in general have an advantage in the process of learning, as a key component of changing behaviour is the 'unlearning' that needs to take place in order to adopt the new information, culture or practice; older organisations may have so much more to unlearn in the process. This learning that takes place is a process that does not happen in isolation, rather it is one that involves varied members of staff to provide contribution and explanation, creating a collective social learning environment.

Finally, it is important to note the impact that different views related to internalisation can create; it was noted that North American organisations may evaluate a relationship (potential or current) based upon the tangible benefits from a business partnership, while in contrast Asian countries will be more focused upon building relationships, and Europeans may find themselves between the two on the continuum. In considering where a company may sit, it is vital to remember that they are impacted by cultural differences.

**Case Study**

GV (the name has been changed) is a leading firm in the tourism sector that provides exclusive, unique, and exciting experience to its customers in many countries. In order to develop its services and products, the business acquired a local company which used to provide very similar products and customer experience. The main goal of Global Voyage was to merge resources of both companies and operations, and to subsequently continue operations under the brand 'Global Village'.

The reality shows that integration was not as smooth as the company was originally hoping for. Very soon both the acquired and acquiring organisations realised that there were a number of cultural differences which made interactions between both teams difficult, which often resulted in clashes. The elements that contributed to the issues experienced by both teams were related to cultural backgrounds, diverse working styles, organisational values, and communication patterns. Some members of staff from the acquired company reported that on a number of occasions they felt that they were being treated as inferior and less valuable members of staff in comparison to members of staff from GV.

What is more, IT platforms used by both companies were not compatible and therefore the companies experienced operational issues with systems, as well as practices found within both companies.

The lack of synchronised and standardised processes and services created a detrimental impact on customer experience where high quality and consistent service was not always provided across different regions.

In order to address the issues experienced, cultural training sessions were provided on number of occasions for all members of staff, with the training focused on sensitivity and awareness of cultural differences between the different parties involved. In this way the company was able to help with building collaboration, respect, and understanding between different teams. In this way it was possible to achieve higher levels of cohesiveness within and among teams and a sense of belonging and shared objectives.

Gradually GV was able to upgrade and integrate operational IT systems by employing a phased approach. In this way the company was able to synergise IT systems and operations that resulted in a smooth information flow. The outcome was very positive, resulting in increased efficiency as members of staff had quicker access to information. The improved platforms and systems allowed enhanced workflow and coordination.

At the same time the company addressed the issues related to local customers' preferences (market segmentation) and provided tailored products and services. Diversified holiday packages were designed and provided in the different geographical areas the company operated within, focusing on regional uniqueness, while brand identity was preserved at the same time. As an outcome, customer experience was maintained at a high level and addressed tourists' diverse preferences.

**Questions for Discussion**

- How were localised and standardised services and processes balanced during the PMI?
- How would you explain and help the teams understand cultural differences?
- How did developing cultural sensitivity and dynamic capabilities affect business performance?
- What were the effects of introducing integrated systems and platforms?

***Big M&As in the Tourism Industry***

- Acquisition of Starwood Hotels & Resorts Worldwide Inc. by Marriott International in 2016.
- Acquisition of P&O Princess Cruises in 2002 by Carnival Corporation.
- AccorHotels acquisition of Fairmont Raffles Hotels International in 2016.
- Acquisition of Orbitz Worldwide, Inc. by Expedia Inc. in 2015.

## References for the Case

Caiazza, R. and Volpe, T., (2015) M&A process: A literature review and research agenda. *Business Process Management Journal*, *21*(1), pp. 205–220.

Cartwright, S. and Cooper, C.L., (1993) The role of culture compatibility in successful organizational marriage. *Academy of Management Perspectives*, *7*(2), pp. 57–70.

**Key Terms and Definitions**

**CQ; Cultural Intelligence**: an individual's ability to adapt to different cultural contexts (Earley and Ang, 2003).

**Contextual Intelligence**: the ability to understand the limits of their knowledge and to transfer that knowledge to foreign markets (Khanna, 2015); a key skill for SMEs to enhance to be able to compete in the international environment.

**Dynamic Capability**: in order to address rapidly changing environments, dynamic capabilities addresses the firm's ability to integrate, build, and reconfigure both internal and external capabilities (Teece, 2018).

**Foreign Direct Investment (FDI)**: a type of cross-border investment involving purchasing a lasting interest in an enterprise located in a different country.

**In-group & out-groups**: ways of creating an 'us' and 'them'; an ingroup is one with which you identify strongly, while an outgroup is one with which you don't due to a plethora of reasons from language, dress codes, festivals, pageants, traditions and rituals (Giles and Giles, 2013).

**Kinship**: How an individual distinguishes another person and assigns them to a particular kin-group (Stone, 2002); individualists do so by focusing on genetical kinships while collectivists focus on culturally expected obligations.

## References


Ang, S., Van Dyne, L., Koh, C., Ng, K.Y., Templer, K.J., Tay, C. and Chandrasekar, N.A., (2007) Cultural intelligence: Its measurement and effects on cultural judgment and decision making, cultural adaptation and task performance. *Management and Organization Review*, *3*(3), pp. 335–371.

Berndt, R., Fantapié Altobelli, C. and Sander, M., (2023) International market entry strategies. In: *International Marketing Management*. Berlin, Heidelberg: Springer Gabler. https://doi.org/10.1007/978-3-662-66800-9_7.

Berscheid, E. and Hatfield, E., (1969) *Interpersonal Attraction* (Vol. 69, pp. 113–114). Reading, MA: Addison-Wesley.

Bowman, C. and Ambrosini, V., (2003) How the resource-based and the dynamic capability views of the firm inform corporate-level strategy. *British Journal of Management*, *14*(4), pp. 289–303.

Browaeys, M.J. and Price, R., (2008) *Understanding Cross-Cultural Management*. New York: Pearson Education.

Brown, D.E., (1991) *Human Universals*. New York: McGraw-Hill.

Brown, K.W. and Ryan, R.M., (2003) The benefits of being present: Mindfulness and its role in psychological well-being. *Journal of Personality and Social Psychology*, *84*(4), p. 822.

Byrne, D., (1971) *The Attraction Paradigm*. New York: Ac.

Caputo, A., Ayoko, O.B., Amoo, N. and Menke, C., (2019) The relationship between cultural values, cultural intelligence and negotiation styles. *Journal of Business Research*, *99*, pp. 23–36.

Charoensukmongkol, P. and Pandey, A., (2021) Trait mindfulness and cross-cultural sales performance: The role of perceived cultural distance. *Canadian Journal of Administrative Sciences/Revue Canadienne des Sciences de l'Administration*, *38*(4), pp. 339–353.

Chen, M.X. and Lin, C., (2020) Geographic connectivity and cross-border investment: The belts, roads and skies. *Journal of Development Economics*, *146*, p. 102469.

Chiu, Y.B. and Lee, C.C., (2020) Effects of financial development on energy consumption: The role of country risks. *Energy Economics*, *90*, p. 104833.

Covas, L. and Pirlog, A., (2019) The importance of national culture dimensions on intercultural competence development of future managers. *Cross-cultural Management Journal*, *21*(2), p. 111.

De Wit, B. and Meyer, R.J.H., (1998) *Strategy-Process, Content, Context: An International Perspective* (2nd edition). London: Thomson Business Press.

Earley, P.C. and Ang, S., (2003) *Cultural Intelligence: Individual Interactions across Cultures*. Stanford, CA: Stanford Business Books.

Earley, P.C., Ang, S. and Tan, J.S., (2006) *CQ: Developing Cultural Intelligence at Work*. Stanford, CA: Stanford University Press.

Earley, P.C. and Mosakowski, E., (2004) Cultural intelligence. *Harvard Business Review*, *82*(10), pp. 139–146.

Earley, P.C. and Peterson, R.S., (2004) The elusive cultural chameleon: Cultural intelligence as a new approach to intercultural training for the global manager. *Academy of Management Learning & Education*, *3*(1), pp. 100–115.

Evans, D.S., Hagiu, A. and Schmalensee, R., (2008) *Invisible Engines: How Software Platforms Drive Innovation and Transform Industries* (p. 408). Cambridge, MA: The MIT Press.

Fink, G., Neyer, A.K. and Kölling, M., (2006) Understanding cross-cultural management interaction: Research into cultural standards to complement cultural value dimensions and personality traits. *International Studies of Management & Organization*, *36*(4), pp. 38–60.

French, R., (2015) *Cross-Cultural Management in Work Organisations*. Kogan Page Publishers.

Gardner, H. and Laskin, E., (1995) *Leading Minds: An Anatomy of Leadership*. New York: Basic Books.

Giles, H. and Giles, J., (2013) *Ingroups and Outgroups*. SAGE Publications, Inc. https://doi.org/10.4135/9781544304106.

Gudykunst, W.B., Ting-Toomey, S. and Chua, E., (1988) *Culture and Interpersonal Communication*. Newbury Park, CA: Sage Publications, Inc.

Hall, E., (1976) *Beyond Culture*. New York. Doubleday.

Hansen, M.H., Perry, L.T. and Reese, C.S., (2004) A Bayesian operationalization of the resource-based view. *Strategic Management Journal*, *25*(13), pp. 1279–1295.

Hills, M.D., (2002) Kluckhohn and Strodtbeck's values orientation theory. *Online Readings in Psychology and Culture*, *4*(4). https://doi.org/10.9707/2307-0919.1040.

Hofstede, G., (1980) Motivation, leadership, and organization: Do American theories apply abroad? *Organizational Dynamics*, *9*(1), pp. 42–63.

Hofstede, G., (1991) Cultures and organizations: Software of the mind. In: *Airaksinen*, House, R.J., Hanges, P.J., Javidan, M., Dorfman, P.W. and Gupta, V. (eds. 2004) (pp. 1–25). London and New York: McGraw Hill.

Hofstede, G., (2006) What did GLOBE really measure? Researchers' minds versus respondents' minds. *Journal of International Business Studies*, *37*, pp. 882–896.

Holden, N., (2002) *Cross-Cultural Management: A Knowledge Management Perspective*. London: Prentice Hall Pearson Education.

House, R.J., Hanges, P.J., Javidan, M., Dorfman, P.W. and Gupta, V. (eds.), (2004) *Culture, Leadership, and Organizations: The GLOBE Study of 62 Societies*. Thousand Oaks, CA: Sage Publications.

Hampden-Turner, C.M. and Trompenaars, F., (2008) *Building Cross-Cultural Competence: How to Create Wealth from Conflicting Values*. New Haven: Yale University Press.

Jha, A., Kim, Y. and Gutierrez-Wirsching, S., (2019) Formation of cross-border corporate strategic alliances: The roles of trust and cultural, institutional, and geographical distances. *Journal of Behavioral and Experimental Finance*, *21*, pp. 22–38.

Karami, M. and Tang, J. (2019) Entrepreneurial orientation and SME international performance: The mediating role of networking capability and experiential learning. *International Small Business Journal*, *37*(2), pp. 105–124.

Khanna, T. (2015) A case for contextual intelligence. *Management International Review*, *55*(2), pp. 181–190.

Kluckhohn, F.R. and Strodtbeck, F.L., (1961) *Variations in Value Orientations*. Evanston, IL: Row, Peterson.

Kolb, D., (1984) *Experiential Learning as the Science of Learning and Development*. Hoboken, NJ: Prentice Hall.

Lau, C.M. and Ngo, H.Y., (2001) Organization development and firm performance: A comparison of multinational and local firms. *Journal of International Business Studies*, *32*, pp. 95–114.

Lian, Z., Sun, W., Xie, D. and Zheng, J., (2021) Cultural difference and China's cross-border M&As: Language matters. *International Review of Economics & Finance*, *76*, pp. 1205–1218.

McCarthy, I.P., Lawrence, T.B., Wixted, B. and Gordon, B.R., (2010) A multidimensional conceptualization of environmental velocity. *Academy of Management Review*, *35*(4), pp. 604–626.

Malik, A., Sinha, P., Budhwar, P. and Pereira, V. (2023) Managing legitimacy in a cross-border post-merger integration context: The role of language strategies. *The International Journal of Human Resource Management*, *34*(21), 4144–4174.

Manik, H.F.G.G., Indarti, N. and Lukito-Budi, A.S. (2023) Examining network characteristic dynamics of kinship-based families on performance within Indonesian SMEs. *Journal of Enterprising Communities: People and Places in the Global Economy*, *17*(1), pp. 72–97.

Marquardt, M.J. and Horvath, L., (2001) *Global Teams: How Top Multinationals Span Boundaries and Cultures with High-Speed Teamwork*. Palo Alto, CA: Nicholas Brealey Publishing.

Mathews, J.A., (2006) Dragon multinationals: New players in 21st century globalization. *Asia Pacific Journal of Management*, *23*, pp. 5–27.

Maznevski, M.L. and Chudoba, K.M., (2000) Bridging space over time: Global virtual team dynamics and effectiveness. *Organization Science*, *11*(5), pp. 473–492.

Maznevski, J.J.D.M.L. and Di Stefano, J.J., (2000) Creating value with diverse teams in global management. *Organizational Dynamics*, *29*(1), pp. 45–63.

McClelland, D.C., (1961) *The Achieving Society*. Princeton, NJ: Van Norstrand Co.

Mintzberg, H. and Westley, F., (2000) Sustaining the institutional environment. *Organization Studies*, *21*(1), pp. 71–94.

Murdock, G.P., (1945) The common denominator of culture. In: *The Science of Man in the World Crisis* (p. 145), Linton, R. (ed.). New York: Columbia University Press.

Parboteeah, K.P. and Cullen, J.B., (2003) Social institutions and work centrality: Explorations beyond national culture. *Organization Science*, *14*(2), pp. 137–148.

Parkin, R., (1997) *Kinship: An Introduction to Basic Concepts*. Oxford: Blackwell.

Quer, D., (2021) Location decisions of Chinese firms in the global tourism industry: The role of prior international experience and diplomatic relations. *Journal of Hospitality and Tourism Management*, *46*, pp. 62–72.

Randolph, G.B. (2000) Collaborative learning in the classroom: A writing across the curriculum approach. *Journal of Engineering Education*, *89*(2), 119–122.

Porter, M.E., (1980) *Competitive Strategy: Techniques for Analyzing Industries and Competitors*. New York: Free Press.

Raisch, S. and Hotz, F., (2010) Shaping the context for learning: Corporate alignment initiatives, environmental munificence, and firm performance. In: *Strategic Reconfigurations: Building Dynamic Capabilities in Rapid-Innovation-Based Industries*. Wall, S., Zimmermann, C., Klingebiel, R. and Lange, D. (Eds.) (pp. 62–85). Cheltenham: Edward Elgar.

Sagiv, L. and Schwartz, S.H., (2000) Value priorities and subjective well-being: Direct relations and congruity effects. *European Journal of Social Psychology*, *30*(2), pp. 177–198.

Sagiv, L. and Schwartz, S.H. (2007) Cultural values in organisations: Insights for Europe. *European Journal of International Management*, *1*(3), 176–190.

Saner, R., Yiu, L. and Søndergaard, M., (2000) Business diplomacy management: A core competency for global companies. *Academy of Management Perspectives*, *14*(1), pp. 80–92.

Sastry, S. (2015) Optimising intercultural synergies: A dynamic model of value creation in post-merger integration contexts. Anglia Ruskin University. Thesis. https://hdl.handle.net/10779/aru.23760999.v1.

Schwartz, S.H., (1992) Universals in the content and structure of values: Theoretical advances and empirical tests in 20 countries. In: *Advances in Experimental Social Psychology*. Schwartz, S.H. and Zanna, M. (Eds.) (Vol. 25, pp. 1–65). Orlando, FL: Academic Press.

Schwartz, S.H., (1994) Beyond individualism/collectivism: New cultural dimensions of values. In: *Individualism and Collectivism: Theory, Methods, and Applications*, Kim, U. et al. (eds.) (pp. 85–119). Thousand Oaks, CA: Sage.

Schwartz, S.H., (2021) A repository of Schwartz value scales with instructions and an introduction. *Online Readings in Psychology and Culture*, *2*(2), 9.

Scott, W.R., (2002) The changing world of Chinese enterprise: An institutional perspective. In: *The Management of Enterprises in the People's Republic of China.* Tsui, A.S. and Lau, C.M. (Eds.) (pp. 59–78). Boston, MA: Springer US.

Stahl, G.K., Maznevski, M.L., Voigt, A. and Jonsen, K., (2010) Unraveling the effects of cultural diversity in teams: A meta-analysis of research on multicultural work groups. *Journal of International Business Studies*, *41*, pp. 690–709.

Staples, D.S. and Zhao, L., (2006) The effects of cultural diversity in virtual teams versus face-to-face teams. *Group Decision and Negotiation*, *15*, pp. 389–406.

Stone, L. (ed.), (2002) *New Directions in Anthropological Kinship*. New York: Rowman & Littlefield Publishers.

Tajfel, H., (1982) Social psychology of intergroup relations. *Annual Review of Psychology*, *33*(1), pp. 1–39.

Teece, D. (2018) Dynamic capabilities [Online]. Available at: https://www.davidjteece.com/dynamic-capabilities.

Thomas, D.C., Elron, E., Stahl, G., Ekelund, B.Z., Ravlin, E.C., Cerdin, J.L., Poelmans, S., Brislin, R., Pekerti, A., Aycan, Z. and Maznevski, M., (2008) Cultural intelligence: Domain and assessment. *International Journal of Cross Cultural Management*, *8*(2), pp. 123–143.

Ting-Toomey, S., (1999) *Communicating across Cultures*. New York: Guilford Publication.

Trompenaars, F. and Hampden-Turner, C., (1997) *Riding the Waves of Culture: Understanding Diversity in Global Business*. London: Nicholas Brealey International.

Trompenaars, F. and Hampden-Turner, C., (2006) *Riding the Waves of Culture: Understanding Diversity in Global Business*. London: Nicholas Brealey International.

Uğur, N.G. and Akbıyık, A., (2020) Impacts of COVID-19 on global tourism industry: A cross-regional comparison. *Tourism Management Perspectives*, *36*, p. 100744.

Wade-Benzoni, K.A., Hoffman, A.J., Thompson, L.L., Moore, D.A., Gillespie, J.J. and Bazerman, M.H., (2002) Barriers to resolution in ideologically based negotiations: The role of values and institutions. *Academy of Management Review*, *27*(1), pp. 41–57.

Wall, S., Minocha, S. and Rees, B., (2010) *International Business*. Harlow, England: Pearson Education.

Williams, K.Y. and O'Reilly, C. A., (1998) Demography and diversity in organizations: A review of 40 years of research. *Research in Organizational Behavior*, *20*(20), pp. 77–140.

Xie, E., Reddy, K.S. and Liang, J., (2017) Country-specific determinants of cross-border mergers and acquisitions: A comprehensive review and future research directions. *Journal of World Business*, *52*(2), pp. 127–183.

Xu, D. and Shenkar, O., (2002) Note: Institutional distance and the multinational enterprise. *Academy of Management Review*, *27*(4), pp. 608–618.

You, N.N., Lou, Y., Zhang, W., Chen, D., & Zeng, L. (2023) Leveraging resources and dynamic capabilities for organizational resilience amid COVID-19. *South African Journal of Business Management*, *54*(1), p. 3802.

# 14 Entrepreneurship in Tourism and Hospitality Research

## A Bibliometric Analysis

*Sanaz Vatankhah, Vahideh Bamshad and Sadaf Tallia*

### Key points

- Entrepreneurship has been receiving mounting research interest among T&H scholars.
- Entrepreneurship concepts have been used to develop rural tourism.
- Sustainable tourism development can be achieved with entrepreneurial initiatives in T&H context.
- Artificial intelligence and its application for T&H entrepreneurship warrant more investigation.

### Introduction

This chapter introduces a pioneering bibliometric analysis focused on the concept of entrepreneurship in tourism and hospitality (T&H) research. Through an in-depth qualitative and quantitative examination of scholarly articles, this study aims to uncover emerging trends, prevalent patterns, and significant contributions. By providing a comprehensive and systematic evaluation, this research offers unique insights into the evolving landscape of entrepreneurship in T&H research. The method of analysis involves meticulous data collection and rigorous mixed techniques, enabling a comprehensive exploration of the subject matter and facilitating valuable implications for future research and industry stakeholders.

### Background

The intertwining of entrepreneurship and the tourism and hospitality (T&H) sector has created a dynamic and ever-evolving landscape. Entrepreneurship plays a significant role in shaping the trajectory of the T&H industries (Dias et al., 2023), acting as a catalyst for innovation, growth, and adaptation (Elshaer & Saad, 2022; Makandwa et al., 2023). In the face of rapid industry transformation, stakeholders in T&H are confronted with the imperative to make well-informed decisions that drive growth, foster innovation, and sustain competitiveness. McGehee and Kim (2004) conducted a quantitative study to understand the motivations behind agri-tourism entrepreneurship. They applied Weber's formal and substantive rationality theory to analyse 11 motivations for launching agri-tourism businesses

DOI: 10.4324/9781003369967-19

among Virginian farm families. Their findings revealed that these motivations combine economic and social factors, showcasing the potential for lifestyle enhancement, cultural preservation, and economic benefits through agri-tourism.

In the context of family businesses within the tourism sector, Getz and Carlsen (2005) used the entrepreneurship lens to apply a family business development framework, encompassing business, family, and ownership aspects, to examine the nature of these businesses in tourism. This framework helps researchers and practitioners understand family-related dimensions in tourism businesses. Moreover, Getz and Petersen (2005) examined the relationship between entrepreneurial orientation and the types of businesses launched in the tourism industry. Their case studies in Denmark and Canada found that lifestyle and autonomy are significant motivations for entrepreneurs in this sector. On the other hand, Komppula (2014) explored private entrepreneurs' role in enhancing rural tourism destinations' competitiveness. Through qualitative methods and case studies in Finland, the study emphasized that entrepreneurs play a crucial role in destination development by leveraging location-specific resources.

In a similar vein, Sharpley (2002) focused on rural tourism in Cyprus and the challenges associated with diversifying tourism offerings. The study emphasized the importance of marketing strategies, community involvement, and thematic attractions in promoting economic sustainability in rural areas and also analysed the key success factors and strategies for rural tourism development. It appears that community involvement, infrastructure development, heritage preservation, marketing, and ongoing evaluation are critical elements for promoting and sustaining rural tourism experiences (Wilson et al., 2001). Chaos theory and a complexity perspective seem to have been strong frameworks for developing research ideas in T&H research. Russell and Faulkner (1999) applied these theories to investigate the role of entrepreneurs in destination development.

They highlighted how entrepreneurial activities can drive changes in tourist destinations and contribute to their viability amid external challenges. The amalgamation of entrepreneurship and T&H sectors offers a unique lens through which to view the current challenge in T&H research. In this context, the significance of comprehending the state of the art in research within the realms of

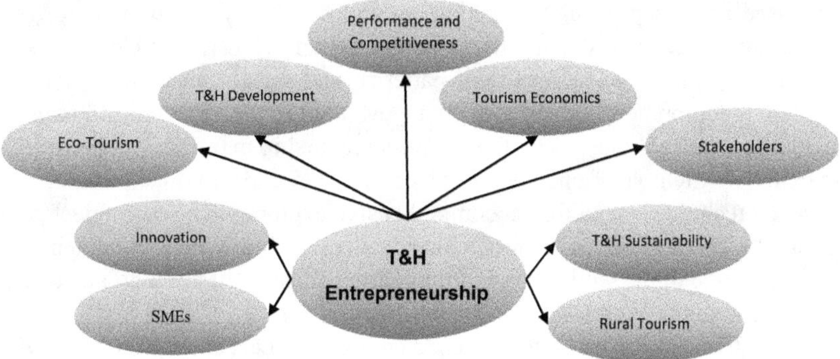


*Figure 14.1* Conceptual model.

entrepreneurship cannot be overstated (Dias et al., 2023). Tourism and hospitality businesses and policymakers can navigate the intricate landscape more effectively by examining the current trends, challenges, and advancements in T&H entrepreneurship research. Figure 14.1 displays a conceptual mapping of key factors related to entrepreneurship in T&H research.

## Method

Bibliometric analysis, an interdisciplinary practice, quantitatively evaluates publication volumes within specific research domains (Zhang et al., 2021). It allows researchers to understand the characteristics and pattern of research activities by analysing scholarly outputs, top-cited publications, countries, most frequent keywords, and the trend of publications. This quantitative exploration of a defined research area identifies influential authors, key journals, and foundational works that have significantly impacted the discourse (Donthu et al., 2021). As a result, bibliometric analysis provides valuable insights for comprehending individual publications and research domains, promoting more in-depth explorations within the literature (Chen et al., 2021). The application of bibliometric analysis in T&H research has been well-established (Chen et al., 2022; Chen et al., 2023; Fauzi, 2023; Gomes et al., 2023; Johnson & Samakovlis, 2019; Koseoglu et al., 2016; Kumar et al., 2023; López-Bonilla & López-Bonilla, 2021). However, there has yet to be a bibliometric study in T&H research to investigate the concept of entrepreneurship as an important dynamic concept in T&H industries (Elshaer & Saad, 2022). Therefore, the current bibliometric study tried to bridge the gap in the literature by addressing the following research questions:

RQ1: Which authors, institutions, countries, and sources are the most influential in applying the concept of entrepreneurship in T&H research?

RQ2: What are the main topics investigated by top-cited T&H literature relevant to entrepreneurship?

Through this exploration, this chapter introduces a pioneering bibliometric analysis focused on entrepreneurship in tourism and hospitality research. Through an in-depth qualitative and quantitative examination of scholarly articles, this study uncovers emerging trends, prevalent patterns, and significant contributions. By providing a comprehensive and systematic evaluation, this research offers unique insights into the evolving landscape of entrepreneurship in tourism and hospitality research. The method of analysis involves meticulous data collection and rigorous mixed techniques, enabling a comprehensive exploration of the subject matter and facilitating valuable implications for future research and industry stakeholders. SCOPUS is one of the well-known bibliometric databases. It has a broad coverage with comparison to WoS (Aghaei Chadegani et al., 2013).

To initiate the secondary data search, a comprehensive exploration of SCOPUS was conducted using the search term "entrepreneur*", ensuring the inclusion of various entrepreneurship variants. The search extended to article titles, abstracts, and keywords. Specifically, the final title search encompassed "entrepreneur*"

AND ("tourism*" OR "hospitality*"), yielding a substantial set of 2,221 documents from SCOPUS. The data collection from the SCOPUS database took place in July 2023. Furthermore, the search results on SCOPUS were meticulously refined to exclusively include journal articles published in English within the domain of *Tourism and Hospitality* (T&H). This meticulous curation resulted in a dataset comprising a total of 671 documents, ready for research analysis. The analysis of these final documents was conducted using the R-Tool of the Bibliometric package, a specialized tool designed for quantitative Bibliometric research (Aria & Cuccurullo, 2017). It was anticipated that a larger volume of SCOPUS data would be retrieved in comparison to WoS for the specified search terms.

## Results

### *Deceptive Analysis of All Documents*

Table 14.1 summarizes information about all documents in research from 1978 to 2023. There were 146 documents written by a single author, while the international co-authorship was 27.91% (Chaparro & Rojas-Galeano, 2021; Elango & Rajendran, 2012).

In addition, Figure 14.2 illustrates the annual scientific production of all documents in the "T&H entrepreneurship research" area. The publications have grown exponentially since 2011 (30 articles), with the maximum number of publications (70 articles) in 2020.

*Table 14.1* Descriptive Breakdown of Included Data

| *Description* | *Results* |
|---|---|
| Main Information about Data | |
| Timespan | 1978:2023 |
| Sources (journals, books, etc.) | 61 |
| Documents | 670 |
| Annual growth rate% | 8.29% |
| Document average age | 7.99 |
| Average citations per doc | 31.45 |
| References | 1 |
| Document Contents | |
| Keywords plus (ID) | 727 |
| Author's keywords (DE) | 2,027 |
| Authors | |
| Authors | 1,314 |
| Authors of single-authored docs | 146 |
| Authors Collaboration | |
| Single-authored docs | 155 |
| Co-authors per doc | 2.49 |
| International co-authorships% | 27.91 |
| Document Types | |
| Article | 671 |

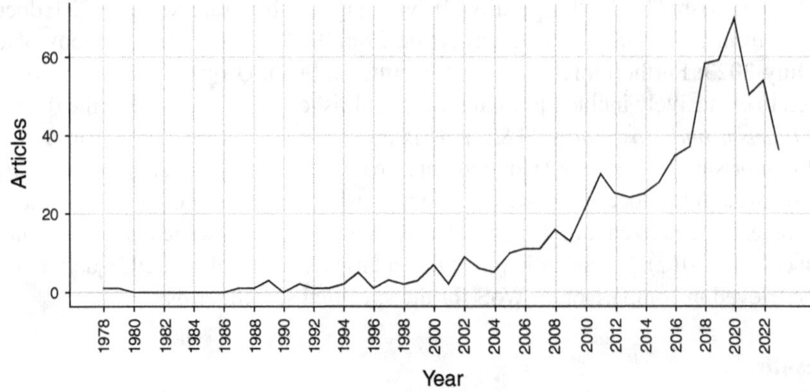


*Figure 14.2* Annual scientific production of all documents in "T&H entrepreneurship research".

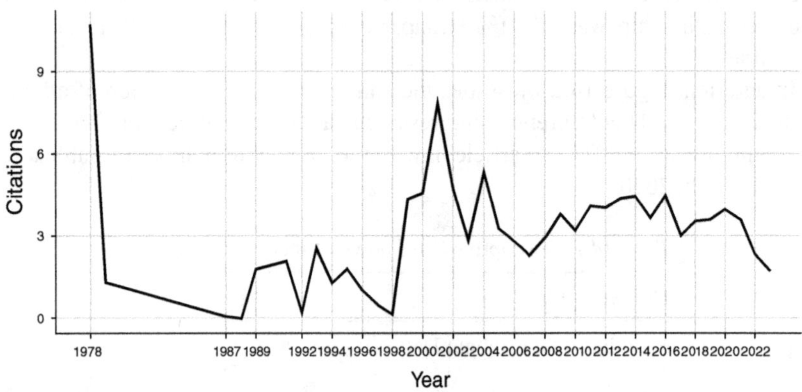


*Figure 14.3* Average article citation per year.

***Average Article Citation Per Year***

The average article citation per year in "T&H entrepreneurship research" are illustrated in Figure 14.3. The first top-cited article in the data set was published by Pizam (1978) entitled "Tourism's Impacts: The Social Costs to the Destination Community as Perceived by Its Residents" was a single-authored document which received 494 citations. Same case for "'Staying within the fence': Lifestyle entrepreneurship in tourism?" (Ateljevic & Doorne, 2000) with a total number of 494 citations. Top cited publications per year have also been assessed. "Post-pandemic recovery strategies: revitalizing lifestyle entrepreneurship" (Dias et al., 2022), received 27 citations per year and ranked first in terms of total citations per year. A review paper by Fu et al. (2019) entitled "The Entrepreneurship Research in Hospitality and Tourism" ranked second with 25.5 citations per year in the list.

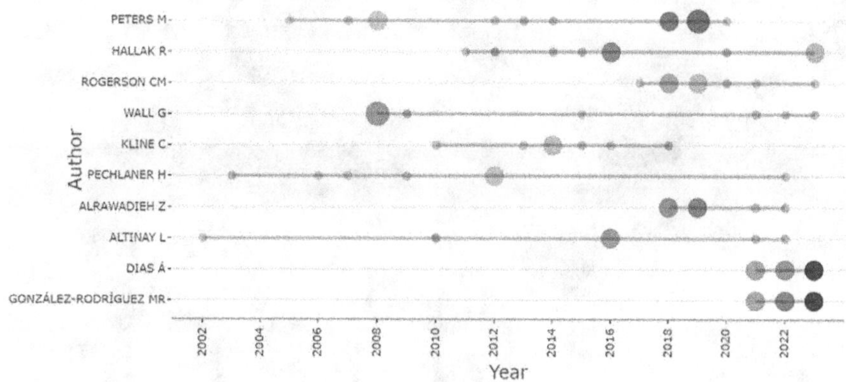


*Figure 14.4* The most influential authors (red line: the author's timeline, bubble size: the number of publications, bubble colour intensity: total citations per year).

### *The Most Influential Authors, Institutions, Countries, and Sources*

Top ten influential authors in the field are illustrated in Figure 14.4. The red line signifies the author's timeline. The bubble size is related to the number of documents published during the years. The bubble colour intensity is also related to the total citations per year (Ale Ebrahim et al., 2020). Top three authors' papers included in the list are "Entrepreneurship in tourism firms: A mixed-methods analysis of Performance Driver Configurations" by Kallmuenzer et al. (2019, total citation of 85 and TCprY of 21); "Innovation, Entrepreneurship, and restaurant performance: A higher-order structural model" by Lee et al. (2016, total citation of 155 and TCprY of 22); and "Tourism, local economic development and inclusion: Evidence from Overstrand Local Municipality, South Africa" by Rogerson and Rogerson (2019, total citation of 33 and TCprY of 7).

University of Surrey and the Hong Kong Polytechnique University with 26 publications jointly ranked first among top ten influential institution followed by University of South Australia (20 publications) and University of Johannesburg (19 publications) as the second and the third respectively. East Carolina University, North Carolina State University, Umea University, University of Central Florida, and the University of Otago are among other influential institutions contributing to "T&H entrepreneurship research". Figure 14.5 illustrates the most influential countries in the research area. The United States of America with 154 publications is the most influential country followed by United Kingdom (153 publications), China (121publications), and Australia (116 publications). Except for China, it appears that the "T&H entrepreneurship research" is dominated by western researchers and scholars, implying a fresh perspective from eastern researchers using fewer western theories and perspectives in "T&H entrepreneurship research".

Source analysis helps to identify the distribution of core journals in the research area. According to the results, 61 journals in T&H research published articles relevant to entrepreneurship. The top ten influential sources in the "entrepreneurship in T&H

*Figure 14.5* Country scientific production in "T&H entrepreneurship research".

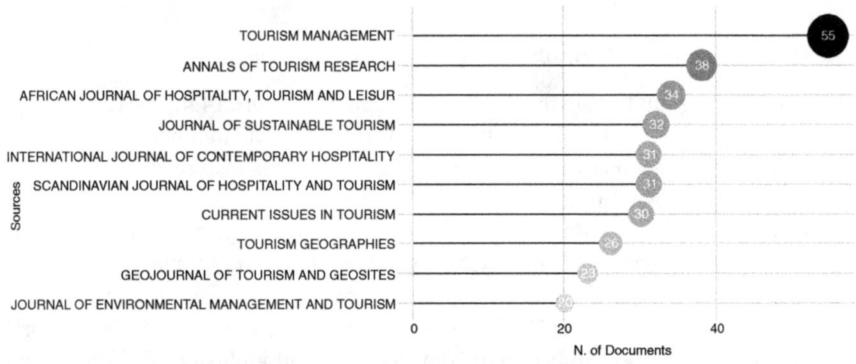


*Figure 14.6* Top 10 influential sources.

research" area are presented in Figure 14.6. Tourism Management as the main journal with the highest CiteScore (22.9) in tourism and hospitality research published 55 papers in this field, followed by *Annals of Tourism Research* and *African Journal of Hospitality, Tourism and Leisure* with 38 and 34 published papers respectively.

***Keyword Growth of the Top Cited Documents***

Figure 14.7 illustrates the authors' keywords trend of all papers published in the "T&H Entrepreneurship research". Except for entrepreneurship, the trendiest topics related to the concept of entrepreneurship in T&H research are innovation, rural tourism and sustainability. The application of these words has been experiencing an ascending growth since 2012. This is in line with the results demonstrated in the word cloud of the authors' keywords as presented in Figure 14.8. The size of the word shows the frequency of its usage as an author keyword. This figure suggests the entrepreneurship is used for sustainable development in tourism destinations. In particular, rural areas are benefiting from innovative entrepreneurial initiatives

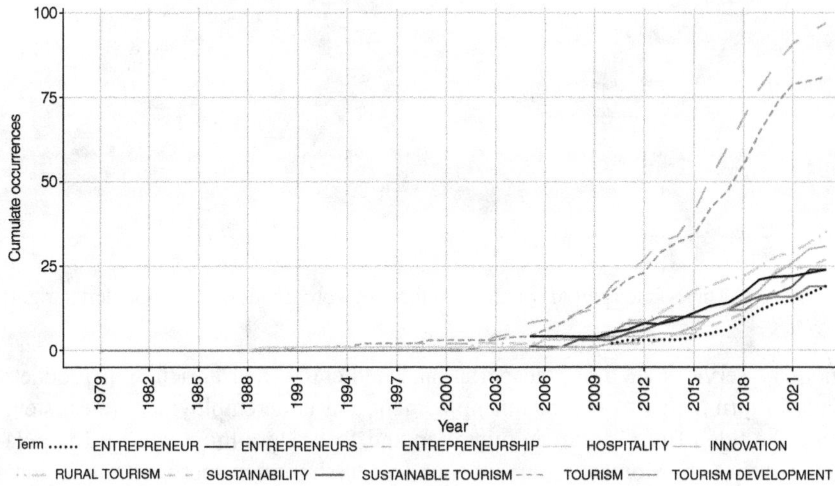


*Figure 14.7* Author's keywords trend.

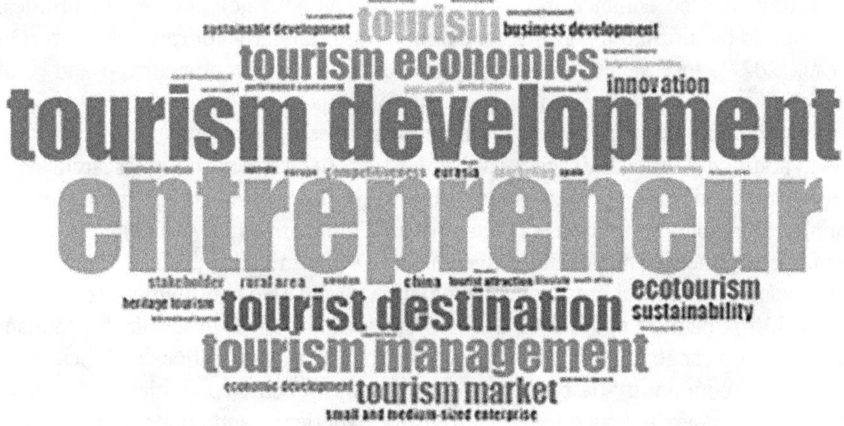


*Figure 14.8* Word cloud of author's keywords.

for tourism development. A more detailed breakdown of the trends in the use of keywords and their connection is represented in Figure 18.9.

As shown in Figure 14.9, rural tourism has always been the focus of entrepreneurship in T&H research. Tourism entrepreneurship holds significant implications for rural development, fostering a transformative impact on local economies and communities (Makandwa et al., 2023). As entrepreneurs invest in creating and promoting tourism ventures in rural areas, a multitude of positive outcomes emerge. Economic diversification occurs as new businesses emerge, offering accommodations, dining, guided tours, and cultural experiences, generating employment opportunities for local residents (Guo et al., 2023). The infusion of tourist spending brings in fresh capital into the community, stimulating the growth of

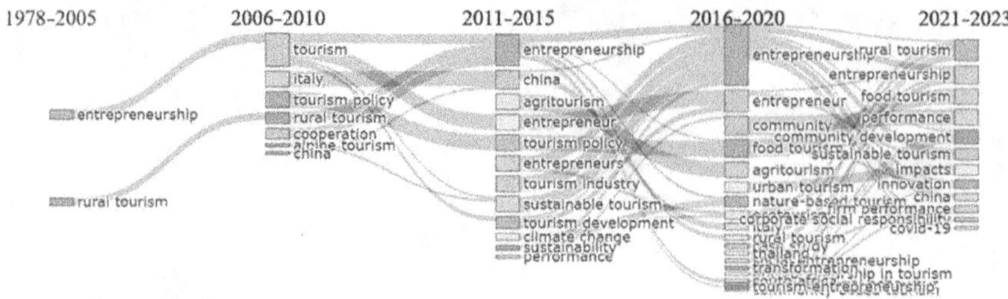


*Figure 14.9* Three-fields plot of countries, author keywords, and sources (from left to right).

ancillary services such as transportation, handicrafts, and agricultural products. This, in turn, helps to reduce unemployment and underemployment rates, stemming the tide of rural-to-urban migration and fostering a more balanced population distribution. Moreover, tourism entrepreneurship often spotlights the region's unique heritage, culture, and natural beauty, instilling a sense of pride and identity among residents while encouraging the preservation of local traditions (Guo et al., 2023; Qu et al., 2022). Collaborative efforts between entrepreneurs and local stakeholders also enhance infrastructure development, including roads, utilities, and public facilities, thus raising the overall living standards for rural residents (Qu et al., 2022). In essence, the convergence of tourism entrepreneurship and rural development presents a promising avenue for sustainable growth, empowerment, and improved quality of life within these often-overlooked areas.

In addition, Figure 14.9 suggests that the "T&H entrepreneurship research" considers the significance of entrepreneurship for tourism sustainability. In fact, entrepreneurship holds significant contributions and implications for sustainable tourism, a vital component of responsible travel that seeks to minimize negative environmental, social, and cultural impacts while maximizing positive contributions to destinations (Lordkipanidze et al., 2005). Entrepreneurial initiatives in sustainable tourism encompass a range of practices, from eco-friendly accommodations and local food ventures to community-based tours and conservation-focused activities (Luu, 2021). These endeavours enhance the overall visitor experience and create economic opportunities for local communities, thus bolstering their engagement in and support for conservation efforts. Entrepreneurs in sustainable tourism often champion practices that respect local cultures, empower marginalized groups, and preserve natural resources (Lordkipanidze et al., 2005). By fostering innovation and collaboration, entrepreneurship enables the development of models that balance economic growth with environmental preservation and social well-being, creating a win-win scenario for both travellers and host communities (Cunha et al., 2020; Lordkipanidze et al., 2005; Lundberg & Fredman, 2012; Nieuwland & Lavanga, 2020).

### *Characteristics of Top Cited Studies*

Bibliometric analysis, while invaluable, should be viewed as a complementary approach rather than a standalone substitute for qualitative peer evaluation (Ale Ebrahim et al., 2019; Donthu et al., 2021). Its application warrants a judicious and cautious approach when evaluating scholarly outputs (Franceschini & Maisano, 2011).

In tandem with bibliometric analysis, qualitative examination of the most frequently cited studies emerges as a pivotal component of this research endeavour, as elucidated by Maghami et al. (2015). Consequently, in the context of this study, the focus was directed towards scrutinizing the most frequently cited documents in the realm of "T&H entrepreneurship research".

This dedicated analysis was undertaken to unveil the distinctive attributes and qualities that have rendered these documents among the most cited articles within the field. By conducting this multifaceted approach, the aim was to provide a more comprehensive and nuanced perspective on the scholarly outputs generated from the perspective of entrepreneurship applied to tourism-related matters.

Accordingly, Getz and Carlsen (2005) published a review paper, McGehee and Kim (2004) conducted a survey study while the majority of articles applied case study methodology (Getz & Petersen, 2005; Komppula, 2014; Russell & Faulkner, 1999).

The first most-cited article is McGehee and Kim's (2004) study which aims to understand the motivations for agri-tourism entrepreneurship through the lens of *Weber's theory of formal and substantive rationality*. Indeed, this study applies the theoretical lens provided by Weber's rationality to the dynamic nature of motivations for agri-tourism entrepreneurship. To bring clarity to this fragmented area, the authors conducted a quantitative methodology and collected data from the Virginia farm families with a survey including 11 motivations for launching an agri-tourism business. The analysis of the data revealed that there are different motivations among Virginian farm families based on various characteristics such as the acres owned, dependence on farming operations, household income, and agri-tourism activity. According to Weber's theory, the motivation behind this kind of business is aligned with the combination of formal (economical) as well as substantive (social) rationality. In other words, motivations for agri-tourism entrepreneurship in Virginian farm families are derived from economic considerations such as additional income and social reasons such as "fully utilize resources, to meet a need in the tourism market, and to educate the consumers". This article is an excellent example of social and economic development as it highlights the unique motivational characteristics of agritourism, such as providing engaging and interactive tourist experiences while supporting local agricultural businesses (McGehee and Kim, 2004). The prominent significant motivational factor identified is lifestyle enhancement. Agri-tourism entrepreneurs revealed interest in the phenomena of growing agricultural and tourist experiences. It revealed a great opportunity for the farmers to preserve their local traditions and cultural heritage and share their experiences and knowledge with others. With this practice, the farmer doesn't have to change the local communities drastically to attract tourists' attention and can reap economic benefits by tapping into the growing market demand. Despite the attractions associated with agritourism, some challenges like dealing with marketing strategies, seasonal attractions, and balancing the demand of agricultural products and tourist activities must be addressed. In conclusion, the authors perceived agritourism entrepreneurship as a viable business opportunity that positively impacts lifestyle, agriculture, community and economic development. The findings emphasize practitioners and academicians investigating the motivational factors and addressing the challenges associated with agritourism entrepreneurship.

The second most-cited article is Getz and Carlsen's (2005) study which aims to depict the state of the art for family businesses in the tourism context. Indeed, tourism offers many opportunities for family businesses in which personal and family needs are dominant to the growth objectives. Recognizing the gap that most of the literature neglects the family-related dimensions of tourism businesses, the authors apply a general framework of family business development for examining the nature of such businesses in tourism. More specifically, this framework contains three main axes in terms of business, family, and ownership. The business axis includes startup, expansion/formalization, and maturity. The family axis includes young business families, entering the business, working together, and passing the Baton. The ownership axis refers to controlling order, sibling partnership, and cousin consortium. Concerning the framework, this paper outlined and discussed the four main themes: small and family business operations, links to entrepreneurship, roles and responsibilities of family members, and destination or community development. Furthermore, the study suggests that these themes can serve as a roadmap for tourism researchers interested in family businesses.

The next study is the study conducted by Komppula (2014) which set forth the goals of understanding the role of entrepreneurs in enhancing the competitiveness of a rural tourism destination. In particular, the study criticizes the literature for overemphasizing the role of DMOs (Destination Management Organizations) and attempts to investigate the private entrepreneurs' perceptions of their own role in the development of the competitiveness of a tourism destination. In terms of methodology, this study adopted a qualitative approach and an intrinsic case study of rural tourism was conducted in Finland. The study analyses one data set of narrative, theme-based, and face-to-face interviews and the other data set of semi-structured interviews with 15 entrepreneurs. The results revealed that the development of tourism destinations is the responsibility of enterprises. In other words, entrepreneurs can strengthen the attractiveness of a destination by exploiting the benefits of location, natural resources, and other unique resources specific to the destination. This study determines the facilitator role for DMOs (e.g., municipalities) and claims that without innovative, committed, and risk-taking entrepreneurs no destination is to flourish.

The study by Getz and Petersen (2005) sits in the fourth position on the most-cited list. The article's main goal is to identify a link between entrepreneurial orientation (profit/growth AND lifestyle/autonomy) and the type of business they tend to launch in the tourism industry. The study analyses two case studies in Denmark and Canada and collected data via questionnaires. The results illustrate that lifestyle and autonomy goals are the predominant motivations for running a business in the tourism and hospitality industry. The entrepreneurial orientation in terms of lifestyle/autonomy, as well as growth/profit, have a direct impact on the type of businesses entrepreneurs tend to launch. Next on the list, Russell and Faulkner's (1999) study established the main objectives of increasing the understanding of the fundamental role of entrepreneurs in destination development. Indeed, this study shifts attention from a traditionally linear view towards a *complexity perspective* and chaos theory, in particular. In other words, the utilization of *chaos theory* and its associated complexity view has been represented as an alternative framework that can be employed in conjunction with the well-established Destination Life Cycle Model. Providing insights into the underlying dynamism of tourism destinations, the authors tend to apply chaos theory to explain the neglected role of entrepreneurs as chaos-makers and as initiators of adaptive responses to the chaos caused by external circumstances.

Analysing a Gold Coast case study, Russell and Faulkner (1999) found that although the destination is suffering from some of the challenges common to tourist resorts at the mature stage of development, it still remains Australia's premium holiday destination. This is attributed to the region's natural assets and the array of man-made attractions. In other words, the study reveals that entrepreneurial activities as actions of movers and shakers brought a sequence of changes to the Gold Coast in terms of the scale and the nature of tourist activity. On the other hand, the findings illustrate that confronting external challenges, the viability of tourist destinations such as the Gold Coast might be due to the presence of entrepreneurs and their entrepreneurial activities. The article sheds light on the emergence of transformation tourism areas for sustainability. The prime insight of this article is to shed light on the challenges and complexities associated with ingraining the concept of entrepreneurship and the tourism industry. It gives intuition towards connecting individual entrepreneurial capabilities for the development of the tourism industry through innovative, risk-taking and sustainable tourism practices. In summary, this article provides scholarly insights into connecting entrepreneurial instincts for the development of destination tourism by Chaos theory. It encourages policymakers and industrial practitioners to embrace entrepreneurship innovation and adaptability during the life cycle of tourism development.

The article "Rural Tourism and the Challenge of Tourism Diversification: The Case of Cyprus" by Sharpley(2002) deals with rural tourism and the challenges Cyprus faces with varying tourism offerings. The author intends to explore the potential of rural tourism to achieve economic sustainability. The core context of this article is to evenly utilize different rural regions to gain economic benefits and preserve those areas' cultural and natural heritage. The instinct was to move the tourist experience from the coastal regions to other parts of the rural areas and spread economic benefits. The author employed the case of Cyprus to examine the challenges faced by tourism diversification in rural parts of the world. Highlighting the complexities of integrating tourist experiences with mass tourism structures and marketing challenges. Thereby suggesting a different approach to consider rural areas' unique attributes and resources to reap economic benefits. However, such benefits can only be exploited with effective branding and promotion strategies highlighting authentic tourism experience, cultural heritage etc. The author emphasizes the importance of marketing strategies and engaging local communities for their contribution and involvement (Sharpley, 2002). In addition, developing thematic tourism attractions will promote seasonal activities in rural tourism. In summary, the article emphasizes the importance of a holistic and sustainable approach to rural tourism development in Cyprus through effective planning brand promotion stakeholders and community support and thematic parks development.

The next article on the list titled "“Staying Within the Fence”: Lifestyle Entrepreneurship in Tourism", authored by Ateljevic and Doorne (2000) examines the concept of lifestyle entrepreneurship in the tourism industry. The prominent factor of this article is to promote the motivational factors and challenges faced by entrepreneurs in aligning their personal interests, values and desired lifestyles with their business development. The article begins with the integration of work and leisure activities of an entrepreneur's lifestyle in the context of promoting the tourism sector. The authors delve into motivational factors like autonomy, creativity and flexibility that drive entrepreneurs to pursue business development in the tourism sector. It also emphasizes the importance of social networks, collaboration and knowledge sharing within the lifestyle of the entrepreneurial community for providing unique and personalized experiences for customers (Ateljevic & Doorne, 2000). Moreover, the article addresses the challenges entrepreneurs face in managing work-life balance, financial uncertainty, and the authenticity and reliability of their business sustainability. This situation can be done through support networks, governmental policies and community collaboration to develop entrepreneurship in the tourism sector. In conclusion, in this article, the author discusses the benefits of cojoining lifestyle entrepreneurship and destination tourism experiences, which bring more authentic and niche experiences to specific market segments. This promoting the policymakers and academics seems to support organizations that nurture and sustain lifestyle entrepreneurship in the tourism sector.

The article titled "Factors for Success in Rural Tourism Development" by Wilson et al. (2001) analyses the key success factors and strategies for rural tourism development. The article begins by emphasizing rural tourism's importance in promoting cultural heritage, community development and economic diversification. This can be further supported through governmental community and individual participation. People can benefit from community resources, knowledge, expertise and destination marketing in creating unique tourist experiences. Engaging residents in planning the scene and making an implementation process promotes a sense of ownership and support for tourism initiatives (Wilson et al., 2001). This article focuses on important factors for promoting destination tourism like infrastructure and amenities attracting and satisfying tourists. Local heritage can be used as a primary attraction for visitors to ruler areas. Additionally, marketing and promotion are envisaged as an important factor for destination tourism. Furthermore, the author addressed the ongoing monitoring and evaluation of tourism activities for value addition and enhancing the tourist experience. In conclusion, this article focuses not only on developing destination tourism rather highlights important internal and external factors for the promotion and sustainability of rural tourism experiences to reap economic benefits.

These studies collectively contribute valuable insights into the complex and dynamic interplay between entrepreneurship and the tourism industry, shedding light on motivations, challenges, and opportunities for entrepreneurs in various contexts.

## Discussion and Future Research Direction

The tourism literature pays increasing attention to the entrepreneurship aspects of the field such as entrepreneurial motivations and orientations (Getz & Petersen, 2005; McGehee & Kim, 2004), family businesses (Getz & Carlsen, 2005), and disruptive role of entrepreneurs in developing various tourism-related dimensions (Komppula, 2014; Russell & Faulkner, 1999). Looking more closely into the articles and the areas that they tend to cover; it becomes clear that topics of entrepreneurship and tourism are compatible. These topics have been developed by addressing the effects of entrepreneurial individuals and their orientations on various dimensions of tourism such as dynamism of tourism destination. Linking individual-level research to more collective settings such as teams, some entrepreneurial lens such as entrepreneurial teams, their diversity, and dynamism can properly adopt to tourism context. Indeed, this research stream is reinforced by the notion that entrepreneurial teams often take entrepreneurial activities of start-ups and businesses as opposed to lone entrepreneurs (Gartner et al., 1994). Applying this notion to the tourism and hospitality literature, Kwapisz (2021) state that the entrepreneurial team's characteristics might influence the profitability of leisure-based start-ups in the tourism industry. So, with respect to this notion that tourism start-ups are also created and managed by teams rather than individuals, investigation of founding teams' diversity and dynamism can pave the way for a better understanding of tourism business growth as well as tourism destination development.

From the viewpoint of theories that are applied in the reviewed articles, it has been found that some have attempted to implement theoretical lens such as *complexity theory* which can explain the dynamism of entrepreneurship. However, only a fragmented and sceptical understanding is known about how entrepreneurship phenomenon can influence the tourism context. In this line, application of the *nexus theory* as one of the most fundamental theories in entrepreneurship context might shed light into tourism issues. The nexus theory claims that within entrepreneurial processes, not only entrepreneurs can exert an influence on opportunities but also entrepreneurs might be affected by those opportunities (Davidsson, 2015). Indeed, this theory redirects scholarly attention from unilaterally adopting a one-sided perspective to embracing a two-sided view, asserting that achieving a more robust application of the entrepreneurship lens is contingent upon scholars considering the dual aspects comprehensively (Casson, 2005; Davidsson, 2012). Tourism through the lens of nexus theory, it can be noticed that prior studies have mainly adopted one-sided view and mainly focused on the effects of entrepreneurs on tourism opportunities. In other words, although studies have attempted to highlight the role of entrepreneurs, their motivation, and orientations in tourism dynamism, interestingly less attention has been paid to how tourism opportunities, their nature, and characteristics can affect entrepreneurial activities in one region.

Finally, the notion that the tourism industry is mostly affected by the external environment (Zhong et al., 2011) might prompt scholars to not only focus on entrepreneurs but also to consider the external enablers and barriers (Vatankhah et al., 2023) for providing a more realistic picture of the phenomenon. In this regard, one of the frameworks that aptly explains the external factors within the entrepreneurship literature is the "external enabler framework" (Kimjeon & Davidsson, 2022). This comprehensive framework clarifies that external enablers can be categorized based on their characteristics, scope (i.e., spatial, temporal, sectoral, socio-demographic), and onset (i.e., gradualness, predictability, evolution). It also explains how the external environment can influence businesses through various mechanisms such as compression, conservation, risk reduction, etc. So, investigating the role of entrepreneurs in the development of tourism destinations is not the whole story. Future tourism research should employ more integrated frameworks that explore the role of entrepreneurs and the influence of environmental changes such as technological, demographic, socio-cultural, etc. In all, this bibliometric study reveals that the ties between entrepreneurship and tourism studies are rather superficial in terms of the applied theories and the variables explored to understand their effects on the tourism phenomenon.

This chapter analysed 671 documents collected from SCOPUS databases based on the search results. Therefore, it might be some documents on other well-known databases, which should have been included in this study. The other limitation is only top cited articles are considered with English as a prime language thus ignoring other articles. Finally, this research work did not include publications other than research articles.

## Conclusion

The discourse emerging from the discussion underscores the nascent but promising relationship between entrepreneurship and the tourism domain. The increasing attention paid to entrepreneurship in the tourism literature signifies a growing recognition of its significance. Explored aspects such as entrepreneurial motivations, family businesses, and the disruptive role of entrepreneurs contribute to understanding the intricate interplay between entrepreneurship and tourism. The analysis of the literature also reveals that substantial compatibility exists between the realms of entrepreneurship and tourism. By examining the effects of entrepreneurial individuals and their orientations on various dimensions of tourism, researchers have illuminated the dynamism within tourism destinations. Moving beyond the individual level, an entrepreneurial lens directed towards teams, their diversity, and dynamism holds the potential to provide a nuanced understanding of entrepreneurship in the context of tourism start-ups.

The application of theoretical frameworks such as *complexity theory* has offered insights into the dynamism of entrepreneurship, but a comprehensive understanding of how entrepreneurship influences tourism remains fragmented. The introduction of the *nexus theory*, which considers the reciprocal influence of entrepreneurs and opportunities, could illuminate the two-sided nature of the relationship between entrepreneurship and tourism. This shift in perspective would encapsulate how entrepreneurs shape tourism and how tourism opportunities shape entrepreneurial activities. Furthermore, recognizing the profound influence of the external environment on the tourism industry, scholars should extend their focus beyond entrepreneurs to encompass external enablers and barriers. The "external enabler framework", which elucidates the diverse characteristics, scopes, and onsets of external enablers, provides a comprehensive perspective on how the external environment shapes businesses and entrepreneurship.

This bibliometric analysis underscores the need for a more robust and integrated approach to understanding the interaction between entrepreneurship and tourism. The research landscape reveals a growing yet surface-level exploration of this relationship, prompting further investigation into the reciprocal influence of entrepreneurship and tourism opportunities. As the tourism industry remains intricately intertwined with external changes, a comprehensive understanding necessitates the incorporation of holistic frameworks that encompass not only entrepreneurial roles but also the influence of multifaceted environmental shifts. In conclusion, this study uncovers the foundational groundwork that has been laid at the nexus of entrepreneurship and tourism research. While the ties between these domains remain relatively unexplored, their potential for mutual enrichment and contribution beckons for more comprehensive investigations. By embracing integrated frameworks and adopting a two-sided perspective, future research can elevate the discourse, shedding light on entrepreneurship's profound influence on the ever-evolving tapestry of the tourism industry.

**Case Study**

***Local Communities and Tourism***

Tourism is considered one of the most crucial industries that substantially affects local communities' lives and behaviour in many countries worldwide. It significantly impacts nations' social, cultural, economic, and environmental aspects. Consequently, tourism has become a mechanism of community development in many destinations, particularly for rural and/or isolated regions, where locals heavily depend on tourism and its related business as a major source of income. Authors have worked towards developing tourism to prosper growth and economic development. Tourism helps create new jobs, strengthens the local economy, and facilitates infrastructure development. This leads to developing the country's economy and currency and attracting tourists for a continuous income stream. Therefore, the authors focused on the motivational factors that lead to agritourism and how to develop through the lens of entrepreneurship. Mostly, the focus remains on developing rural areas where farmers can find substitute income generation methods to help improve their economic reliance. Much work is done to understand how farmers can promote and attract tourists while preserving their cultural values and promoting job and infrastructure development. Local communities may perceive tourism and hospitality positively due to its potential for employment, income generation (e.g., direct, indirect, and induced income), and improved community infrastructure. This has given rise to the entrepreneurial intention towards developing destination tourism. The intertwining of entrepreneurship, tourism, and hospitality development remains a prime focus in the academic sector. Thinking of novel and unique attractions to promote experiential tourism is a new way of promoting economic growth in rural areas. This promotes the concept of entrepreneurship in agritourism and understanding of different factors that impact the development and understanding of entrepreneurial intention towards the growth and diversification of agritourism and destination tourism. Authors reveal multiple motivational factors which focus on tourism development in rural areas. The prime motivational factors include economic growth, development of a country, creation of new jobs, infrastructure development, promotion of agriculture, alternative sources of income, innovation and creativity, tourism attraction, destination tourism, life cycle, etc. The authors employed multiple theories to promote the relationship between entrepreneurship with tourism and hospitality development. Some studies employed Weber's theory of Formal and Substantive Rationality to understand the dynamic nature of motivations for Agri-tourism entrepreneurship. According to Weber's theory, tourism development is persuaded to align the combination of formal (economical) and substantive (social) rationality. Other authors focused on understanding the fundamental role of entrepreneurship towards destination tourism development. This led to the rise of complexity perspective and chaos theory. The implications of chaos theory and its associated complexity give rise to a new dimension towards destination tourism that can be employed with the well-established destination lifecycle model. The role of entrepreneurs as chaos makers and initiators of adaptive responses to the chaos caused by external circumstances. External factors impacting destination tourism, including environment, economy, technology, research and development, were also identified. However, on the other hand, some may negatively perceive tourism because of its negative socio-cultural impacts and environmental costs. The fear of losing the cultural values by mixing the culture of tourists with local tribes. Moreover, some other factors also lead towards the discouragement of tourism development, like pollution, littering, waste and strain on the natural environment by producing more products to fulfil the needs of the tourists and exhausting country resources. Tourism often leads to several income dependencies and may cause fluctuations due to its unpredictable nature. Lastly, there is an influx of increased crime onto the influx of different people. As a result, community attitudes towards entrepreneurship, tourism, and hospitality development constantly and simultaneously reveal positive and negative perspectives. A key point to be considered here is that the interaction between visitors and locals could influence community attitudes and behaviour towards tourism. Tourism offers several opportunities, especially to the ruler areas, to find alternative income sources and expand their growth production scales while preserving and promoting their local cultural values, supporting vulnerable groups, including minorities, youth and women. Such growth activities will serve as new insight for the children to follow and promote the same paradigm. The advantages of agritourism are far beyond the negative aspects. Rural areas can easily reap the benefits of hospitality and tourism development through peaceful and tolerant ways. They can generate additional sources of income while preserving their cultural and local values. This depends on how tourists are aware of or sensitive to local customs and traditions. Furthermore, it should be mentioned that local attitudes towards tourism are a vital criterion in identifying, measuring and analysing the changes generated by tourism. Locals can easily set values where they can promote entrepreneurship and hospitality and tourism development while preserving their culture and entertaining tourists at the same time. Some negative agritourism aspects include social change, changing values, difficulty in preserving religion and culture, control of crime, problems with tourist host relationships, and destruction of heritage. Yet there are multiple ways locals can save their sociocultural aspects while promoting tourism in their area as an alternative source of income. One of the most important ways that support sustainable tourism practices while preserving the environmental and cultural aspects is by adhering to local, national, and international laws. If the locals show persistent attitudes and behaviour towards environmental sustainability, the tourists endeavour not to damage the environment and comply with the rules. Therefore, local attitudes towards tourism are crucial to determining their responses to tourism development policies and planning within their countries. Nonetheless, as lofty as sustainable tourism goals may be, sometimes it's difficult to achieve. Situations impacting the culture and environment in a local area remain appalled. Most tourists rarely visit a destination site with the intention of causing harm either to the residents or the environment. Yet often, damage is inflicted accidentally, unconsciously and usually with visitors never realizing the harm they have wrought. This might happen due to the lack of understanding of different sociocultural aspects. Therefore, conscious acts on the part of the local entrepreneurs are deemed necessary towards the promotion of tourism and hospitality development in ruler areas.

**Answer the Following Questions**

Q1 Discuss the contribution of entrepreneurial intention to tourism growth and development. Give examples from the country of your choice.

Q2 Explain the negative effects of destination tourism on the social-cultural environment from a theoretical perspective.

Q3 Provide FOUR suggestions on how to enhance local community attitudes towards tourism.

**Key Terms and Definitions for the Key Constructs**

*Bibliometric analysis*: Bibliometric analysis is a quantitative research method used to evaluate and analyse patterns, relationships, and trends within academic publications, primarily scholarly articles, books, and citations. It involves the examination of citation patterns, authorship, publication frequency, and impact factors to assess the significance and influence of research within a specific field.

*Complexity theory*: Complexity theory, also known as complex systems theory, is an interdisciplinary framework that studies complex and nonlinear systems. It explores how simple interactions among components or agents within a system can give rise to emergent properties and behaviours that are not easily predictable. Complexity theory is often applied to understand phenomena in science, economics, sociology, and other fields.

*Eco-tourism*: Eco-tourism, short for ecological tourism, is a type of sustainable tourism that emphasizes nature conservation and environmental responsibility. It involves visiting natural areas, such as national parks or wildlife reserves, with the goal of appreciating and preserving the environment while minimizing negative impacts.

*Entrepreneurship*: Entrepreneurship refers to the process of identifying, creating, and pursuing business opportunities with innovation, creativity, and a willingness to take risks. Entrepreneurs are individuals or groups who initiate and manage ventures, often with the goal of achieving growth, profitability, and societal impact.

*Innovation*: Innovation is creating and implementing new ideas, methods, products, services, or processes to bring about positive change, improvement, or advancement. It involves the application of creativity and original thinking to address challenges, meet needs, or seize opportunities. Innovation can occur in various domains, including technology, business, science, and social sectors. It often leads to increased efficiency, competitiveness, and the development of novel solutions that benefit individuals, organizations, and society. Innovation can take the form of incremental improvements or ground-breaking breakthroughs.

*Local community engagement*: Local community engagement involves actively involving and collaborating with the residents, organizations, and stakeholders of a particular geographic area or community. It often pertains to initiatives, projects, or decision-making processes that affect the community's well-being, development, and sustainability.

*Rural tourism*: Rural tourism is a subset of tourism that takes place in rural or countryside areas rather than urban destinations. It typically involves visitors seeking authentic experiences in rural settings, including activities such as farm stays, nature exploration, cultural interactions, and participation in local traditions.

*SMEs* (Small and Medium-sized Enterprises): SMEs are businesses characterized by their relatively small size and limited financial resources compared to larger corporations. The specific criteria for categorizing businesses as SMEs may vary by country, but they typically include factors such as the number of employees, annual revenue, or total assets. SMEs play a crucial role in economic development and innovation.

*Sustainable tourism*: Sustainable tourism refers to an approach to tourism that aims to minimize its negative impact on the environment, society, and economy while maximizing the benefits for all stakeholders. It involves responsible travel practices that conserve natural resources, protect cultural heritage, and support local communities, thereby ensuring that tourism can be enjoyed by present and future generations.

*Tourism economics*: Tourism economics is a branch of economics that focuses on analysing the economic aspects of the tourism industry. It examines the economic impact of tourism activities, including factors such as tourism expenditure, job creation, GDP contributions, and the influence of tourism policies on regional or national economies.

*Tourism stakeholders*: It refers to individuals, groups, organizations, or entities that have a vested interest in the tourism industry. These stakeholders can have various roles and objectives related to tourism, including planning, development, promotion, management, or consumption of tourism products and services. Tourism stakeholders include government agencies, local communities, tour operators, hotels, airlines, travel agencies, tourists, and non-governmental organizations (NGOs). The interactions and cooperation among these stakeholders play a crucial role in shaping the tourism sector.

## References

Ale Ebrahim, S., Ashtari, A., Zamani Pedram, M., & Ale Ebrahim, N. (2019). Publication trends in drug delivery and magnetic nanoparticles. *Nanoscale Research Letters*, 14, 1–14.

Ale Ebrahim, S., Ashtari, A., Zamani Pedram, M., Ale Ebrahim, N., & Sanati-Nezhad, A. (2020). Publication trends in exosomes nanoparticles for cancer detection. *International Journal of Nanomedicine*, 15, 4453–4470.

Aria, M., & Cuccurullo, C. (2017). Bibliometrix: An R-tool for comprehensive science mapping analysis. *Journal of Informetrics*, 11(4), 959–975.

Ateljevic, I., & Doorne, S. (2000). 'Staying within the fence': Lifestyle entrepreneurship in tourism. *Journal of Sustainable Tourism*, 8(5), 378–392.

Casson, M. (2005). The individual–opportunity nexus: A review of Scott Shane: A general theory of entrepreneurship. *Small Business Economics*, 24, 423–430.

Chadegani, A. A., Salehi, H., Yunus, M. M., Farhadi, H., Fooladi, M., Farhadi, M., & Ebrahim, N. A. (2013). A comparison between two main academic literature collections: Web of Science and Scopus databases. arXiv preprint arXiv:1305.0377.

Chaparro, N., & Rojas-Galeano, S. (2021). Revealing the research landscape of Master's degrees via bibliometric analyses. arXiv preprint arXiv:2103.09431.

Chen, W., Ahmed, M. M., Sofiah, W. I., Isa, N. A. M., Ale Ebrahim, N., & Hai, T. (2021). A bibliometric statistical analysis of the fuzzy inference system - Based classifiers. *IEEE Access*, 9, 77811–77829. doi:10.1109/ACCESS.2021.3082908.

Chen, S., Tan, Z., Chen, Y., & Han, J. (2023). Research hotspots, future trends and influencing factors of tourism carbon footprint: A bibliometric analysis. *Journal of Travel & Tourism Marketing*, 40(2), 131–150.

Chen, S., Tian, D., Law, R., & Zhang, M. (2022). Bibliometric and visualized review of smart tourism research. *International Journal of Tourism Research*, 24(2), 298–307.

Cunha, C., Kastenholz, E., & Carneiro, M. J. (2020). Entrepreneurs in rural tourism: Do lifestyle motivations contribute to management practices that enhance sustainable entrepreneurial ecosystems? *Journal of Hospitality and Tourism Management*, 44, 215–226.

Davidsson, P. (2012, July). Entrepreneurial opportunity and the entrepreneurship nexus: A reconceputalization. In *Academy of Management Proceedings* (Vol. 2012, No. 1, p. 13290). Briarcliff Manor, NY: Academy of Management.

Davidsson, P. (2015). Entrepreneurial opportunities and the entrepreneurship nexus: A re-conceptualization. *Journal of Business Venturing*, 30(5), 674–695.

Dias, Á., Patuleia, M., Silva, R., Estêvão, J., & González-Rodríguez, M. R. (2022). Post-pandemic recovery strategies: Revitalizing lifestyle entrepreneurship. *Journal of Policy Research in Tourism, Leisure and Events*, 14(2), 97–114.

Dias, Á., Silva, G. M., Patuleia, M., & González-Rodríguez, M. R. (2023). Developing sustainable business models: Local knowledge acquisition and tourism lifestyle entrepreneurship. *Journal of Sustainable Tourism*, 31(4), 931–950.

Donthu, N., Kumar, S., Mukherjee, D., Pandey, N., & Lim, W. M. (2021). How to conduct a bibliometric analysis: An overview and guidelines. *Journal of Business Research*, 133, 285–296.

Elango, B., & Rajendran, P. (2012). Authorship trends and collaboration pattern in the marine sciences literature: A scientometric study. *International Journal of Information Dissemination and Technology*, 2(3), 166–169.

Elshaer, I. A., & Saad, S. K. (2022). Entrepreneurial resilience and business continuity in the tourism and hospitality industry: The role of adaptive performance and institutional orientation. *Tourism Review*, 77(5), 1365–1384.

Fauzi, M. A. (2023). A bibliometric review on knowledge management in tourism and hospitality: Past, present and future trends. *International Journal of Contemporary Hospitality Management*, 35(6), 2178–2201.

Franceschini, F., & Maisano, D. (2011). Regularity in the research output of individual scientists: An empirical analysis by recent bibliometric tools. *Journal of Informetrics*, 5(3), 458–468.

Fu, H., Okumus, F., Wu, K., & Köseoglu, M. A. (2019). The entrepreneurship research in hospitality and tourism. *International Journal of Hospitality Management*, 78, 1–12.

Gartner, W. B., Shaver, K. G., Gatewood, E., & Katz, J. A. (1994). Finding the entrepreneur in entrepreneurship. *Entrepreneurship Theory and Practice*, 18(3), 5–9.

Getz, D., & Carlsen, J. (2005). Family business in tourism: State of the art. *Annals of Tourism Research*, 32(1), 237–258.

Getz, D., & Petersen, T. (2005). Growth and profit-oriented entrepreneurship among family business owners in the tourism and hospitality industry. *International Journal of Hospitality Management*, 24(2), 219–242.

Gomes, S., Lopes, J. M., & Ferreira, L. (2023). Looking at the tourism industry through the lenses of industry 4.0: A bibliometric review of concerns and challenges. *Journal of Hospitality and Tourism Insights* (ahead-of-print). https://doi.org/10.1108/JHTI-10-2022-0479.

Guo, Y., Zhu, L., & Zhao, Y. (2023). Tourism entrepreneurship in rural destinations: Measuring the effects of capital configurations using the fsQCA approach. *Tourism Review*, 78(3), 834–848.

Johnson, A. G., & Samakovlis, I. (2019). A bibliometric analysis of knowledge development in smart tourism research. *Journal of Hospitality and Tourism Technology*, 10(4), 600–623.

Kallmuenzer, A., Kraus, S., Peters, M., Steiner, J., & Cheng, C. F. (2019). Entrepreneurship in tourism firms: A mixed-methods analysis of performance driver configurations. *Tourism Management*, 74, 319–330.

Kimjeon, J., & Davidsson, P. (2022). External enablers of entrepreneurship: A review and agenda for accumulation of strategically actionable knowledge. *Entrepreneurship Theory and Practice*, 46(3), 643–687.

Komppula, R. (2014). The role of individual entrepreneurs in the development of competitiveness for a rural tourism destination–A case study. *Tourism Management*, 40, 361–371.

Koseoglu, M. A., Rahimi, R., Okumus, F., & Liu, J. (2016). Bibliometric studies in tourism. *Annals of Tourism Research*, 61, 180–198.

Kumar, S., Kumar, V., Kumari Bhatt, I., Kumar, S., & Attri, K. (2023). Digital transformation in tourism sector: Trends and future perspectives from a bibliometric-content analysis. *Journal of Hospitality and Tourism Insights* (ahead-of-print). https://doi.org/10.1108/JHTI-10-2022-0472.

Kwapisz, A. (2021). Team aspects of leisure-based entrepreneurship. *Leisure Studies*, 40(4), 529–544.

Lee, C., Hallak, R., & Sardeshmukh, S. R. (2016). Innovation, entrepreneurship, and restaurant performance: A higher-order structural model. *Tourism Management*, 53, 215–228.

López-Bonilla, J. M., & López-Bonilla, L. M. (2021). Leading disciplines in tourism and hospitality research: A bibliometric analysis in Spain. *Current Issues in Tourism*, 24(13), 1880–1896.

Lordkipanidze, M., Brezet, H., & Backman, M. (2005). The entrepreneurship factor in sustainable tourism development. *Journal of Cleaner Production*, 13(8), 787–798.

Luu, T. T. (2021). Green creative behavior in the tourism industry: The role of green entrepreneurial orientation and a dual-mediation mechanism. *Journal of Sustainable Tourism*, 29(8), 1290–1318.

Lundberg, C., & Fredman, P. (2012). Success factors and constraints among nature-based tourism entrepreneurs. *Current Issues in Tourism*, 15(7), 649–671.

Makandwa, G., de Klerk, S., & Saayman, A. (2023). Culturally-based community tourism ventures in Southern Africa and rural women entrepreneurs' skills. *Current Issues in Tourism*, 26(8), 1268–1281.

Maghami, M. R., Asl, S. N., Rezadad, M. E., Ale Ebrahim, N., & Gomes, C. (2015). Qualitative and quantitative analysis of solar hydrogen generation literature from 2001 to 2014. *Scientometrics*, 105, 759–771.

McGehee, N. G., & Kim, K. (2004). Motivation for agri-tourism entrepreneurship. *Journal of Travel Research*, 43(2), 161–170.

Nieuwland, S., & Lavanga, M. (2020). The consequences of being 'the Capital of Cool'. Creative entrepreneurs and the sustainable development of creative tourism in the urban context of Rotterdam. *Journal of Sustainable Tourism*, 29(6), 926–943.

Pizam, A. (1978). Tourism's impacts: The social costs to the destination community as perceived by its residents. *Journal of Travel Research*, 16(4), 8–12.

Qu, M., McCormick, A. D., & Funck, C. (2022). Community resourcefulness and partnerships in rural tourism. *Journal of Sustainable Tourism*, 30(10), 2371–2390.

Rogerson, C. M., & Rogerson, J. M. (2019). Tourism, local economic development and inclusion: Evidence from Overstrand Local Municipality, South Africa. *Geo Journal of Tourism and Geosites*, 25(2), 293–308.

Russell, R., & Faulkner, B. (1999). Movers and shakers: Chaos makers in tourism development. *Tourism Management*, 20(4), 411–423.

Sharpley, R. (2002). Rural tourism and the challenge of tourism diversification: The case of Cyprus. *Tourism Management*, 23(3), 233–244.

Vatankhah, S., Bamshad, V., Altinay, L., & De Vita, G. (2023). Understanding business model development through the lens of complexity theory: Enablers and barriers. *Journal of Business Research*, 155, 113350.

Wilson, S., Fesenmaier, D. R., Fesenmaier, J., & Van Es, J. C. (2001). Factors for success in rural tourism development. *Journal of Travel research*, 40(2), 132–138.

Zhang, B., Rahmatullah, B., Wang, S. L., Zhang, G., Wang, H., & Ebrahim, N. A. (2021). A bibliometric of publication trends in medical image segmentation: Quantitative and qualitative analysis. *Journal of Applied Clinical Medical Physics*, 22(10), 45–65.

Zhong, L., Deng, J., Song, Z., & Ding, P. (2011). Research on environmental impacts of tourism in China: Progress and prospect. *Journal of Environmental Management*, 92(11), 2972–2983.

# Index

Note: **Bold** page numbers refer to tables and *italic* page numbers refer to figures.


Abi-Hanna, N. 29
accessible tourism 112–113
achievement *vs.* ascription 255
Action Research (AR) 11, 12, 14, 15, 17, 18; activity 14–15; CLATC 18; cycle 11, *11*; implementation 11, **11**; social innovation 11–12; tools 14, 15
Adeyinka-Ojo, S.F. 36
agri-tourism entrepreneurship 276–277, 285, 292
Ahmad, Ijaz 2
AirBnB 195–196
Air Travel Organisers' Licensing Scheme (ATOL) 56
Alkelani, W. 219–221
allocentrism 108, 218
Almeyda-Ibáñez, M. 153
Altman, I. 172
Amadi, Robert 2
Amatulli, C. 117
Anabila, P. 31
An, D. 31, 32
Anderson, A.A. 8
Anderson, A.R. 15
Ang, S. 260
Arroyo, Carmen 2
artificial intelligence (AI) 76, 79, 195, 197, 276
Arvanitis, Pavlos 2
Assaker, G. 37
Asset-Based Community Development (ABCD) model 19, 21
Association of British Travel Agencies (ABTA) 49, 50, 54, 55, 58, 59
Ateljevic, I. 288
Athwal, N. 117
attribution bias 224
augmented reality (AR) 81, 131–133, 137–139, 141–143, 157, 160
autonomous style 258
autonomous vehicle 192, 197, 201

Bamshad, V. 2
Banyte, J. 37
Barber, N. A. 116
Baxter, J. 171
Bennett, M. J. 218, 219, 224, 225
Berdychevsky, L. 34
Berry, J. W. 211, 224
Berscheid, E. 260
big data 76, 81, 195
Bilgihan, A. 151
Black and Minority Ethnicity (BAME) 10, 13, 14, 17
Blackden, P. 34
Blain, C. 152
blue spaces 91–92, 100
Boddewyn, J. J. 32
Boisen, M. 171
Bratianu, C. 222
Braun, E. 149, 171
Brexit: affecting tourism industry, examples of 56–58; ATOL 56; case study 60–61; future research directions 58–59; holiday packages 55–56; hospitality and tourism 54; parameters evaluated after 53–54; rising prices might discourage visitors 56–57; solutions and recommendations 58; staff shortages 57; trade 53; travel industry after 55; for travel industry, implications of 50–53; unemployment 53; visitors to UK 57–58
Browaeys, M.J. 255
Brown, P. 220

Brundtland Report 68
Buhalis, D. 69, 154
building AR reality 138–139
Business Continuity Management Plan (BCMP) 236–237
Byrne, D. 260

Carlsen, J. 277, 285, 286
Casais, B. 171, 175
Cassinger, C. 180
Cetin, G. 151
Chang, K. 173
chaos theory 277, 286–287, 292
charismatic/value-based style 258
chemical process industry (CPI) companies 238, 240, 241
Chen, A. 117
Chen, G. M. 217
Chen, R. 150, 165
Chief Executive Officers (CEO) 50, 54, 57
Childers, T. L. 30
China National Tourism Administration 38–40
Choi, B. K. 151
Christy, T. P. 33
Chu, T. Q. 32
circular economy 74, 98, 99, 101, 195
Civil Aviation Authority (CAA) 56
Clark, H. 8
Clays Advice and Training Centre (CLATC) (case study): ABCD framework 21; 2023 Budget **22**; challenges and recommendations 22–23; co-operative society 19, **20**; future training and project proposals 22; '3 LIQUD' 21; Meputa Oruaka (Handiwork) concept 18–19; objectives 18; ongoing projects and work progress 19–21; seed business loan 19, **20–21**; seminars 19, **19**; workshops 19, **19–20**; *see also* South-East Nigeria
Cohen, E. 211
communication: content *vs.* context 216–217; contextual process 213, 214; in cross-cultural encounters 212, 220–221, 223; and cultural diversity 212, 213–214; defined 213; intercultural effectiveness (*see* intercultural communication); is culture 216, 220; low context and high context 225; multicultural communication competence 219–220; non-verbal 216; symbols 213–214; transactional interaction 213, 214
community-based organizations (CBOs) 8, 9
community-led tourism initiatives 175–176
complexity theory 277, 286, 289, 290, 292
consumer personal factors 108–109
consumer psychological factors 107–108
consumer social factors 109–110
Contextual Intelligence 260, 261
Conversational Constraints 220, 221, 224
cooperative societies 8–9, 14
corporate social responsibility (CSR) 75, 121
Covas, L. 254
COVID-19-related problems 49–50
Crabolu, G. 72
Crisis Management Plan 236
critical decision-making *234*; celebrating all successes 245; championing cultural honesty 244; corporate 235–236; corporate risk register development 243; digitalisation 246; directors' enhanced roles 243; incident recall 245; reinforcing commitments to safety 245; senior management and C-suite executive coaching 244; skilled staff attraction and retention 244; talent pool development 244; technical staff mentoring/coaching 243
Croft, Joss 57
cruise tourism 6
cultural and artistic hubs 174
cultural diversity: app-based technology 224–225; awareness of contexts 223; and communication (*see* communication); cross-cultural interactions 212; cultural diffusion 215; cultural variabilities, awareness of 225; culture, defined 212, 214; global culture 224; intercultural competence 212; intercultural leadership 222–223; intergroup relationships 212; overcome perceptual barriers 223; research model *213*; self and other 217–219; tourism and resentment to tourists (case study) 226; universalities of culture 215–217; unlearning process (culture shedding) 224
cultural home 212, 221
Cultural Intelligence (CQ) 260–261, 263
cultural tourism authenticity 151
Customer Relationship Management (CRM) 150, 154
cyber resilience 242, 248

Dahl, D. W. 35
data protection 54–55
Deale, C. 116
De Lange, D.E. 16
destination authenticity: cultural tourism authenticity 151; increased visitors' engagement 150, 152; methods 151; online destination brand 151; on tourist satisfaction 151; on tourists' decision-making processes 150–151; and VRM 151–152
destination branding: case study 162–164; defined 149; DMOs 150, 152–154; on DMOs 152–154; elements 152; future research directions 160–161; post-industrial city 149; social media and digital platforms 153; strategic branding approach 153; VRM 154–155; youth travel 149–150
destination image: classification 149; defined 34–35; impact on 36–37
Destination Life Cycle Model 287
destination management organisations (DMOs) 34, 69, 71, 72, 149, 150, 152–160, 286
destination managers (DMs): future research suggestions 37–38; shock advertising 28–31; solutions and recommendations 37; tourist behaviours 29; tourist misbehaviour 34
Deutsch, M. 109
Diaspora, South-East Nigerians in 14, 17
digitalization 99, 102, 191, 199
digital trends: accessibility 200; AI 197; airlines 196–197; autonomous mobile robots 198; autonomous or self-driving vehicles 192; background of 192–193; carbon neutrality 196; digital transformation 198–199; electric vehicles 192; emerging technologies 193, *193*; 5G networks 192; GPS technology 192; last-mile delivery 198; MaaS (*see* Mobility as a Service (MaaS)); OTR visibility 198; Rasch method 200–201; real-time visibility solutions 197–198; sustainable transportation systems 199–200; time slot management 198; user-centric innovation 199; value co-creation 198–199
distance learning 246
Dolnicar, S. 117
Doorne, S. 288
drones 79, 198
Drucker, Peter: *The Practice of Management* 7–8
dynamic capabilities processes 262–265
dynamic workforce *234,* 248

Earley, P.C. 260
e-bikes 197
Echtner, C. M. 34
eco-friendly initiatives and sustainability 178–179
ecotax 100
eco-tourism 79
education and training *234*
Ekwugha, M. 2
electric vehicles (EVs) 192
embeddedness *vs.* autonomy 256
Emergency Response Plans 236
entrepreneurial tourism 2; external (corporate) 2; internal (organizational) 2
entrepreneurship: agri-tourism entrepreneurship 276–277; context of family businesses 277; future research direction 289–290; local communities and tourism (case study) 291–293; rural tourism 277; theoretical frameworks 290–291; and T&H sector (*see* tourism and hospitality (T&H) sector); in tourist destinations 277–278
Environmental, Social and Governance (ESG) 161
environmental sustainability: concept of sustainable tourism 68; policies and practices 67
Equiano, Olaudah 16
Estimated Time of Arrival (ETA) predictions 197, 198
Ethnocentrism 218
EU General Data Protection Regulation (GDPR) 54
European Tourism Indicator System for Sustainability (ETIS) 72
Evans, A. I. 2
experiential learning 263–265; coordination/integration 264; dynamic capabilities processes 264, 265; learning 264; learning styles 265; outcome of 'thinking' processes 263–264; reconfiguration 264; repetitive activities and experimenting 264
extended reality (XR) 137–138, 140, 143
"external enabler framework" 290–291

face 220, 222
Fam, K. S. 31, 32

Fan, D. X. 212
Faulkner, B. 277, 286, 287
Fermani, A. 107
Filieri, R. 109
Fink, G. 254
5G networks 192, 195
Font, X. 72
foreign direct investment (FDI) 252, 259, 266, 268
Fu, H. 280
Fullerton, R. A. 41

García, C. 97
Gardner, H. 261
gastronomic excellence 179
Generation Z 120, 155, 156, 160
Geographic Information System (GIS) 200
George, B. P. 153
Gerard, H. B. 109
Getz, D. 277, 285, 286
Gibson, H. J. 34
gig economy (case study): challenges and opportunities 204–205; economic impact of 204; legal challenges in 202–203; tax windfall for UK government 203; workforce trends and future 203–204; *see also* transport
Global Future Council for Sustainable Tourism 197
'Global Leadership and Organizational Behavior Effectiveness' (GLOBE) 257–258
Global Positioning System (GPS) 138, 192
global tourism 1, 6–7, 77, 78, 92, 149, 155
Gomez-Lopez, R. 76
Go, M. F. 174
Good *vs.* Bad Human Nature 257
Gössling, S. 79
Govers, R. 171, 174
green marketing: adoption of 76–77; environmental certifications 76; and tourism management 77
Gregori, G. L. 111
Grellier, J. 91
Gross Value Added (GVA) 54
Group of 20 (G20) tourism 50
Grün, B. 117
Gudykunst, W.B. 256
Guirdham, M. 220, 222, 225
Guirdham, O. 220, 222, 225
gyroscopes 138

Hall, E. T. 31, 214, 216, 220, 225, 256
Hallmann, K. 35, 41
Hampden-Turner, C.M. 254, 255
Hanks, L. 116
Hansen, M.H. 262
Haptic technology 133
Harries, E. 8
Haspeslagh, Philippe 266
Hatfield, E. 260
Heckler, S. E. 30
Heilig, Morton 147
heritage conservation and restoration 177–178
Herrero, A. 76
hierarchy *vs.* egalitarianism 256
Hofstede, G. H. 31, 215, 221, 254, 257, 259
Holm, E. D. 178
House, R.J. 257, 258
Howard, J. A. 30
Huang, S. 38
Hughes, K. 34
humane style 258
hydric stress 92, 95, 98

Igbo Apprenticeship System (IAS) 6
immersive technologies: AR (*see* augmented reality (AR)); case study 143–146; challenges 133–134; COVID-19 pandemic 131; future research 141–143; international travel 137–138; mixed reality (MR) 133; on society 132; solutions and recommendations 140–141; tourism and travel 134–136; VR (*see* virtual reality (VR))
incongruity 30, 37
individualism *vs.* communitarianism 255
Ind, N. 178
in-group & out-groups 257
inner- *vs.* outer-directed 255
innovation 265, 267, 276; balance of 234–235; digital 192–193; entrepreneurial 1, 177, 182, 287; social (*see* social innovation); technological 197, 198; user-centric 199
Insch, A. 181
intercultural communication: appropriateness of loudness 220; assertive communication 221–222; clarity 221; in collectivist cultures 221; in cross-cultural setting 220–221; empathy 221; person-centred messages 221; relationship orientation 221; sharing information 222; *see also* communication

internal branding: and place attachment 171; through resident involvement and entrepreneurship 171–174, *173*
internalisation strategies: integration types 266; intercultural resources 266; international diversity 267; process of globalisation 267
Internet of Things (IoT) 79, 81, 195, 197, 198

Jarrillo, J.C. 15
Jemison, David 266
Jerez-Jerez, María Jesús 2
Jha, A. 252
Jiménez-Barreto, J.
Jimenez Jimenez, Isabel Dolores 2

Kallmuenzer, A. 281
Kallström, L. 177
Kangjuan, L. 175
Kantenbacher, J. 72
Kaushal, V. 151
Kaushik, A. K. 151
Kent, M. 92
Kim, H.H. 115
Kim, K. 276, 285
Kim, M. S. 223
Kim, S. 31, 32
kinship system 259, 268
Kirzner, I.M. 15
Kisasembe, R. 39
Kletz, Trevor 245
Kluckhohn, C. 214
Kluckhohn, F.R. 254, 257
Knezevic Cvelbar, L. 117
Koester, J. 212, 213
Kolb, D. 264, 265
Komppula, R. 277, 285, 286
Kroeber, A. L. 214
Kumar, V. 151
Kunz, H. 32
Kwapisz, A. 289

Laskin, E. 261
Latour, M. S. 35
Lau, C.M. 259
Law, R. 115, 154
learning organisation 233, *234*, 247
Lee, C. 281
Lee, M. S. W. 41
Lee, T. J. 151
leisure-oriented tourism 7
Levinson, S.C. 220
Levy, S.E. 152
lifestyle entrepreneurship 288
Line, N.D. 116
Lings, I. 150, 154
linked travel agreement (LTA) 56
local community engagement 177, 183–184
logic models 8, 13, 17
Loi, K. I. 34
Low, S. M. 172
Lui, F. 31
Lustig, M. W. 212, 213
luxury and sustainability 116–117
Lv, X. 151

Machová, R. 36
Magala, S. 174
Maghami, M. R. 285
management literature 258–262
maritime industry 15–17
maritime tourism 12–17, *13*
Martinez, P. 72, 76
mastery *vs.* harmony 256
Mathews, J.A. 263
Matulevičienė, M. 31
Mayer, J. 28
McClelland, D.C. 257
Mcdermott, V. 221
McGehee, N. G. 276, 285
Melubo, K. 39
Méndez Pérez, Ester 2
Meng, F. 39
Meputa Oruaka (Handiwork) concept 18
millennials 117, 120, 149, 150, 155–158, 160, 164, 238, 239, 248
Miller, G. 72
mindfulness 261
Mintzberg, H. 263
mixed reality (MR) 133, 137, 140, 143, 144, 147
Mobility as a Service (MaaS): digital foundation of 194; and digital innovations, intersection of 193; environmental considerations 194, 195; personalization 194; priorities and trends in digital transport 196–201; regulatory considerations 195–196; smart cities 195
Monteiro, P. 171, 175
Moriizumi, S. 221
Mosakowski, E. 260
M-PESA 14
multicultural communication competence 219–220
multinational companies (MNCs) 232; challenges 232–233; environmental

crisis 232; foundations of success 233–234
Murdy, S. 150, 154

Navio-Marco, J. 2
Near Field Communication (NFC) 79
Netto, A. 6
neutral *vs.* affective cultures 255
nexus theory 289, 290
Ngo, H.Y. 259
Ng, S.L. 39
Nwala, Uzodinma 16

Oham, C. 2
Ojiagu, N.C. 2
Olivia, P. 30, 35, 68
organisation: balance of innovation and exploitation 234–235; challenges 237–238, 241, 243; changes 237–238; closing skills gap 240; conceptual framework *234*; conspiracy theories 242; Covid-19 pandemic 237; critical corporate decision-making 235–236; cybersecurity 242; emergency plans and contingencies 236; foundations of success 233–234; global workforce and technical skills shortages 239–240, *240*; integrity 234; issues 238–242; strategic workforce planning 241; survival and growth 232–233; training and development 238
organisational principles 235
organisational resilience *234*
Organisation for Economic Co-operation and Development (OECD) 90
over-the-road (OTR) shipments 198

Package Travel and Related Travel Arrangements Regulations 2018 56
Package Travel Directive (PTD2) 55, 56
Paiuc, D. 222
Palomo Martinez, J. 72
Paris Climate Accord 197
Park, R. E. 227
participative style 258
Pearce, P. L. 34, 36
Peng, N. 117
Pérez-Zabaleta, A. 2
Perry, W. 219
Petersen, S. 170
Petersen, T. 277, 286
Peterson, R.S., 260
Pettersson, A. 29
Pflaumbaum, C. 29
Phau, I. 32
Phippard, Simon 55
Pike, S. 34, 150, 154
Pirlog, A. 254
Pizam, A. 280
place branding: importance of residents' involvement 170; and reputation 171
Plog, S. C. 108
PMI: dynamic capabilities 262–263; experiential learning 263–265; future research directions 267–268; GV (case study) 269–270; internalisation strategies 266–267; management literature 258–262; values 255–256
Pope, N. K. L. 30
post-Brexit tourism: and consequences in travel industry *51*; EU nations and markets for UK, in 2018 **51,** 51–52; implications of Brexit for travel industry 50–53; parameters evaluated after Brexit 53; *see also* Brexit
power imbalance 223
*The Practice of Management* (Drucker) 7–8
Pratisto, E. H. 139
Prendergast, G. 32
Price, R. 255
pro-poor tourism 7
Punj, G. 41
Purwa, Agi Agung Galuh 2

Qastharin, Annisa Rahmani 2
Quer, D. 252

Radio Frequency Identification (RFID) 79
Rasch method 200–201
reefs (case study): adopting circular printing process 83; biomimicry and organic design 83; in climate change mitigation 82; Coastruction's vision for conservation 83; economic importance of 82–83; importance of 82; seamless integration with environment 83–84; *see also* sustainable tourism
Resident Entrepreneurship Acts: in community-led tourism initiatives 175–176; in cultural and artistic hubs 174; in eco-friendly initiatives and sustainability 178–179; in gastronomic excellence 179; in heritage conservation and restoration 177–178; leading trends and create opportunities 176–177;

Mountain Haven (case study) 183–185; promoting destination through 183; social impact and community development 176
responsible tourism 67, 93, 94, 100, 161
Reynolds, P.D. 16
risk assessment *234,* 235–236, 241, 243
risk management 233, *234,* 237, 241
Ritchie, B. J. R. 34
Ritchie, J. R. B. 152
Ritchie, R. J. B. 152
robots 79, 198
Rodriguez-Oromendia, A. 2
Rogerson, C. M. 281
Rogerson, J. M. 281
rural tourism 74, 276, 277, 283, 286–288
Russell, R. 277, 286, 287

Sabri, O. 31
Sagiv, L. 256
Sastry, S. 2, 253
Sawang, S. 31, 37
Schänzel, H.A. 36
schema 30, 213, 261
Schumpeter, J. 15
Schwartz, S.H. 255–257
Schwarz, N. 220
self and other: acceptance stage 219; adapting and integrating new aspects of culture 217; cultural identity 217; defense stage 218; denial stage 218; integration stage 219; intercultural sensitivity 217–218
self-driving vehicles 192
self-protective style 258
sequential *vs.* synchronic 255
Sevilla-Sevilla, C. 2
Sharpley, R. 277, 287
Shenkar, O. 258, 268
Sheth, J.N. 30
Shi, H. 154
shock advertising 28, 29; ambiguity 30; barriers to the effective application of 36; components of 29, 35–36; cultural dimension of 31–33; distinctiveness 30; role of culture in 36; transgression of norms and taboos 30–31
Simpson, Julia 50
skills gap 50, 58, 62, 234, *234,* 239, 240, *240,* 246
Skorupa, P. 29
Small and Medium-sized Enterprises (SMEs) 262, 267, 277
smart tourism: Bliss holiday (case study) 121–123; evolution of 121; external factors 118–120; ecological and social problems 120; economic stability 118; online eco-system 119; political stability and safety concerns 119; factors influencing consumer decision-making 107–112; personal factors 108–109; post pandemic transformation 111–112; psychological factors 107–108; social factors 109–110; sustainable tourism preferences 110–111; future research directions 120–121; health and mobility 107; impact on individuals' daily behaviours 107; industry challenges 114–118; luxury and sustainability 116–117; travel mobility and technology 114–116; primary objectives 106; tourism industry offerings 112–114; accessible tourism 112–113; urban tourism 113–114; wellness tourism 114
social distance 137, 227
social enterprise, in South-East Nigeria 8–10
social entrepreneurship: activity 14–15; business mission 11; enabler 14; inputs 15; intermediate outcomes 14; long term goal 14; social mission 11; societal issues 10
social impact and community development 176
social innovation 10–12; Action Research 11, *11,* **11,** *11*; M-PESA 12, *12,* 14, 17
sojourner 211
Song, S. 110
South-East Nigeria: accountability line 8; BAME (*see* Black and Minority Ethnicity (BAME)); CLATC (case study) (*see* Clays Advice and Training Centre (CLATC) (case study)); global tourism 6–7; IAS 6; maritime tourism 13, *13,* 15–17; population 5; slavery and emigration 5; social enterprise in 8–10; social innovation 10–12; theories for tourism industry 7; Theory of Change 7–8; tourism, defined 6; travelling 6
space tourism 6
specific *vs.* diffuse cultures 255
Spencer-Oatey, H. 212
stakeholders: capitalism system (IAS) 6; collaboration and participation 77, 80, 81; customers 76; destinations for 34, 36, 69, 72; industry 100–101, 119, 197,

276, 278; internal and external 69, 170, 171, 181, 262; local 154, 155, 284; Theories of Change 7; tourism 38, 93, 100, 112
Starosta, W. J. 217
State-Owned Company for the Management of Innovation and Technological Tourism (SEGITTUR) 73–75; Costa Rica 74; Formentera 73–74; Smart Tourist Destinations (DTI) 73, *73*; Sydney 74–75
statistics & facts 54
"Stay Away" campaign of 2023 218
stereotyping 222–224
Stevenson, H.H. 15
Stoica, I. S. 2
Strodtbeck, F.L. 254, 257
Stuart, M. 181
subcultures 216
Sultana, Saira 2
sustainability: defined 68; and eco-friendly initiatives 178–179; environmental (*see* environmental sustainability); and green marketing 75–77; and luxury 116–117; at travel destinations 69, 71–72; water–tourism 93–94
sustainable development 69, 71, 76, 77, 116, 117, 163, 195, 282
Sustainable Development Goals (SDGs) 8, 9, 16, 68, 70, 93, 94, 161
sustainable tourism 106; advent of onlife-world **78,** 78–79; benchmarking 72; collaboration for long-term development 70–71; communication of initiatives 69–70; defined 68, 93; examples of good practices 71; global greenhouse gas emissions 67; importance of reefs (case study) 82–84; meta-challenges **77**; new technologies 77–79; promoting responsible practices 70–71; role of sustainability and green marketing 75–77; SEGITTUR 73–75; at travel destinations 71–72

Tallia, Sadaf 2
Tanzer, Mark 58, 63
team-oriented style 258
Teruel, L. 68
Thelander, Å. 180
Theory of Change *7,* 7–8
Thomas, D.C. 261
Ting-Toomey, S. 261
Torres-Delgado, A. 72
tourism and hospitality (T&H) sector: author's keywords trend 282–284, *283, 284*; average article citation per year 280, *280*; bibliometric analysis 278–279, 291; characteristics of top cited studies 284–288; deceptive analysis of documents **279,** 279–280, *280*; and Entrepreneurship 276–278, *277*; most influential authors, institutions, countries, and sources *281,* 281–282, *282*
tourism, defined 6
tourism stakeholders 38
tourist misbehaviour 29, 34, 36–40
Trade and Cooperation Agreement (TCA) 48, 55
transport: airlines 196; defined 191; digital transformation 201; digital trends 192–193, *193*; future research directions 201–202; gig economy (case study) (*see* gig economy (case study)); MaaS (*see* Mobility as a Service (MaaS)); people or goods 191
travel mobility and technology 114–116
Travel Simply Tours (case study): diversifying recruitment sources 61; join trade organizations 62; market expansion 62; remote work and flexibility 62; results 62; retention methods 62; solution 61; training programmes 62; *see also* Brexit; post-Brexit tourism
Trompenaars, F. 254, 255
Tsaur, S.-H. 38
2030 Agenda for Sustainable Development 68, 70, 81, 93, 101

Uber 192, 195, 196, 201–203
Uchinaka, S. 175
United Nations (UN) 6, 12, 16, 90, 93, 122
universalism *vs.* particularism 255
urban tourism 107, 109, 113–114, 117, 121, 123
Urbinati, S. 175
Urwin, B. 33
user-generated content (UGC) 158–159

value co-creation 198–199
Vatankhah, S. 2
Venkat, R. 29
Venter, M. 33, 175
Vézina, R. 30, 35, 68
videohaptic mixed reality 147

Viñals, M.J. 68
virtual communications 246
virtual reality (VR) 131–133; benefits in travel 136; future of 136; tours 139–140; travel experiences 135–136
virtual tourism 6, 18, 137, 141–143
Virvilaitė, R. 31
visitor relationship management (VRM) 150, 154, 155, 161, 162, 164; challenges in building 157; concept of "living like a local" 155; defined 154; destination authenticity on destination branding 154, *155*; DMOs 155, 157; experiential marketing 157; importance of 157; information search process 156; millennials and Generation Z 156; Smiling West Java Application 164; tourism 154
Volgger, M. 38

Waller, D. S. 31
Wall, S. 266
Wang, T. 39
water consumption: factors 94; in hotels 94–95, **95**; in tourist destinations 94–98, **95,** *96*
water footprint 90, 96; blue 96; components 97; direct and indirect use 96, *96,* 97; green 96; grey 96
water use: and consumption in tourist destinations 94–98, **95,** *96*; future research 101; goal of sustainable tourism 90; mitigation measures, impacts of tourism on 98–100; solutions and recommendations 98–100; in tourist destinations, Benidorm (case study) 101–102, *102*; value of water 90; water footprint 90; water resources, impacts of tourism on 98; water–tourism nexus 90, 91–93; water–tourism sustainability 93–94
Weber, M. 285, 292
wellness tourism 107, 114
Wen, J. 39
West Java 162–164
West Java Tourism Center (WJTCC) 164
Westley, F. 263
Wilson, S. 288
Withdrawal Agreement 48, 55
Women Organizations 9
World Bank 7
World Economic Forum 79, 239
World Tourism Organization (UNWTO) 68, 69, 72, 90
World Travel Market (WTM) 49, 50
World Travel & Tourism Council (WTTC) 50, 67, 79
Wu, Y. 151
Wypych, L. 2

XBorders 248
Xu, D. 258

Yoo, M. 110
You, N.N. 262
Youth Mobility Plan (YMS) 59
youth travel 149–150
Yukako, N. 28

Zaim, I. A. 2
Zenker, S. 149, 170
Zouganeli, S. 154

For Product Safety Concerns and Information please contact our EU representative GPSR@taylorandfrancis.com
Taylor & Francis Verlag GmbH, Kaufingerstraße 24, 80331 München, Germany

www.ingramcontent.com/pod-product-compliance
Lightning Source LLC
Chambersburg PA
CBHW052119230326
41598CB00080B/3874